Excavation Systems Planning, Design, and Safety

About the Author

Joe M. Turner, PE (Santa Rosa, California), a registered civil engineer in several states, is founder and former owner of J.M. Turner Engineering, which specializes in excavation shoring system engineering and excavation safety planning. Over the past 17 years the company has provided excavation engineering services throughout the United States.

Excavation Systems Planning, Design, and Safety

Joe M. Turner, PE

McGraw Hill

New York Chicago San Francisco
Lisbon London Madrid Mexico City
Milan New Delhi San Juan
Seoul Singapore Sydney Toronto

The McGraw·Hill Companies

Library of Congress Cataloging-in-Publication Data

Turner, Joe M.
Excavation systems planning, design, and safety / Joe M. Turner.
p. cm.
Includes index.
ISBN 978-0-07-149869-2 (alk. paper)
1. Excavating machinery. I. Title.
TN277.T87 2008
624.1'52—dc22

2008026990

McGraw-Hill books are available at special quantity discounts to use as premiums and sales promotions, or for use in corporate training programs. To contact a special sales representative, please visit the Contact Us page at www.mhprofessional.com.

Excavation Systems Planning, Design, and Safety

1 2 3 4 5 6 7 8 9 0 DOC/DOC 0 1 4 3 2 1 0 9 8

ISBN 978-0-07-149869-2
MHID 0-07-149869-9

This book is printed on acid-free paper.

Sponsoring Editor
Larry S. Hager

Acquisitions Coordinator
Alexis Richard

Editorial Supervisor
David E. Fogarty

Project Manager
Harleen Chopra, International Typesetting and Composition

Copy Editor
Patti Scott

Proofreaders
Ragini Pandey and Megha Roy Chaudhary

Indexer
WordCo Indexing Services Production

Production Supervisor
Richard C. Ruzycka

Composition
International Typesetting and Composition

Art Director, Cover
Jeff Weeks

To my wife, Elizabeth, and daughters,
Abigail McEntee and Sarah Maxam

Contents

Preface

This book represents a culmination of knowledge that I have gained through 10 years of working as a construction engineer for contractors constructing pipelines and water and wastewater treatment projects and 16 years of developing an engineering firm that provided shoring design to contractors and shoring manufacturers. During the construction years I estimated excavation projects, planned the work, and worked in the trench with the workers. In the engineering years I was fortunate to be able to work with hundreds of underground contractors, the largest and the smallest, all over the United States. This wide range of exposure to level of knowledge and approach to excavation work and safety that these contractors had was priceless. During that entire period I never worked on the "other side of the street" for owners and civil design firms, but I was continually working with them on excavation work issues. Later in my career I worked with lawyers to help them understand excavation issues. Throughout this career I accumulated my basic knowledge and understanding of how all parts of the excavation industry functioned and interacted. All of us involved in the industry pulled basic excavation and safety technology information from wherever we could find it. We searched it out, adapted it to our needs, and used it to construct safer and more efficient excavations. I saw across the board that new entrants to the excavation industry faced a huge task in trying to acquire basic knowledge of the industry. I saw the construction side looking for better and more efficient ways to do things and the civil design side looking for ways to accept what they were proposing to do based on current state-of-the-art technology and accepted practices. At the end of a career it is easier to look back and see where it all came from and how it all fits together. It is not so easy to write it down. This book is an effort to assemble what I learned in one place and add some perspective so that everyone involved can better understand excavation work.

This book was written with the purpose of providing a basic introduction to principles, industry practices, safety, and legislation in excavation work. It is specific to producing safe, stable excavations through planning and is not about mechanical excavating equipment such as excavators and production in that sense. I firmly believe that fundamental information and background is important to gaining confidence and competence on any subject, and that on the subject of excavation work any person with a high school education and basic algebra skills can read and understand the material presented here. There is something here for every level of background and experience. The material is directed toward six major groups of individuals within the excavation industry: workers in the trench; contractors and their estimators and engineers; shoring manufacturing and distribution companies; civil design side and construction design side engineering firms; excavation safety professionals; and the legal profession. An overarching theme throughout the book is division of responsibility and the relationships among these groups.

Producing excavations, excavation safety, and shoring are all synonymous with excavation work. It is hard to talk about one without talking about the others. This book's scope is close to that of OSHA Subpart P—Excavations and the book stops where the standard leaves off, except that it presents detailed design and tabulated data information on manufactured shoring systems which were not envisioned at the time OSHA wrote the standard. This book offers depth and discussion in the form of commentary, design, engineering, use, and safety issues with all manufactured worker protection systems within the standard. It speaks in general about major engineered shoring systems such as pile and plate, sheet pile, and tieback systems, but does not go into specific engineering for those systems. The engineering provided on the Subpart P systems is intended to help bridge the gap between the construction and manufacturing side use and the civil design side review and inspection of those systems.

Chapters 1 and 2 provide background to the excavation industry and fundamental engineering concepts used throughout the book. Chapter 3 discusses the relationship between the construction side and the civil side of excavation work, the excavation planning process, and industry standards.

Chapter 4 is about the work that goes on prior to start-up of excavation production operations. Outside force damage on existing subsurface installations is becoming a major focus and engineering discipline because of the severity of accidents associated with it. There is a section on damage from surface loads and support of exposed utilities. There is a section on trench plates and traffic covers over open excavations.

Chapters 5, 6, and 7 are about the soil, soil loading, surcharge, and overall stability of the excavation.

Chapters 8 and 9 look at worker protection systems covered in OSHA Subpart P: open cuts; timber shoring; trench jacks; and shoring shields. Additionally these chapters cover manufactured shoring systems that were not contemplated at the time the standard was written such as aluminum shoring boxes and slide rail shoring. In these chapters there is emphasis on how manufactured shoring is engineered and tabulated so that engineers who review submitted shoring systems can see how the manufacturers arrived at their data tabulation. Emphasis is also placed on shoring equipment selection criteria and safety issues so that the contractors' estimating and engineering staff can do more design and planning in-house prior to taking the plan to a shoring design engineer.

Appendices 1 and 2 are the Subpart P—General Requirements and Appendix A—Soils with commentary. These sections present basic material that should be integrated into competent person training seminars and offer insight for the contractor and legal profession.

We require that excavation workers receive competent person training in excavation work. There is a similar requirement for professionals involved in the work that states they shall not produce work that they have no experience with. The material covered in this book could be considered competent person training topics for the professional.

This book is intended to serve as a first introduction to excavation work for any person at any level of education, age, and experience who is new to excavation issues in his or her work.

For those engineers and construction managers interested in focusing their practice around excavation-related construction engineering, I would like to offer some encouragement. The exciting thing about this work is that engineering design, start to finish, usually takes days to weeks, not months and years. The design process is much different than for permanent projects. The concept is originated by contractors, and the approval process is many times limited to the project design engineer, again taking a very short time. In most cases the project is funded and already under construction, with the overall project cost being in the millions while the shoring design cost is insignificant in comparison to the project cost and the shoring cost. This means that the shoring design engineer's experience and ability to make things happen fast is far more important than the design fee. Construction of the design is usually started within weeks and sometimes completed and removed within weeks. There is a shortage of engineering firms specializing in construction engineering. In my engineering practice we found that because of a lack of experienced engineers in this field we could not meet the demand, no matter how many engineers we put to the task. As a result our hourly rates were 25 to 50 percent higher than those in other similar disciplines such as building structure design. We also found it necessary to operate in many states rather than citywide or countywide because large excavation contractors operate regionally. This leads to demands on time at home and frequent travel, which is appealing to some engineers and an extreme disadvantage to others. These dynamics are the main factors that separate construction engineers from project design side engineers. The contractor is the client of the construction engineering firm, and it is not recommended to work both sides of the street.

This book is by no means a complete compilation of knowledge on the subject, and the opinions expressed are just one view. Where specific excavation-related stamped engineered work is a requirement, experience and knowledge beyond that presented in this book are necessary.

Joe M. Turner, PE
www.excavationsafety.com

Acknowledgments

The excavation safety and shoring and industry has been promoted, innovated, and influenced by many talented individuals; a few of these people I have had the privilege to know. Through my conversations with them I have gained much of the knowledge put forth in this book. Naming some of these giants of the industry whom I have known would surely leave out others of equal importance whom I have not met but who have equally influenced me. I would like to gratefully acknowledge their effort and selfless contribution to the excavation safety industry.

In preparing this book I would like to thank my good friend Mark Mathews for his advice and help with research; my wife, Elizabeth, and my daughters, Sarah and Abigail, for their editing work and encouragement. I would like to thank another good friend and author, Jeff Thomas, who advised me on issues involved in technical writing. I would also like to thank another good friend and colleague in the shoring industry, Jimmy Conway. Jimmy's enthusiasm and thirst for knowledge about excavation safety and shoring issues helped inspire me to write this book. The drawings and photography work in this book would not have been possible without the technical skill of Dustin Maxam and the drafting skills of Shaun Jones and Gary Nguyen.

The research and publications that the National Technical Information Service, the federal Occupational Safety and Health Administration, state OSHA administrators, and Naval Facilities Engineering Command (NAVFAC) have made available also have been invaluable to me in preparing this book.

CHAPTER 1

Introduction to Excavation Safety and Shoring Design

1.1 Introduction

The focus of this book is excavation safety and excavation work productivity. There was the time when these two concepts were seen as being at odds with each other; however, over the past 60 years a fundamental concept has changed that view. The concept is that excavation safety is just another aspect of what it takes to produce the end product. Like planning, environmental protection, digging, shoring, bedding, pipe laying, and backfill, excavation safety must be produced to deliver a pipeline project. The project cannot be completed without it. The more effectively each ingredient is produced, the more productive and profitable the project is. Failure to adequately provide any of these elements amounts to project failure; however, today by far the most costly and damaging failure is seen as failure to produce effective safety. The industry, worldwide, has come to accept that the value of human life is greater than the value of what is constructed inside the excavation and that in all cases excavation work can and must be produced without injury to workers.

It is hard to talk about excavation safety without talking about the Occupation Safety and Health Administration (OSHA) and shoring. Both of these terms are synonymous with excavation safety. Due to its dual nature, the word *shoring* is also closely associated with excavation production. Excavations cannot be planned and produced without some consideration of soil stability. In an unstable excavation the production work cannot be performed, and workers in that environment are not safe. Shoring stabilizes and protects simultaneously. In the first half of the 19th century, increased excavation productivity in the form of moving less dirt and avoiding damage to surrounding facilities, not digging faster (that is the realm of excavation equipment technology), drove the development of shoring technology. In the latter half of the 19th century, an increased concern for worker safety became a second driving force for shoring technology development. They both complement each other; however, faster and more efficient ground stabilization is most likely still the major mover of shoring technology development. Aside from stable ground issues there are other aspects of worker safety associated with excavation work, the major one being existing subsurface facilities. In this area protection of the lives of workers and the surrounding public is foremost with cost of damage to facilities being secondary. OSHA, as it relates to workers, and the Common Ground Alliance (CGA), as it relates to everyone involved, over the last half of the century have elevated this important issue to the forefront of excavation safety.

Awareness, knowledge, and technology are the tools used to produce safety and productivity in excavation work. The intent of this book is to help the reader gain access to those tools.

1.2 Excavation Work Categories

In the context of excavation work, safety and shoring are only one aspect. Excavation work in its entirety is anything involving mechanical movement of the earth's surface. Categorically it can be broken down by sequence of work, as in Fig. 1.1. There are four major categories; *surface work*—cut and fill, then construct production work; *below surface work*—cut, construct production work, then cover it up; *mining and tunneling*—cut then restore; and *agricultural*—cut, plow, and plant. A single project usually can originate in any one of these categories and involve all of them. There are unique safety issues associated with each category. This book focuses on excavation and safety issues surrounding *below surface work*.

Below surface work is the category of excavation work that OSHA Subpart P was written for. In 1926.650(a) OSHA states the scope and application. This subpart applies to all open excavations made in the earth's surface. Excavations are defined to include trenches.

OSHA 1926.651 applies to general excavation safety; however, 9 of 12 citations apply to work that involves excavated holes or trenches. Only three citations—Exposure to Vehicular Traffic, Exposure to Falling Loads, and Warning Systems for Mobile Equipment—apply to general excavation work. OSHA 1926.652 and all the remaining appendices apply to worker protection from cave-in. The entire 29 CFR Part 1926 Safety and Health Regulations for Construction apply to all construction operations; however, OSHA Subpart P is specific to when excavation work causes there to be a hole in the ground.

In terms of the process, excavation work and productivity fall into the categories of machinery productivity, sequencing of materials

Surface work Cut, fill, construct	**Belowground work** Cut, construct, and cover	
Transportation facilities	**Foundation**	**Pipeline**
Roads	Buildings	Urban service lines
Railroads and light transit	Bridges	Pressure pipe
Airport	Retaining walls	Storm water
Shipping ports	**Buried structures**	Gravity sewer
Land development	Basement	Gas service
Subdivision	Belowground parking structures	Electrical service
Commercial site development	Treatment plant structures	Pressure transmission
Land shaping	Industrial process structures	High-pressure gas and oil
Waterway and erosion control	Pump station	Slurry
Land stabilization	Utility vault and manhole	Water
Canal and viaduct		Sewer
Dams	**Temporary**	Electrical transmission
	Trenchless access pits	Duct banks
	Environmental cleanup	Fiber optic
	Exploratory excavations	Direct buried cable
	Archaeological	

Mining and tunneling Cut and restore	**Agriculture** Cut, plow, and plant
Open pit mining	**Farming**
Cave mining	**Grazing**
Drilling	**Forestry**

FIGURE 1.1 Excavation work subclassifications.

equipment and workforce (Fig. 1.1), and logistics of the trench or excavation. There are safety issues involved with each category; however, the focus here is the logistics of the trench, the configuration, and how to safely and efficiently get it into that condition and keep it that way until the production work is completed.

1.3 Basic Excavation Terminology

A few terms that will be used throughout this book can use some up-front clarification because they are used so often and sometimes interchangeably.

Excavation OSHA defines excavation as meaning any human-made cut cavity, trench, or depression in the earth's surface, formed by earth removal. This is a good definition until it becomes necessary to talk about an excavation that is not a trench such as excavation of a large rectangular hole. Sometimes this excavation is referred to as a structure excavation or open cut. Within the excavation industry the word *excavation* generally means a cut that is not a trench. Here it further excludes surface work that does not have a potential for cave-in on workers, swimming pool digging (excluded from OSHA excavation requirements), and tunneling work except for tunnel temporary entrance and exit structures.

Trench OSHA defines a trench or trench excavation as a narrow excavation (in relation to its length) made below the surface of the ground. In general, the depth is greater than the width, but the width of a trench (measured at the bottom) is not greater than 15 ft (4.6 m). If forms or other structures are installed or constructed in an excavation so as to reduce the dimension measured from the forms or structure to the side of the excavation to 15 ft (4.6 m) or less (measured at the bottom of the excavation), the excavation is also considered to be a trench.

One of the reasons that OSHA has defined a trench in this way is so that a distinction about access and egress of the excavation can be made. In a trench there must be access to a ladder within 25 ft of travel. In an excavation there needs to be a way to get into and out of it; otherwise, an excavation the size of a city block would need a lot of ladders.

OSHA purposes aside, there is a linear aspect to a trench, and it is also associated with pipe and utility work. The word is most often used in that context.

Excavation industry Every enterprise involved with excavation work.

Shoring industry Enterprises involved with shoring work including manufacturers and distributors of shoring equipment, the contractor, and OSHA and other safety-related organizations when focused on excavation work.

Production work There are two separate aspects to work in an excavation—constructing the excavation and constructing the work that caused the excavation to be opened in the first place or that which is to be buried inside the excavation. The latter is considered production work. Piles driven in the bottom of an excavation that are permanent and support the structure are *production piles*. Bedding, pipe, formwork, etc. are all production work.

The owner and project design engineer focus all their attention on the production work and not on the logistics of the excavation. The contractor is primarily responsible for the safety and success of the excavation work. From a safety aspect it is important to note that when a design engineer is focused on the production work, which is her or his primary function, excavation safety issues are secondary to that. Even though there has been some effort to place greater liability for accidents on the project designers, no amount of regulation or legal action can change this fundamental dynamic of "ownership of responsibilities" between the engineers and the contractors. This is not to say that the engineers do not have an obligation to design a safe constructible project and bring everything within their realm to bear on worker safety, just that the contractors should do their own due diligence because the buck stops with them, not the other way around.

1.4 Excavation Industry Makeup

The excavation industry is a conglomeration of many different endeavors. The makeup and diversity of the work render an expert in one area totally inexperienced in another. A pipe layer who knows how to install waterlines can be clueless about gravity sewer lines. Welders who work on high-pressure cross-country gas lines may work an entire lifetime on them and would be useless on a gas service utility crew. Contractors who build water treatment plants are not usually good at building pipelines. In true economic fashion, everyone in the industry focuses on what he or she does best. Figure 1.2 is definitely an incomplete listing of the people involved in the industry, but it does give an indication of the diversity. Three commonalities that run through all these niches are the need to protect existing facilities, stabilize excavations, and protect workers inside them.

Contractor
Office
- Project manager
- Estimator
- Engineer

Field
- Superintendent
- Engineer
- Foremen
- Equipment operators
- Teamsters
- Surveyors
- Laborers
- Carpenters
- Pipe fitters
- Electricians

Subcontractors

Project owner
Government agencies, military, private
Project engineers
- Geotechnical engineers
- Civil
- Mechanical
- Environmental

Interface personnel
- Municipal
- Agency

Supply
- Equipment
  - Mechanical excavating
  - Mechanical lifting
  - Shoring manufacture
  - Shoring distribution
  - Pump and power
  - Small tools
- Production materials
  - Aggregate
  - Pipe
  - Concrete and precast

Peripheral
- Underground contractor associations
  - Local
  - National
- Banking
- Insurance
- OSHA
- CGA
- Legal
  - Claims
  - Accident litigation

FIGURE 1.2 Makeup of underground excavation industry.

These three issues affect every person involved in this community. The collective knowledge, focus, and commitment of resources by this community are what produce safety and productivity within the industry.

1.5 Excavation Safety Regulation

Throughout the history of humankind, worker safety has always been a concern to the society at large. To some extent accidents in the pursuit of industry have always been seen as a necessary evil. The concept that industry is responsible to the worker for the cost of injury did not take root until the early 18th century, and it started to be taken seriously in the United States and Canada from around 1908 to 1915 as evidenced by the adoption of worker compensation laws by several states and provinces. The driving force for adoption of these laws was, and still is today, the "in your face" nature of industrial accidents. The press coverage following bad accidents and the accumulation of statistics related to recurring incidents cause the public to question the benefits they receive from industry at the expense of the labor force, and the employers face lawsuits demanding that they financially compensate victims and pay a price in fines and restrictions to their ability to do business the same as before. Workers compensation insurance provides compensation to the worker for injury on the job and also prevents the worker from suing the employer for liability of the accident. This allows employers to factor the cost of the insurance into their job costs and eliminate the unknown cost risk of the lawsuit. As it still is today, the compensation to the employee is never enough to offset the damage done by the accident; as the numbers of accidents increase, the employers' cost of insurance rises, and the public pays an increased cost for the product—so it becomes in everyone's financial interest to prevent accidents. Today's worker safety rules were born out of this atmosphere and this time period.

Workers compensation laws enacted by state legislatures created industrial relations commissions that administered the insurance, studied and investigated causes of injury in employment, and enacted regulations that carry the force of law to prevent accidents. By 1925 most states and provinces had issued General Construction Safety Orders. The very first orders recognized and stated that success hinged on the recognition and commitment to safety by the employer. Education and accident reporting were also seen as important elements. In the area of excavation work there was a conceptual division between excavations and trenches. Excavations were thought of as building foundation excavations, and the danger was associated with a single wall of earth collapsing on workers and the foundations of surrounding structures falling into the excavation. The solution was sheeting and angled timbered strut bracing or cross lot bracing.

Bracing not exceeding 8-ft horizontal spacing and allowable stress timber design values were specified. Trench Construction Safety Orders stood outside the general orders and defined the trench to have a depth greater than its width. A depth of 4 ft was the maximum unshored trench wall height. The maximum horizontal spacing of shoring was 8 ft, and maximum vertical spacing was 4 ft. There was a requirement for placing excavated materials at least 1 ft from the edge of the trench, and there was a ladder rule for trenches over 5 ft deep and access within 25 to 200 ft depending on the state issuing the order. Very few changes to these orders occurred until after 1950 when trenches came back into the realm of general safety orders. One code under a heading of "All Excavations" stated, "No employer shall cause or permit his employees to work in or adjacent to any excavation until a reasonable examination of the same has been made to determine that no conditions exist exposing them to injury from unstable ground." This looks a lot like the precursor to today's requirement that all excavations be inspected by a competent person to determine that worker protection is not required. In that same code there was reference to protection while installing shoring and sloping and benching. In 1940, a survey was undertaken under the auspices of the Work Projects Administration (WPA) of different western state trench safety codes and practices toward a goal of standardization. When one looks at the language that is used today in the excavation standard, it becomes obvious that it has been carried over from the original work. Much of what was relevant and worked then is still important and works today.

The *Occupational Safety and Health Act of 1970* established two separate agencies to deal with worker safety. Under the Department of Labor, the Occupational Safety and Health Administration (OSHA) was established for the purpose of developing and enforcing workplace safety and health rules. Under the Department of Health and Human Services, the National Institute of Safety and Health (NIOSH) was established with the mission of providing research, education, and training in the field of occupational safety and health. This sounds like two overlapping agencies, but in actuality they complement each other. To separate the two, keep in mind that OSHA is charged with establishment and enforcement of the rules and NIOSH provides research, education, and training. Together their common purpose is to leave no stone unturned when it comes to ways to protect workers in the workplace. When they first came on the scene in 1970, OSHA was viewed by contractors and their workers as policemen and they were unaware of NIOSH. Through the years both agencies have worked to change the image to one of their being part of the excavation community working to prevent accidents in any way possible. In fact they are the only ones within that community who come to work for the sole purpose of preventing death and injury to workers; there is no profit motive and no parallel purposes.

In 1979 OSHA focused on existing excavation safety regulations with the goal of adopting an updated and improved version. The National Bureau of Standards had conducted a survey of current trenching practice and concluded that due to the rapid and changing nature of soil it was not practical to have a registered engineer design the sloping and shoring and that it was necessary to have standards that could be understood and implemented by field supervisory personnel. They also recognized the need for a soil identification system that could be used by field personnel. With the help of NIOSH they drafted and put out for comment the original version of our current excavation safety requirements. The draft was reviewed by representatives from affected parties, including engineers, labor unions, contractors, shoring manufacturers, and safety officials. The process was accomplished through a series of four workshops in Atlanta, Ga., Boston, Mass., Dallas, Tex., Milwaukee, Wis., and San Francisco, Calif., and took 3 years to complete. On October 31, 1989, the final revised draft resolution 29 CFR Part 1926 Subpart P was issued. Except for a change in 1994 to add guardrails on walkways over trenches and elimination of a requirement for barricading at remote locations, the standard remains unchanged today.

In 2002 OSHA initiated a regulatory review of Subpart P as required by government resolution. The purpose of the review was to determine if it has been effective, efficient, and is up-to-date with current knowledge and technology. Written comment from industry professionals was solicited and reviewed, and cost and efficiency studies were performed. The conclusion was that "in relation to increased construction activity, fatalities have been reduced by more than 40%." The standard is fairly clear and at this time is well understood and usable although there are some areas that can be improved on; it is cost-effective; and fatalities result from violation of the standard.

On the issue of cost-effectiveness the report goes on to say, "The cost of protective systems has decreased by about 10% in real dollars between 1990 and 2001," and "Newer types of protective systems, including various 'trenchless' technologies, slide rail systems, and modular trench boxes, are being used with increasing frequency. Although the costs of these newer systems have not been examined for this study, it is reasonable to assume that each enjoys a net cost advantage over the older methods." Although a study has not been performed, a reduction in the cost of excavation production due to development of these systems can also be attributed to safety regulation. This reduction is likely to be far more than the 10 percent reduction impact listed in the report.

Reasons for noncompliance with the standard were lack of commitment on the part of contractors and their supervisors, lack of education and training, and cost pressure. Although there were no proposed changes resulting from this review, the federal government is working on some areas that will most likely result in future changes.

- Underground installations have become a focus of attention due to the increase in serious accidents. Increasingly crowded and aging utility corridors have become the scene of devastating excavation-related accidents that have resulted in death to not only workers but also the public. Major issues that are not clearly addressed in the current standard are definition of the effort required to identify, method of exposing, and notification requirements when lines are damaged.
- Provisions for hazardous atmospheres are not clear as to when a trench becomes a hazardous atmosphere or confined space. Future rule making on this subject will affect the excavation rules.

On both of these issues and many other trenching and excavation rules, states have already made significant changes.

State OSHA Offices

The federal government mandates that all states enforce federal OSHA requirements. They also encourage states to establish their own job safety and health programs. To date 26 states and territories have their own OSHA programs that are modeled after federal OSHA rules. The federal government provides up to 50 percent of the funding for these offices. The federal OSHA rules are a minimum standard, and usually state offices adopt the federal standards and then make amendments to them that result in higher standards. Some advantages to having state OSHA offices are as follows:

- The state has control of the program and can take direct actions to effect changes to it.
- Citations are violations of state statutes that can be handled in state court versus the federal court system.
- Local businesses have greater power in dealing with their own state legislature.
- States can partner with local associations and businesses to innovate worker safety programs.

Most changes made by state programs result from local accidents and enforcement problems. There has also been an effort to make the language used in the federal regulations easier for business and workers to understand. When changes are proposed, they have to clear the bar that says they have not weakened the federal standard and have not imposed an unnecessary financial burden on the affected businesses.

State OSHA offices do not have jurisdiction over federal reservations. Work on military bases, federal offices, and Indian reservations is still regulated by federal OSHA even though the locations are within a state.

Federal OSHA and state OSHA rules only apply to protection of employees and only apply when workers are present. Excavations of any depth or configuration can be made as long as workers do not enter and the stability of the walls does not affect workers in the vicinity.

Office of Pipeline Safety

The Office of Pipeline Safety (OPS) operates within the U.S. Department of Transportation's Pipeline Hazardous Materials Safety Administration (PHMSA). OPS is the primary federal regulatory agency responsible for ensuring the safe, reliable, and environmentally sound operation of U.S. energy pipelines. It develops and implements pipeline safety regulations at the federal level, and it shares regulatory responsibility with the states. OPS and its state partners oversee more than 2 million mi of pipelines.

External force damage during excavation work near these facilities is the leading cause of incidents on these pipelines. Unfortunately in this case the word *incident* is a euphemism for widespread death and destruction. The risk associated with excavation work in the vicinity of these lines cannot be understated. Excavation workers and the public need greater education in this area, and consequently OPS has put considerable effort into accident prevention. The OPS has initiated public awareness programs about leak detection, notification, and emergency procedures; invested in and promoted Common Ground Alliance (CGA) as the leading damage prevention association, which in turn developed "Best Practices" for preventing damage to pipelines and other underground facilities and "Dig Safely" for the purpose of educating pipeline excavators about preventing damage; developed the National Pipeline Mapping System (NPMS); and initiated education and training programs for pipeline operators.

Public Safety and Excavation Work

Although the major thrust of public concern and written safety rules is directed toward protection of workers, some of the regulations have the added benefit of protecting the public. Requirements for underground facility location, barricading and protection of workers from traffic, continuation of existing drainage, stability of existing structures, barricading of remote sites, etc. all enhance public safety; however the bulk of public protection requirements are governed by the need to protect the project owner from liability. Basic legal principles such as a property owner's right to full lateral support of her or his land, the public's right to safely access public facilities and the individual's own property, are an engineer's major obligation toward public safety in the work he or she designs, and the public building codes work to ensure public safety. Insurance and contract requirements are written to address risks specific to each project.

precision that a good operator can achieve make the operator and the machine appear superhuman. This is well and good when it comes to productivity, but when it comes to unexpected objects such as persons or buried lines in a normally clear workspace, it can spell disaster. Severe injury can result from head impact with a moving excavator bucket, even with a hard hat on.

Grade checkers and laborers working with excavator operators to locate buried lines should be in constant communication with the equipment operator. This is a case where technology has definitely outpaced OSHA rule making. The closest thing OSHA has in Subpart P to a requirement about this is the following:

> *1926.651(f) Warning system for mobile equipment*. When mobile equipment is operated adjacent to an excavation, or when such equipment is required to approach the edge of an excavation, and the operator does not have a clear and direct view of the edge of the excavation, a warning system shall be utilized such as barricades, hand or mechanical signals, or stop logs. If possible, the grade should be away from the excavation.

Today on an excavation project most people have a telephone with them. Headset communication systems are also used extensively. In excavation work the communication between the excavator and the lead ground worker is every bit as critical as it is with crane operators. There should be a regulation requiring an active communication link between the two; hand signals are no longer sufficient.

Since the speed of excavation equipment is faster, the linear progress of pipeline excavations is faster. The hourly cost of the pipeline production operation is staggering, to the point where every effort is made to speed it up and also to keep it from being delayed. The pipeline excavation and laying operation moves as a freight train that cannot stop for obstacles, so that the obstacle-clearing process has to be fine-tuned. When there is no room for error, there has to be a safety valve. Existing line location work, traffic planning, and equipment and material access planning are far more important in these circumstances.

To enhance safety under these conditions, the amount of planning should be stepped up, the operation should be worked up to speed slowly so that bugs can be worked out of the system, and there should be a safety officer within the company in addition to the competent person, separate from the person in charge of the pipeline productivity, with the authority to shut down the operation if there is an uncleared safety issue.

A 108-in × 12-ft reinforced concrete pipe (Fig. 1.6) weighs approximately 45,000 lb, or 22.5 tons; if it rolls over someone, it will flatten her or him like a gingerbread cookie. If it swings even slightly on the end of a crane cable when a worker is between it and a hard spot, it will crush heads, chests, pelvises, hands, or anything else that gets in the way. Pipe has to be picked off a truck and strung out along a pipeline. It then has to be picked again and set inside a trench that

Figure 1.6 Concrete pipe-laying operation.

usually has shoring elements in the way. This is a short list of risks inherent with these operations:

- Hoisting equipment tip over during truck unloading
- Dropping pipe on a person while it is being picked
- Rolling off of inadequate blocking when pipe is being strung
- Hoisting equipment tip over during setting operation
- Dropping pipe on a person while it is being set
- Swinging into and damaging shoring installation
- Crushing workers inside the excavation between the pipe and hard places

The size, mass, and physics of a large pipe being tethered on a long cable are something that is not encountered by a person in everyday life, so people do not have any inherent safety sense about it. Every new person working around large equipment and materials should have related training before being allowed to work with them. This should be more than a tailgate safety meeting and should be project-specific. There should be a minimum experience requirement for the equipment operator and at least one member of the ground crew. Large pipeline projects are usually multimillion-dollar affairs while in comparison the money allocated for safety on them is puny. There should be specifications on these projects that require more money and effort to be put into safety planning than is currently being spent. The OSHA Subpart P regulation regarding this issue reads as follows:

> *1926.651(e) Exposure to falling loads.* No employee shall be permitted underneath loads handled by lifting or digging equipment. Employees shall be required to stand away from any vehicle being loaded or unloaded to avoid being struck by any spillage or falling materials. Operators may remain in the cabs of vehicles being loaded or unloaded when the vehicles are equipped, in accordance with 1926.601(b)(6), to provide adequate protection for the operator during loading and unloading operations.

This is a starting point; however, it should be understood by all concerned that there is much more to it.

Confined Spaces and Hazardous Atmosphere

In Subpart P 1926.651(g) Hazardous Atmospheres, OSHA has eight citations, the first of which refers to additional requirements in general construction safety standards and none of which use the term *confined space*. These citations are discussed individually in the Subpart P commentary at the end of this book; however, general discussion follows here.

The subjects of confined space and hazardous atmospheres are so closely related that they should almost be discussed as one topic. The essence of the subject is egress and accumulation of toxic substances. Understanding the dynamics of accidents related to this subject is helpful. Unique to excavation work is the idea that excavations by themselves can become confining and a major purpose of excavation work—installing pipes and manholes—creates confined space. The other unique circumstance is that most hazardous gases are heavier than air, and hazardous fluids flow like water; both flow downhill, and excavations, manholes, and pipelines create a hole for the fluids to flow into. Also being heavier than air, hazardous gas when breathed into the lungs displaces air and is too heavy to exhale, creating disastrous results for those persons exposed. If the hazardous gas is lighter than air, it is most likely to head into the atmosphere and harass the public at large; however, if it is created in the bottom of the hole by a construction process such as welding and painting, the gases can flow uphill through pipes into manholes literally miles away. The third factor that enters into the picture is that the percentage of excavation work that is subject to this problem is small. It is hard to keep the focus when it rarely applies.

One key to success in preventing this class of accidents is clear identification of the circumstances that can create the problem. This is similar to the issue of worker protection from cave-in, where even in the case of a trench less than 5 ft deep a competent person has to determine that the excavation does not require sloping or shoring. There should be a requirement that a competent person make a determination about the classification of confined space and potential for development of hazardous atmospheres. It is understood that this is contained in the definition of a competent person; however, it should be more specific just as it is with protection from cave-in. This allows focus and action to be directed to areas where there is a potential problem. OSHA 1926.651(g)(1)(i) uses the phrase ". . . or a hazardous atmosphere exists or could reasonably be expected to exist. . . ." This catches all, but it is blunted by the fact that most all excavations do not fall into that category.

There is no question that inside pipes and manholes there is a potential problem. Soils reports and project specifications are required to identify potential contaminated soils. Work in industrial areas, refineries, and the like can produce gases that can roll into swales and ditches, so these areas require extreme caution whereas a new subdivision on virgin land that is not connected yet to anything has no potential. As with all construction activities, this hazard identification process needs to be performed and acted on prior to physical construction activities.

The other aspect of this topic is that it is complicated. There is excellent, easy-to-use detection equipment readily available; however, the person using it must be trained. The idea of effecting a rescue with poorly trained, inexperienced workers is scary, although at times necessary. If possible, go to those who do this the best, the fire department and emergency rescue teams.

Confined space and hazardous atmospheres are a broad subject that should be taken seriously. The excavation competent person cannot confine his or her knowledge to Subpart P requirements and be effective. In excavation work the focus for hazardous atmospheres should be on identification of the potential problems and then taking it to persons with complete knowledge of the subject for solutions.

Induced Hazards in Excavation Work

It is not known who coined the term *induced hazards*, but OSHA uses it and defines it as follows: "Induced hazards arise and are induced from a multitude of incorrect decisions and actions that occur during the construction process." One of the reasons for induced hazards is that everyone is looking at the small picture and no one is looking at the big picture. Decisions that make sense for production and budget do not always make sense from a safety standpoint. In construction it usually requires several individuals, each with a different focus and perspective, to complete any given production element. There may be several individuals from several different organizations looking at safety aspects, with each person thinking that the other person in the loop will catch errors and omissions that he or she makes. Persons in the field figure that management took care of the risky stuff. There is usually no clear designation of

CHAPTER 2

Engineering Structural Principles for Shoring Design

2.1 Introduction

The design of structures that resist the forces of soil is a large part of developing worker protection systems. When the first OSHA standards were being developed, the idea of having a registered engineer on every excavation project was considered and discussed at the regional workshops. The specific issues concerning this were that it required a trained engineer to identify soil and to perform engineering calculations to determine the loading on the shoring system, design shoring elements, or certify that premanufactured shoring elements were structurally adequate to resist the soil. The solution to the soil issue was development of Appendix A, Soil Classification. This classification system was developed for use by a trained, competent person with the concept that if she or he went through the steps correctly, it could be concluded that the soil was correctly identified. The system is fairly prescriptive and does not require broad background knowledge or education. Soil loading falls out in degree of severity with the determinations of type A, type B, and type C soil. Structural adequacy is then developed through tables and charts for timber or manufacturers' tabulated data for manufactured equipment.

Given the depth and soil type, most tabulated protection systems can be developed without calculations and the background knowledge of an engineer. A 20-ft depth limitation on a competent person's determination of sloped excavations was set due to concerns about bottom stability and the consequences of slope failure in deeper excavations. There is still confusion about whether the 20-ft limitation applies to manufactured shoring equipment applications because manufacturers tabulate for depths greater than 20 ft. Since depth limitations and acceptable conditions for use are set by the tabulating manufacturer, it appears that this equipment can be used to whatever depth it is tabulated for, and this is typically the way it is used. The problem is that the tabulated data usually state a requirement that the competent person determine whether stable bottom conditions exist; however, Appendix A gives the competent person no method for determining this.

This solution to the engineer-at-the-site problem was brilliant, not just because it took the registered engineer off the payroll of the contractor but more importantly because it involved the workers in developing the solution. Moreover, it requires those workers in the trench to understand the soil and the protection system. Involvement in the plan by those implementing it is a key element to the success of the plan. This is the major reason why engineering structural design principles are presented here. Structural design concepts are basic and simple to learn. Anyone with the mathematical ability to add, multiply, divide, and do basic algebraic manipulation and the conceptual ability to draw can learn structural engineering basics. Even without mathematics the basic and intuitive concepts of compression, tension, shear, bending, deflection, force, geometric properties, and stress are better understood with a simple course in engineering basics. For those who were exposed to it in high school or college and have long since tried to forget it, this section will help to refresh memories. The basic engineering presented here is directed strictly toward understanding applications in soil mechanics and shoring design.

There are plenty of trained people who are not structural engineers in the excavation industry and who are required to understand these principles to do their jobs. Contractors' estimators and project managers need to be able to develop shoring systems in order to bid and plan the work prior to bringing in a registered engineer to finalize a plan. Shoring design engineers can do their best work when the contractor is heavily involved and understands the structural limitations of the system. The competent person with a basic understanding of engineering principles is more capable of evaluating applied shoring systems; e.g., he or she should know that excessive deflection is an indication of bending failure, and that a couple of dings in the side of the shoring box near the end do not have much effect on bending, but in the middle of the box it does. Engineers on the project design side who do not perform structural designs still need to be able to review plans developed by the contractor and her or his engineers. Much of the shoring equipment design calculations presented later in this book are there so that the design-side engineers can see how the manufacturers engineered the equipment and developed the tabulated data. None of this is intended to suggest that a registered engineer is not necessary where it has been determined that a registered engineer is required to develop and stamp an engineering application. There is a vast array of engineering work involved with excavation planning and shoring that falls between selection of protection systems from tabulated data and engineer stamped plans.

2.2 Forces

A force can be thought of as weight acting in a direction. The weight of a person who is standing, measured in pounds, is exerting a vertical force on the floor. The person who pushes on a wall is exerting a horizontal force on the wall. Isaac Newton's third law of motion proclaimed that for every action there is an equal and opposite reaction. This directly leads to the concept that for every force there is an equal and opposite force, and further to the conclusion that the summation of forces in any direction is equal to zero. In structures these forces are referred to as *action* and *reaction*. The person standing on the floor is the action force, and the force of the floor resisting is the reaction force. Forces are either moving (dynamic) or fixed to one location (in equilibrium). In structures everything is considered to be in equilibrium and analyzed that way, referred to as static analysis versus dynamic analysis, such as in the case of a car in motion impacting an object.

Force and direction of a force can be broken down into components using the pythagorean theorem, which states that in a right triangle the square of the hypotenuse equals the sum of the squares of the two sides, or $c^2 = a^2 + b^2$ (Fig. 2.1). Usually the most important component directions are vertical and horizontal. Vertical is the direction of gravity and is considered perpendicular to the earth's level surface. Other terms sometimes used by engineers for force direction are *longitudinal*, in the long direction of something; and *transverse*, perpendicular to the long direction.

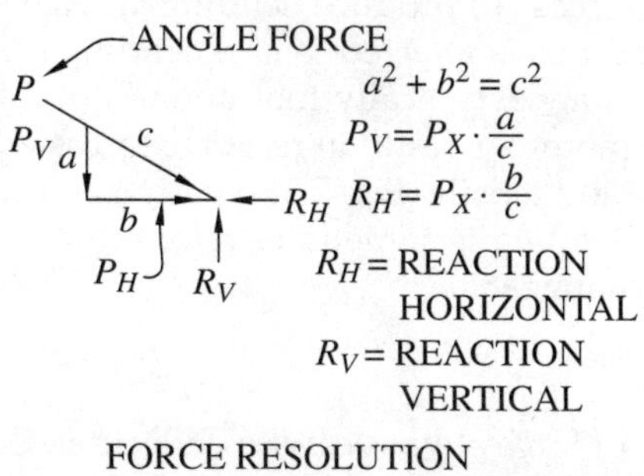

FIGURE 2.1 Resolution of an angled force into horizontal and vertical forces.

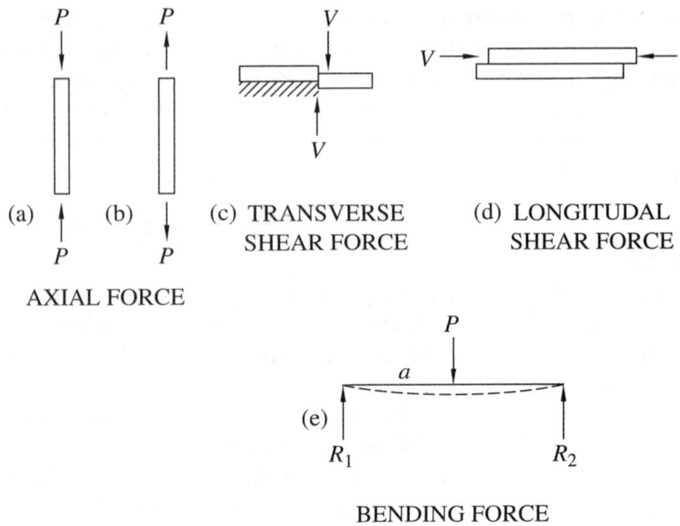

FIGURE 2.2 Force and effect on structural member.

Forces fall into major types—axial, shear, bending, and bearing—depending on the effect they have on the structural member (Fig. 2.2). One force can have more than one effect on a member; for instance, a bending force also creates shear at the reactions, and an axial force can crush a member or if the member is long and thin, it can cause buckling which is attributable to bending forces. The easiest way to conceptualize these forces is to grasp an object such as a ruler or pencil and try to bend it, compress it end to end, and shear it off. The force and direction that you have to apply to get the result will indicate which type of force is being applied.

Bending Force

Bending forces have another dimension to them that is distinct from axial and shear forces. To produce bending, a force has to be applied at a distance, and this is referred to as a bending moment with units of force times distance, typically foot-pounds (ft·lb) or inch-pounds (in·lb). Bending moments result in resistive forces within a beam that are compressive and tensile (Fig. 2.3). They also produce shear forces at the reaction. Bending is sometimes referred to as flexure, and the words are interchangeable.

Deflection

The application of force always causes deflection. Compressive axial force causes a member to get shorter, tensile axial force causes a member to get longer, shear force causes the beam particles to move

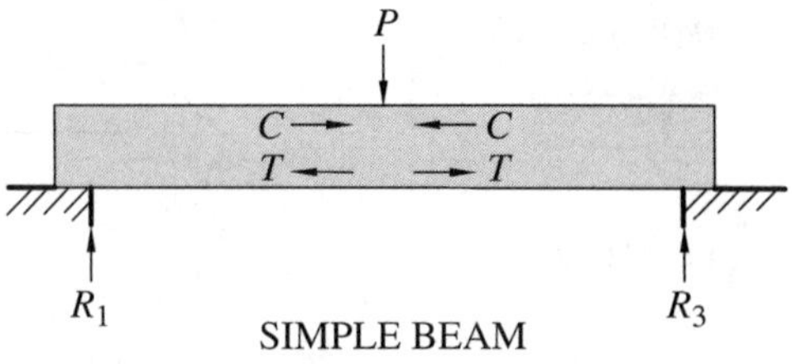

FIGURE 2.3 Simple beam compression at top and tension at the bottom.

laterally compared to adjacent particles, and bending forces cause the beam to curve away from the force. The measured distance that a beam moves as a result of a bending force or the increased length of a tension member is called the *deflection*. The Greek letter delta δ (lowercase) or Δ (uppercase) is used to refer to deflection.

Calculating Reactions for Simple Beams

To determine the internal forces, axial, shear, bending, and deflection, in a beam, all the external forces must be known. All forces in shoring design are distributed over an area. However, for calculation purposes we choose to look at them in two different forms, as point loads (force concentrated at a single point) [Fig. 2.4(a)] and distributed loads (force per unit length) [Fig. 2.4(b)]. A wheel or person standing is seen as delivering a point load because the contact area is small. Distributed loads can be summarized in point load form by multiplying by the length that it is distributed over and concentrating it at the center of action. In shoring design the forces on a beam are usually the result of soil, water, and surcharge loading. These forces are distributed most often in a rectangular, triangular, or trapezoidal form [Fig. 2.4(a) to (c)].

The forces and moments on a beam are derived from the surrounding conditions, and the resisting reactions to those forces are calculated. The reactions may be located at any point along a beam and in any direction, although they are usually horizontal or vertical, and there is no limitation to the number of reactions. Beams with more than two reactions are continuous [Fig. 2.5(a)].

Depending on the number of reactions and movement limitations that the reaction provides, a beam is considered simple, which is to say solvable by simple mathematics, or indeterminate, or unsolvable by simple hand calculations. For the purposes of this section a two-point pin reaction or a one-end fixed cantilever beam are simple [Fig. 2.4 (a) through (f)]. All timber and most steel applications in shoring can be broken down into these two cases. There are published beam formula tables and easy-to-use computer beam programs available for the rest.

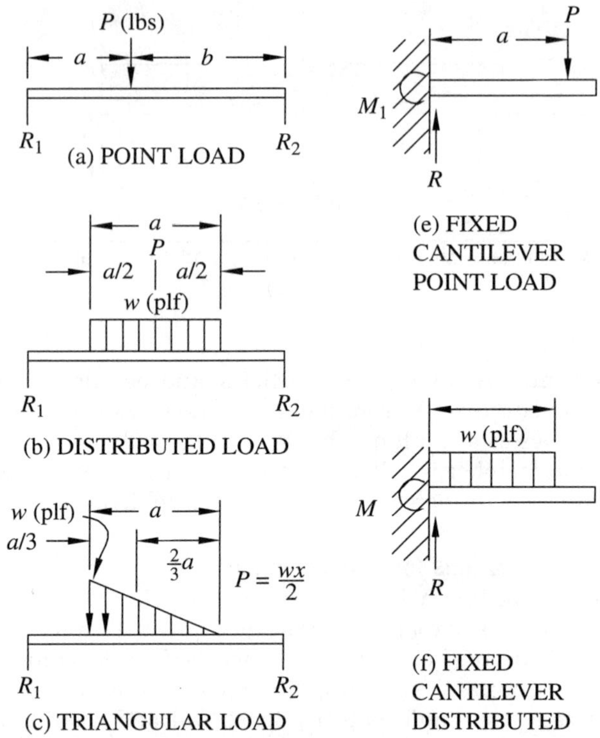

FIGURE 2.4 Beam load configurations.

Reactions limit the movement of the beam. Possible movement is directional (movement away from a point) and rotational (movement around a point). Pinned connections [Fig. 2.5(b)] limit movement but do not limit rotation. Fixed connections [Fig. 2.5(c)] limit movement and rotation. Not normally used in shoring calculations are roller-type connections that allow pinned movement in a horizontal direction and spring-type connections that allow for pinned connection movement in a vertical direction.

There are two basic equations based on Newton's third law that are used to solve for beam reactions:

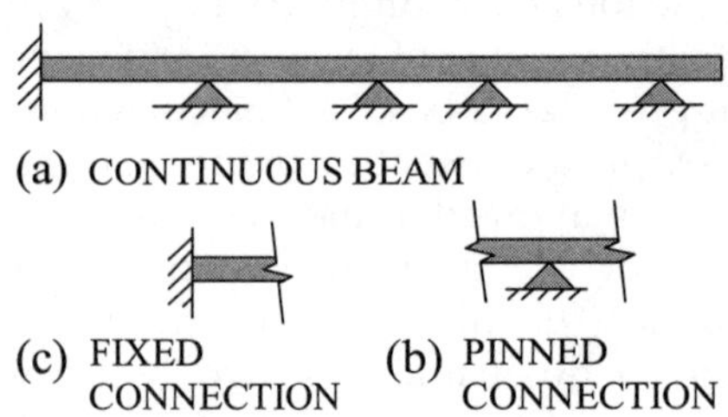

FIGURE 2.5 Beam reaction conditions.

1. The summation of forces in the horizontal and vertical directions equals zero:

$$\sum F_H = 0 \qquad \text{and} \qquad \sum F_V = 0 \tag{2.1}$$

2. The summation of moments about any point equals zero:

$$\sum M_a = 0 \tag{2.2}$$

A moment is simply a force multiplied by a distance from a point. A directional system has to be kept to simplify the thought process during the calculation. Assume that all vertical forces downward are positive and forces upward are negative, horizontal left is negative and horizontal right is positive, and moments causing rotation in a clockwise direction are positive and the opposite rotation is negative. By multiplying all forces in a system, including the reactions, by the distance from any point and adding them, the total will equal zero if the system is in equilibrium (not in motion).

Simple beams have two unknown forces, the reactions. One of the reactions can be solved for by summing the moments about a reaction because the distance to the reaction being summed about is zero and multiplication by zero causes the reaction to drop out of the equation. The easiest way to understand and learn this is to go through the calculations. Figure 2.6, sometimes referred to as a free-body diagram, shows a simple beam supporting some point loads and a distributed load. In this book we use the abbreviation k = kip = 1000 pounds. Also klf = kips per lineal foot. The summation of moments is taken about R_1, and R_2 is solved for. Once R_2 is known, the fact that the summation of vertical forces equals zero can be used to solve for R_1.

$$\sum M_{R_1} = 0$$

$$(R_1 \times 0) + (8 \times 6.9\text{ k}) + \left(\left(8 + 5 + \frac{6}{2}\right)(6 \times 10\text{ klf})\right)$$

$$+\ (24 \times 2\text{ k}) + (30 \times 15\text{ k}) - (24 \times R_2) = 0$$

$$55.2\text{ k}\cdot\text{ft} + 960\text{ k}\cdot\text{ft} + 48\text{ k}\cdot\text{ft} + 450\text{ k}\cdot\text{ft} = 24 \times R_2$$

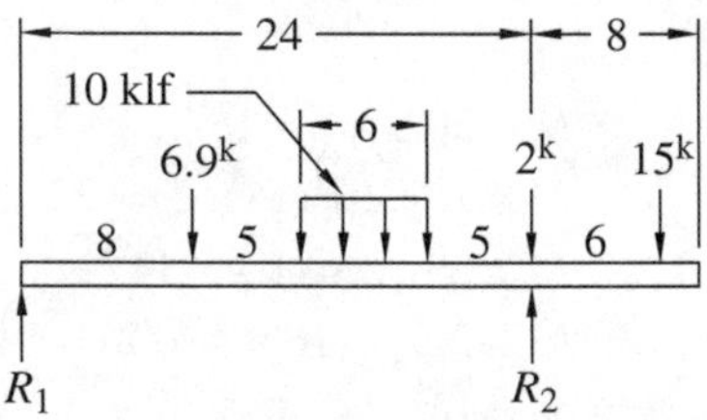

FIGURE 2.6 Simple beam free-body diagram.

$$\frac{1513.2}{24} = R_2$$

$$R_2 = 63 \text{ k}$$

$$\sum F_v = 0$$

$$6.9 \text{ k} + (6 \times 10 \text{ klf}) + 2 \text{ k} + 15 \text{ k} - 63 \text{ k} - R_1 = 0$$

$$R_1 = 20.9 \text{ k}$$

These calculations can be checked by calculating $\Sigma M_{R_2} = 0$ and getting $R_1 = 20.9$ k.

Shear and Moment Diagrams in Simple Beams

The purpose of solving beams is that the reactions can be designed to resist the loads and the beam can be designed to resist the internal shear and moments. The internal shear and moments change along the length of the beam. Usually the shear is largest at the reactions, and the moment is largest in the middle. Determining the exact location and quantity of the maximum shear and moment is required for the beam design. After calculation of the reactions for the beam shown in Fig. 2.6, a shear and moment diagram can be developed. Figure 2.7(b) shows a shear force diagram (VFD). Often V is used to denote shear. Figure 2.7(c) shows a moment force diagram (MFD). Both of these diagrams can be simply derived from the free-body diagram, seen in Fig. 2.7(a). To get the shear, plot the forces and moments shown on the free-body diagram. In Fig. 2.7(b):

Step 1 $R_1 = 20.9$ k
Step 2 20.9 k – 6.9 k = 14 k
Step 3 14 k – 10 k/ft for 6 ft = –46 k
Step 4 –46 k – 2 k = –48 k
Step 5 –48 k + 63 k = 15 k
Step 6 15 k – 15 k = 0

The maximum shear is –48 k at R_2 and 20.9 k at R_1.

To develop the moment diagram, plot the area of the shear diagram. In Fig. 2.7(c):

Step 1	at 8 ft	$M = 20.9 \times 8 = 167.2 \text{ k} \cdot \text{ft}$
Step 2	at 13 ft	$M = 167.2 + (14 \times 5) = 237.2 \text{ k} \cdot \text{ft}$
Step 3	at 14.4 ft	$M = 237.2 + (1.4 \times 14/2) = 247 \text{ k} \cdot \text{ft}$
Step 4	at 19 ft	$M = 247 - (46 \times 4.6/2) = 141.2 \text{ k} \cdot \text{ft}$
Step 5	at 24 ft	$M = 141.2 - (46 \times 5) = -90 \text{ k} \cdot \text{ft}$
Step 6	at 30 ft	$M = -90 + (15 \times 6) = 0 \text{ k} \cdot \text{ft}$

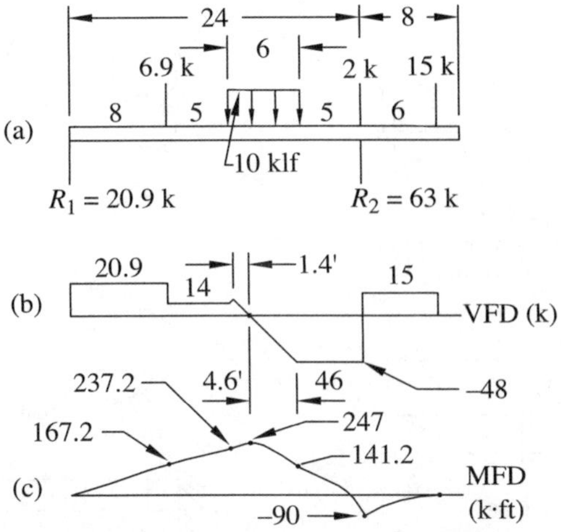

FIGURE 2.7 (a) Free-body diagram, (b) shear force diagram, and (c) moment force diagram.

The maximum moment is 247 k · ft at 14.4 ft from R_1, and the moment at R_2 = – 90 k · ft. Positive moments cause compression in the top fibers of the beam and tension in the bottom fibers, maximum at the surface and zero at the neutral axis (center of a rectangular or symmetric I beam). Negative moments such as the cantilever moment at R_2 cause tension in the top and compression in the bottom.

Although it is not normally done, a different beam section, e.g., tapered, could be designed for different locations along the beam based on the moment and shear at each location. If beams are to be notched or have holes cut through them, it is best to do so in the area between maximum moment and maximum shear, where the full moment and shear capacity are not needed. Also evident from these diagrams is that if additional load is going to be added to the beam, such as a temporary surcharge load from a crane pad, the best place to do so is toward the reactions because it will add the least amount of bending and shear to the beam,

Developing shear and moment diagrams checks the reaction calculations because they will not balance if the reactions have been calculated incorrectly; they provide a clear picture of everything that is going on inside the beam, and there is no mistake about the location of maximum shear and moment.

Given the moment of inertia and modulus of elasticity of the beam, the deflection diagram, or M/EI diagram, can also be derived by plotting the area of the moment diagram. However the mathematics starts to become tortured, and there are easier ways to approximate the deflection.

Most of the loads on shoring systems fall into the category of standard cases that have generalized formulas. Figure 2.8 is a collection of

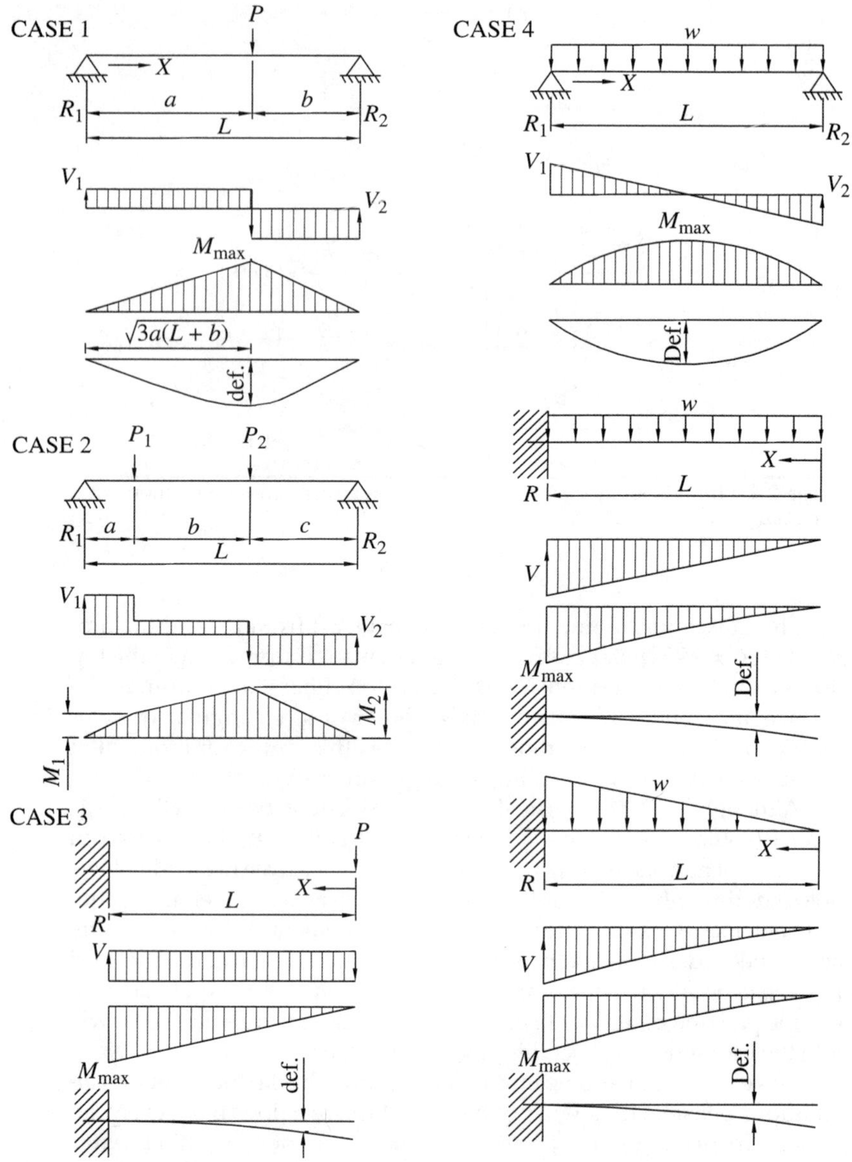

FIGURE 2.8 Simple beam diagrams.

standard beam formulas and diagrams. Beam loading can also be broken down into several different cases that fit the diagrams and then added together to get the total beam moments, shear, and deflection. There are also several excellent easy-to-use structural software programs that calculate and draw the diagrams for simple and indeterminate beams.

2.3 Beam Properties

The basic structural materials of shoring systems are beams made of wood, steel, and aluminum. These materials have three basic types of properties associated with their makeup: strength, material properties; shape, geometric properties; and weight. Wood is made of organic fiber and has different strength properties in the longitudinal and transverse directions. Steel and aluminum are mineral mixtures and have the same strength properties in both directions. Material strength properties are derived from testing. There are many different types of strength properties associated with a particular material; however, the basic properties of interest with structural design are resistance to bending, compression, tension, shear, and deflection. Specifically how large can these types of forces be before the material is loaded beyond its strength to resist? There are two ranges—elastic and plastic—that a material goes through before it breaks apart. If the beam supports the applied forces and returns to its original shape when the load is removed, it was loaded in the elastic range; if it returns, curved, bent, or deformed, it was in the plastic range; and if the material is broken, it exceeded the ultimate strength.

Steel and Aluminum Beams

The test to determine the strength range for steel and aluminum is to take a cylindrical rod and pull it until it starts to deform and then break. By measuring and recording the force divided by the area of the rod, the yield strength and the ultimate strength can be determined. These two strength values, yield strength F_y and ultimate strength F_u, are reported in units of force per unit area, thousands of pounds, or kilopounds (kips or k), per square inch (ksi).

For steel and aluminum all other design strengths are derived from F_y and F_u. Different metal mixtures have different strength properties. Specifications for the mixes and the test procedures are set by the American Society for Testing and Materials (ASTM) and are identified by ASTM numbers. Most steel plate and I beams used in shoring applications are ASTM A-36, min $F_y = 36$ ksi and $F_u = 58$ ksi, and A-572 grade 50, min $F_y = 50$ ksi and min $F_u = 65$ ksi; the most common today is A572. Steel tube is A500 grade B, $F_y = 46$ ksi and $F_u = 58$ ksi. Most common for structural steel pipe is A53, min $F_y = 35$ ksi and min $F_u = 60$ ksi. It is possible to purchase steel that tests higher than the minimum specified; in fact most are stronger because most steel today is recycled from scrap that usually has higher-strength steels in the mixture. If higher strength values are used in design, a mill certification for the material and strength should be acquired and the beams should have identification that ties them to the certification for the life of the beam. Lack of beam material identification usually causes it to be designed with the lowest strength or $F_y = 36$ ksi. It is possible at any time during the life of the beam to have the material tested and recertified. Just cut a small piece of the material, a "cookie," and send it to a laboratory.

There is one other important material property derived from the same test, the *modulus of elasticity E*, reported in units of pounds per square inch (psi) or kips per square inch (ksi). The modulus of elasticity is unit strength divided by unit deformation in the elastic range. For example, if the area of the rod is 1 in² and it is pulled by a 1000-lb force and the deflection is measured to be 0.0000349 in/in of total elongation through the elastic range, the modulus of elasticity is

$$E = \frac{1000 \text{ lb} / 1 \text{ in}^2}{0.0000349 \text{ in/in}} = 28{,}653{,}295 \text{ psi } (29{,}000 \text{ ksi})$$

This number $E = 29{,}000$ ksi is used for common structural steels and is independent of the yield strength. The modulus of elasticity is used in developing allowable design strengths for steel; however, the design engineer uses E mostly in deflection calculations. The weight of a piece of steel 1 in thick × 12 in × 12 in is 40.8 lb, and 1 ft³ of steel weighs 490 lb.

The specifying organization for steel is the American Institute of Steel Construction (AISC). They publish the *Manual of Steel Construction*, 10th edition, which contains the design specifications and all pertinent information about steel shapes manufactured in the United States. Any person with a career in excavation-related work should own or have access to this book.

Structural aluminum used for shoring is almost always 6061-T6, with tension yield stress $F_{ty} = 35$ ksi, tension ultimate stress $F_{tu} = 42$ ksi, and modulus of elasticity $E = 10{,}100$ ksi. Note that the yield strength for aluminum is close to that of steel, but since E is one-third that of steel, the deflection under the same conditions is three times that of steel. The allowable design strength for aluminum is somewhat less also due to the low E value. A piece of 6061 aluminum measuring 1 in × 12 in × 12 in weighs 14.11 lb, and 1 ft³ weighs 169 lb.

The specifying organization for aluminum is the Aluminum Association, Inc., and they publish design specifications and pertinent information about aluminum shapes in *The Aluminum Construction Manual*. If a person is going to do a lot of design work with aluminum through a career, this is also a must-have publication; however, the design specifications for aluminum, as well as for steel, can be found in building code publications.

Wood Beams

Wood has an elastic range that varies with duration, temperature, and moisture content. There is no test that determines the plastic range (over time under a load it will have a permanent set); it just breaks at some ultimate load. The strength varies with each type of force being applied. Conditions of use such as repetitiveness, time, temperature, and wet shape affect the strength. Strength variations in the material are also due to different species and grades within each species. There are several different grading agencies and grading rules that further complicate the certainty of the values. The approach to determining strength design values for wood is to test and publish allowable design values for each species and grade. For example, for Douglas fir larch, No. 1 and better, 2- to 4-in thick, visually graded dimension lumber bending $F_b = 1200$ psi and $E = 1{,}800{,}000$ psi. For all species and grades there are seven different design values depending on the force application. The modulus of elasticity for wood is roughly one-sixteenth that of steel, or more to the point, the deflection of the exact same beam shape and load will give 16 times more deflection. The allowable bending strength for wood is roughly one-twenty-fifth times that of steel. Douglas fir weighs approximately 2.6 lb per board foot (bd ft) (1 in × 12 in × 12 in), and 32 lb/ft^3. See Sec. 9.2.1 Timber Shoring Design by a Registered Engineer for a complete discussion of wood design for shoring applications.

The specifying organization for wood is the American Forest and Paper Association (AF&PA), and they publish *National Design Specification for Wood Construction*. These design specifications and some of the information in that publication can be found in the Uniform Building Code.

Geometric Properties of Beams

All beams that have the same shape, regardless of material, have the same geometric properties. Geometric properties are determined by cross-sectional shape, not length. Basic shapes of beams are rectangular, box (rectangular with a rectangular hole), round (solid or hollow), and web-flanged. Basic geometric properties of all these shapes are area, depth, width, and thickness. From the basic geometric properties the inertial properties—ability to resist bending and deflection, moment of inertia (I in^4), section modulus (S in^3), distance to extreme fiber (c in), and radius of gyration (r in)—can be calculated. Area is important to resisting axial forces. Distribution of area is important to resisting bending forces; moment of inertia, section modulus, and distance to extreme fiber are all related to bending strength. The radius of gyration is associated with length and buckling. Except for square and round sections the inertial properties are different in each major axis, referred as the x-x direction and y-y direction or strong and weak axes. In irregular shapes such as the L shape [Fig. 2.9(d)] and extruded nonsymmetric shapes, there is a third axis, z-z, of minimum moment of inertia.

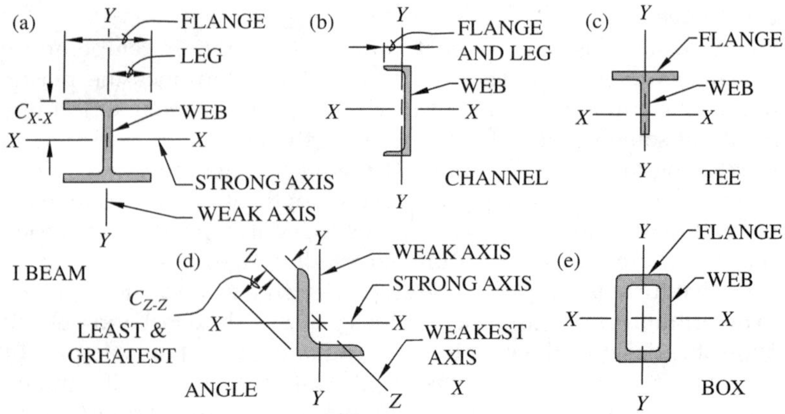

FIGURE 2.9 Beam shapes.

Web-flanged shapes are I beams, channels, and T shapes [Fig. 2.9(a) to (c)]. Usually the web is in the direction of the weak *y-y* axis and centered or close to it, and the flange is in the direction of the *x-x* axis and away from it. Bending is deflection away from the axis. In Fig. 2.9(a) bending about the *x-x* axis would mean that over the length of the beam it deflects up or down, and bending about the *y-y* axis means that it bends sideways to the left or right along the length as one looks at the end. The farther away from the *x-x* axis and the greater the area in the flanges, the stronger the beam is in bending about the axis. The flange is usually not connected on at least one end, and the portion not connected is referred as the leg. The web holds the flanges in place and carries the shear forces. The angle and box [Fig. 2.9(d) and (c)] can be visualized as a flanged beam, using the rule that the flange is the one that is the farthest away from the axis and the web is closest.

For rectangular shapes (Fig. 2.10), the geometric properties can be calculated as follows:

Area $A = bd$ in^2 (2.3)

Distance to extreme fiber $c = \dfrac{d}{2}$ in (2.4)

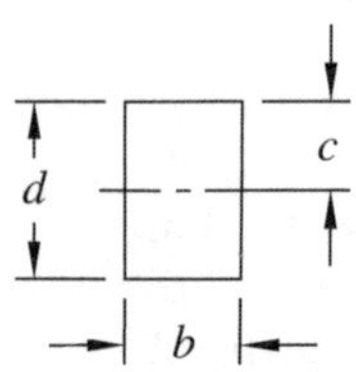

FIGURE 2.10 Rectangular beam.

Moment of inertia $$I = \frac{bd^3}{12} \quad \text{in}^4 \tag{2.5}$$

Section modulus $$S = \frac{I}{c} \quad \text{or} \quad S = \frac{bd^2}{6} \quad \text{in}^3 \tag{2.6}$$

Radius of gyration $$r = \sqrt{\frac{I_{x-x}}{A}} = \frac{d}{\sqrt{12}} \quad \text{in} \tag{2.7}$$

These formulas are all the geometric properties that are needed for wood design. For steel design all the needed geometric properties are calculated and tabulated in the AISC manual, and the same is done for aluminum in the *Aluminum Construction Manual.* None of the geometric properties for these materials are tabulated in building code manuals.

In steel and aluminum, some additional geometric issues are important. Compression parts of beams, specifically flange tips, can buckle before their yield strength is exceeded. The solution to this problem is to proportion them or brace them so that it does not happen before the allowable yield strength is exceeded. This is approached in different ways with steel and aluminum. In steel design beams are classified as compact, noncompact, and slender based on the flange being continuously connected to the web and the width thickness ratios of the flange compression elements, b/t ratio; depth to thickness, d/t; clear distance between flange and web thickness, h/t_w; depth to web thickness, d/t_w; and the outside diameter to thickness, D/t, for circular beams. See Table B5.1, Limiting Width-Thickness Ratios for Compression Elements, in the AISC manual. Practically all shapes listed in the manual in 36 and 50 ksi beams are compact. In aluminum design this is handled with buckling constants and slenderness ratios. Beams are manufactured with these ratios in mind. Manufactured steel beams are designed to be "efficient" by proportioning them to get the most strength from the least amount of weight. In structure design beams are first sized for basic resistance to primary forces, axial or buckling based on length, bending, shear, and deflection, and then checked for localized problems with stability, flange bracing, and web thickness. In preliminary design the way to get around the secondary problems is to pick a beam that is one to two sizes larger than indicated by design. For instance, if the designer is looking for a 14-in-deep I beam for bending purposes with a section modulus of 135 in^3, the closest beam to that is a W14 × 90 with $S = 143$ in^3; a W14 × 99 with $S = 157$ in^3 is most likely to be adequate for shear, braced flange spacing, and end bearing. Sizing up one or two sizes also puts an extra factor of safety on the original design assumptions. Final design of beams should always be checked by a registered engineer.

2.4 Factor of Safety

The term *factor of safety* can be quite confusing when one is looking at the results of a design process. It is generally thought of as some sort of reserve capacity. If a person needs 5 gal of gas to drive to a destination and has 10 gal of gas in the tank, the factor of safety is 2. The other 5 gal of gas covers the possibility that the person might get lost on the way, get caught in a traffic jam that burns up extra gas, have engine problems that burn gas faster than expected, or encounter a multitude of other predictable problems. It is still possible to run out of gas before the person gets there. If the driver is just going out to lunch, the factor of safety against failure of getting there is totally adequate; and if the trip is to the hospital for imminent delivery of a baby, the factor of safety may not be enough. A full tank of gas, say, a factor of safety of 4, would not be enough to protect against the extreme possibility of getting a hole in the gas tank. The factor of safety says something about the overall possibility of getting there but nothing about the possibility of any one of the particular events listed happening.

In engineering a primary distinction relates to whether the factor is applied to the material, the loads, or both. Two different design philosophies stem from this—*allowable strength design* (ASD) and *load factor and resistance design* (LFRD).

In ASD the safety factors are applied to the material strength properties, and the loads are not usually factored or a factor of 1 is used. In construction design a factor of 1.5 is usually considered a minimum, so that the failure material stress is at least 1.5 times greater than the calculated stress. In LFRD, design factors greater than 1 are applied to loads depending on predictability, and resistance factors less than 1 are applied to material strength. When the factored loads are less than the factored load resistance, the structure is said to be safe.

Both ASD and LFRD yield similar results, and most of the time they yield the same results. ASD has been around longer and is generally preferred for steel, timber, and aluminum, mostly because that is the way it was taught in colleges. Concrete design uses LFRD. LFRD is thought to produce a more efficient design, is now the primary method taught in college, and is used as the primary design method outside the United States. This book uses ASD because it involves less factoring and applies to the material side only. On the load side the minimum loads are often specified or easily calculated.

2.5 Allowable Stress Design

Applied force causes stress. In our lives we talk about it in an abstract and unquantified way until it becomes too much, at which point we say we are stressed out and some people snap mentally under the pressure. In engineering we define and quantify it. Stress is force applied to an area. The Greek letter σ (sigma) is generally used to denote stress, and it is defined as

$$\sigma = \frac{P}{A} \tag{2.8}$$

where σ = stress, psi, psf
P = force, lb, k
A = area, in^2, ft^2

In structural engineering stress is usually denoted by a capital F or lowercase f with a subscript denoting the type of stress it is. The capital F is used when the stress is stated, and the small f when it is the result of a calculation. For example, the allowable bending stress for a particular beam may be stated to be F_b = 20 ksi, and the calculated stress would be $f_b = x$ ksi. The following nomenclature for allowable stress is used in steel and aluminum design:

f computed stress, ksi
F_a axial compressive stress, ksi
F_b allowable bending stress, ksi
F_p allowable bearing stress, ksi
F_t allowable tensile stress, ksi
F_v allowable shear stress, ksi
F_y minimum yield strength, ksi
F_u minimum breaking strength, ksi

Similar nomenclature is used in wood construction except the units are pounds per square inch.

Allowable stress design is a three-step process:

1. Develop the loads and structure configuration (free-body diagram). The beam loading is intended to be what is anticipated or reasonably expected without factors of safety cranked in. Maximum wind loads, traffic wheel loads, etc. are expected and predicted to occur during the life of the structure; perfect storms or oversize loads on bridges are not expected and are in the realm of the factor of safety. As is evidenced by structural failures in well-designed structures, the system is not perfect. Every design situation is different, and it is the responsibility of the designer to root out worst-case loading conditions.
2. Determine the allowable stress based on some factor of safety as a percentage of the yield stress F_y or ultimate strength F_u. In ductile materials F_y is usually used, and where brittle failure would be expected such as with bolts and cables the factor is based on F_u. The allowable stress is the maximum anticipated

stress after considering all possible load cases. Material strength variations, generalizations in the model, calculation accuracies, load variations, variations in natural forces such as earthquake and wind, load duration, and consequences of failure are some of the considerations that go into developing the allowable stresses. The allowable stress specifications are usually developed by the associations dedicated to promoting the particular material and then adopted or further altered by specifying agencies and engineers. Safety and efficient use of the material are most at odds during the process.

3. Calculate the stresses in the beam and compare them to the allowable stress in the beam.

$$f_{\text{calculated}} \leq F_{\text{allowable}}$$

The process is often iterative, or repetitive, by rearranging the loading configuration or the beam size until the calculated stress is close to the allowable so that the lightest-weight and least expensive beam is found.

Calculated Stress in Beams

The following formulas are used to compute each different type of stress in beams:

Bearing stress

$$f_p = \frac{P}{A} \tag{2.9}$$

where P = axial load and A = cross-sectional area to which the axial load is applied.

Bending stress at extreme fiber

$$f_b = \frac{Mc}{I} = \frac{M}{S} \tag{2.10}$$

where M = bending moment
c = distance to extreme fiber
I = moment of inertia
S = section modulus

Average vertical shear

$$f_v = \frac{V}{A} = \frac{V}{dt_w} \tag{2.11}$$

where V = shear
A = section area for solid shapes
d = beam depth
t_w = web thickness for shaped beams

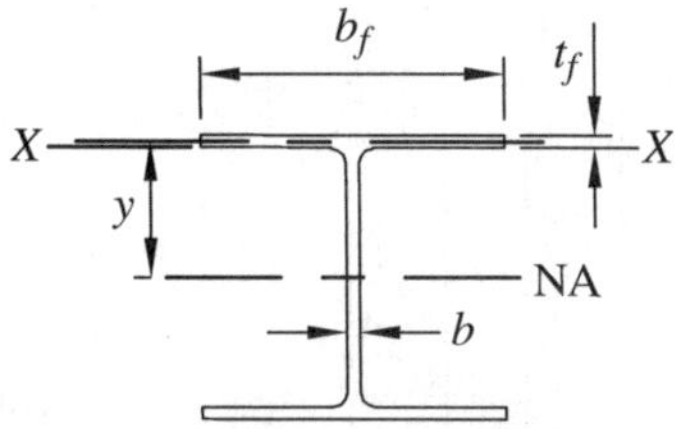

$Q_{x\text{-}x}$ = FIRST MOMENT OF AREA ABOUT x-x

$Q_{x\text{-}x} = b_f \times t_f \times y$

y = DISTANCE FROM NA TO CG OF $b_f t_f$

FIGURE 2.11 First moment of area.

Horizontal shearing stress at any section (see Fig. 2.11)

$$f_{vh} = \frac{VQ}{Ib} \tag{2.12}$$

where V = shear
Q = moment of area outside section about neutral axis
I = moment of inertia
b = width of section

Calculated Stress and Buckling in Columns

Columns are beams with the forces concentrated perpendicular to the long axis of the beam. If the force is centered on the center of the section, it is concentrically loaded; and if it is not, it is eccentrically loaded. The distance away from the center is the eccentricity e, and the eccentricity times the force is the eccentric moment $P \times e$.

Column stress

$$f_a = \frac{P}{A} + \frac{Mc}{I} = \frac{P}{A}\left(I + \frac{ec}{r^2}\right) \tag{2.13}$$

where P = axial load
A = area
$M = Pe$
e = eccentricity
c = distance to extreme fiber
I = moment of inertia
r = radius of gyration

Stability of columns is based on ability to resist buckling. In addition to cross-sectional properties, buckling resistance varies with the length and the end condition. To reduce the length/end condition variable

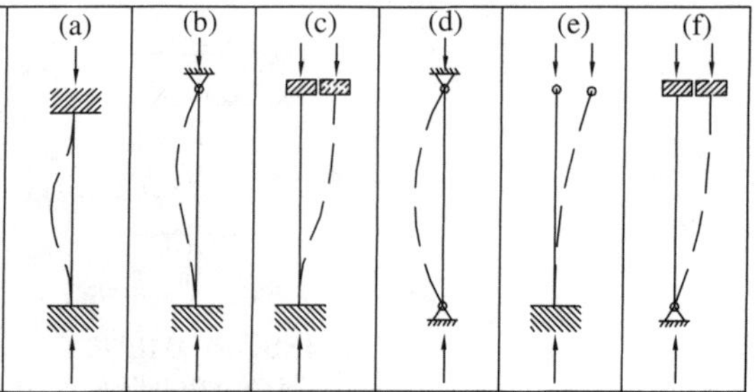

	(a)	(b)	(c)	(d)	(e)	(f)
Buckled shape of column is shown by dashed line						
Theoretical k value	0.5	0.7	1.0	1.0	2.0	2.0
Recommended design value when ideal conditions are approximated	0.65	0.80	1.2	1.0	2.10	2.0

End condition code	
	Rotation fixed and translation fixed
	Rotation free and translation fixed
	Rotation fixed and translation fixed
	Rotation free and translation fixed

FIGURE 2.12 Column effective.

to one item, the concept of effective length is used. Figure 2.12 gives K values to be multiplied by the column length to obtain an effective length.

2.6 Construction Engineering Design

Following the three-step process outlined above, a structural system to support temporary loads encountered during the construction process can be developed. Subsequent chapters in this book discuss in detail all aspects of the process and use it as a basis for understanding excavation work.

Here we will look at the mechanics of the process. There are two basic types of problem:

1. Defined loading. The geometric and loading conditions are developed and the elements are designed, and sometimes the word *proportioned* is used, to resist the loads. Usually bending is the primary consideration, so the beam is selected by determining a required section modulus and then other stresses are checked.

Example In the case of the loaded beam in Fig. 2.6, the maximum shear and moment were determined to be $V = 48$ k and $M = 247$ k·ft. A steel beam with $F_y = 50$ ksi is to be designed to resist the forces. A 14-in-depth beam is desired due to availability and spatial considerations.

Step 1 Determine the allowable stress. In the final design, check AISC design specifications. For preliminary design using steel, a good starting point for allowable stresses is

$$F_b = 0.60F_y = 0.60 \times 50 = 30 \text{ ksi}$$

$$F_v = 0.40F_y = 0.40 \times 50 = 20 \text{ ksi}$$

Step 2 Solve for section modulus, using

$$f_b = \frac{M}{S}$$

Solve for S. $$S = \frac{M}{F_b} = \frac{247 \text{ k}\cdot\text{ft} \times 12 \; in/1\,ft}{30 \text{ ksi}} = 98.8 \text{ in}^3$$

Step 3 Go to a table of beam properties (Fig. 2.13), and find a beam with a section modulus greater than 98.8 in^3. A W14 × 68 has the following properties:

$$S = 103 \text{ in}^3 \qquad d = 14.04 \text{ in} \qquad \text{and} \qquad t_w = 0.415 \text{ in}$$

Step 4 Check the shear, using

$$f_v = \frac{V}{A} = \frac{V}{dt_w}$$

$$f_v = \frac{48 \text{ k}}{14.04 \text{ in} \times 0.415 \text{ in}} = 8.23 \text{ ksi} \qquad \text{OK}$$

Step 5 Check other design specification requirements. In the estimating or preliminary planning process this may be as far as the design goes. The beam weight can be used to estimate the cost of purchase or rent and the minimum beam section and weight can be used to find available beams. In the final design the beam is checked for all design requirements such as deflection, allowable flange brace length, end bearing, etc. The reaction forces and beam bearing geometry can be used to design the support.

2. Defined elements. The strength of specific elements is given, and then they are positioned and loaded so that their strength is not exceeded. The objective is to determine how much load the element can withstand.

Example A shoring shield wall panel can be considered as a long flat beam. An 8-ft-deep × 20-ft-long shoring shield has a section modulus of 190 in^3. The material is 55 ksi steel. Find the allowable loading in pounds per square foot.

Step 1 Use the allowable bending strength.

$$F_b = 0.6F_y = 0.6 \times 55 \text{ ksi} = 33 \text{ ksi}$$

Step 2 Determine the maximum allowable moment, using

$$f_b = \frac{M}{S}$$

$$M = SF_b = 190 \text{ in}^3 \times 33 \text{ ksi} \times \frac{1 \text{ ft}}{12 \text{ in}} = 522 \text{ k}\cdot\text{ft}$$

Step 3 Find the distributed loading that results in a 522 k·ft moment.
using $M = \frac{Wl^2}{8}$ from Fig. 2.8 beam-loading formulas, distributed load

$$W = \frac{8M}{\ell^2} = \frac{8 \times 522\ \text{k}\cdot\text{ft}}{20^2\ \text{ft}} = 10.44\ \text{klf}$$

And since the shield wall is 8 ft deep,

$$w = \frac{10{,}440\ \text{plf}}{8\ \text{ft}} = 1305\ \text{psf}$$

Step 4 The remaining elements of the shield, end beam, and struts would be designed around this loading.

In shoring design this method is used to design and develop tabulated data for shoring equipment. The psf load is further calculated into allowable depth. In the field the competent person determines the depth and soil condition and then looks for a shield panel with that capacity. An engineered shoring design would develop a soil loading in pounds per square foot and look for a shield with that capacity.

2.7 Steel I Beams Used for Shoring

Steel I beam is used in shoring for piles, wales, and struts on pile and lagging shoring and for wales and struts on sheet pile shoring. Figure 2.13 shows two common beam categories used in the industry, W14 and HP14. Of these the W14 × 120, W14 × 90, and HP14 × 117 are most common in the inventories of shoring suppliers and contractors because of the following:

- Deeper beams take up workspace and increase the excavation quantity.
- Normal strut spacing is 18 to 22 ft because typical pipe lengths are 16 and 20 ft. These beams are very efficient with shoring loads in this range.
- These beams are well suited to bridge false work spans and traffic clearance requirements.
- The beam strength and flange width work well for H-pile and lagging systems. Both 14-in W and HP shapes will fit into a 24-in round drilled hole.

The basic difference between the W and HP shapes is that the W was developed for bending and the H was developed for bearing or axial loads, hence the designation HP for H-pile. One can be substituted for the other as long as the required design parameters are met; however, the W is more efficient (strength compared to weight) for wales, and the HP is more efficient for struts. In the field the HP is distinguishable from the W because the flanges and web have practically

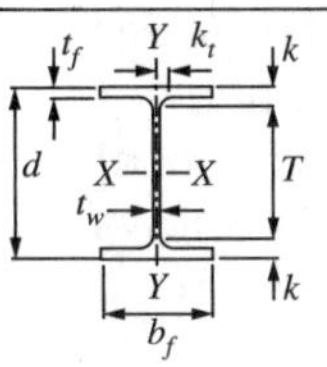

W Shape

Dimensions

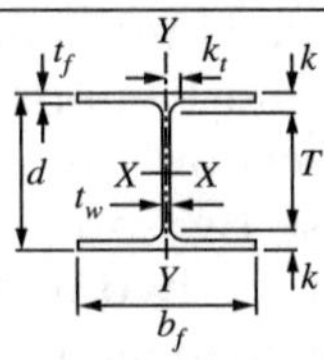

HP Shape

Shape	Area A	Depth d		Web: Thickness t_w		Web: $\frac{t_w}{2}$	Flange: Width b_f		Flange: Thickness t_f		Distance: k k_{des}	Distance: k k_{det}	Distance: K_1	Distance: T	Distance: Workable gage
	in²	in		in		in	in		in		in	in	in	in	in
W14×132	38.8	14.7	$14\frac{5}{8}$	0.645	$\frac{5}{8}$	5/16	14.7	14 3/4	1.03	1	1.63	2 5/16	1 9/16	10	5 1/2
×120	35.3	14.5	$14\frac{1}{2}$	0.590	9/16	5/16	14.7	14 5/8	0.944	15/16	1.54	2 4/16	1 1/2	↓	↓
×109	32.0	14.3	$14\frac{3}{8}$	0.525	1/2	1/4	14.6	14 5/8	0.860	7/8	1.46	2 3/16	1 1/2	↓	↓
×99	29.1	14.2	$14\frac{1}{8}$	0.485	1/2	1/4	14.6	14 5/8	0.780	3/4	1.38	2 1/16	1 7/16	↓	↓
×90	26.5	14.0	14	0.440	7/16	1/4	14.5	14 1/2	0.710	11/16	1.31	2	1 7/16	↓	↓
W14×82	24.0	14.3	$14\frac{1}{4}$	0.510	1/2	1/4	10.1	10 1/8	0.855	7/8	1.45	1 11/16	1 1/16	10 7/8	5 1/2
×74	21.8	14.2	$14\frac{1}{8}$	0.450	7/16	1/4	10.1	10 1/8	0.785	13/16	1.38	1 5/8	1 1/16	↓	↓
×68	20.0	14.0	14	0.415	7/16	1/4	10.0	10	0.720	3/4	1.31	1 9/16	1 1/16	↓	↓
×61	17.9	13.9	$13\frac{7}{8}$	0.375	3/8	3/16	10.0	10	0.645	7/8	1.24	$1\frac{1}{2}$	1	↓	↓
HP14×117	34.4	14.2	$13\frac{1}{4}$	0.805	13/16	7/16	14.9	14 7/8	0.855	13/16	$1\frac{1}{2}$		1 1/16	11 1/4	5 1/2
×102	30	14.0	14	0.705	11/16	3/8	14.8	14 3/4	0.785	6/8	1 3/8		1	↓	↓
×89	26.1	13.8	$13\frac{7}{8}$	0.615	5/8	5/16	14.7	14 3/4	0.72	5/8	1 5/16		15/16	↓	↓
×73	21.4	13.6	$13\frac{5}{8}$	0.505	1/2	1/4	14.6	14 5/8	0.45	1/2	1 3/16		7/8	↓	↓

Properties

Shape	Compact Section Criteria $\frac{b_f}{2t_f}$	Compact Section Criteria $\frac{h}{t_w}$	Axis X-X I	Axis X-X S	Axis X-X r	Axis X-X Z	Axis Y-Y I	Axis Y-Y S	Axis Y-Y r	Axis Y-Y Z	r_{ts}	h_o	Torsional Properties J	Torsional Properties C_w
			in⁴	in³	in	in	in⁴	in³	in	in	in	in	in⁴	in⁶
W14×132	7.15	17.7	1530	209	6.28	234	548	74.5	3.76	113	4.23	13.6	12.3	25,500
×120	7.8	19.3	1380	190	6.24	212	495	67.5	3.74	102	4.2	13.5	9.37	22,700
×109	8.49	21.7	1240	173	6.22	192	447	61.2	3.73	92.7	4.17	13.5	7.12	20,200
×99	9.34	23.5	1110	157	6.17	173	402	55.2	3.71	83.6	4.14	13.4	5.37	18,000
×90	10.2	25.9	999	143	6.14	157	362	49.9	3.7	75.6	4.11	13.3	4.06	16,000
W14×82	5.92	22.4	881	123	6.05	139	148	29.3	2.48	44.8	2.85	13.5	5.07	6,710
×74	6.41	25.4	795	112	6.04	126	134	26.6	2.48	40.5	2.82	13.4	3.97	5,990
×68	6.97	27.5	722	103	6.01	115	121	24.2	2.46	36.9	2.8	13.3	3.01	5,380
×61	7.75	30.4	640	92.1	5.98	102	107	21.5	2.45	32.8	2.78	13.2	2.19	4,710
HP14×117	9.25	14.2	1220	172	5.96	194	443	59.5	3.59	91.4	4.15	13.41	8.02	19,900
×102	10.5	16.2	1050	150	5.92	169	380	51.4	3.56	78.8	4.1	13.31	5.39	16,800
×89	11.9	18.5	904	131	5.88	146	326	44.3	3.53	67.7	4.05	13.22	3.59	14,200
×73	14.4	22.6	729	107	5.84	118	261	35.8	3.49	54.6	4	13.11	2.01	11,200

FIGURE 2.13 Steel Beam dimensions and properties for W14 and HP14. (Redrawn from *AISC Steel Construction Manual*, 13th ed., AISC, Chicago)

the same thickness and the depth and flange width are close to the same. The W shape has a web thinner than the flange, and the flange width can be substantially less than the depth. Examination of the two beams and dimensions in Fig. 2.13 will demonstrate this.

For bending, the beam property that is designed for first is the section modulus, and then shear and deflection are checked. Note that in Fig. 2.13 the W14 × 109 has $S_{x\text{-}x} = 173$ in³ and the HP 14 × 117 has $S_{x\text{-}x} = 172$ in³, essentially the same. The W14 × 109 shape is more efficient because it weighs 8 lb/ft less. For shear the most important dimension is the web thickness t_w, and for deflection the most important property is $I_{x\text{-}x}$. Comparison of these two properties for the W14 × 109 and HP14 × 117 shows that the two beams are quite similar. These two beams are used interchangeably for wales.

When one is designing axial members, struts, the controlling properties are area and radius of gyration $r_{y\text{-}y}$. The W14 × 109 has $A = 32$ in^2 and $r_{y\text{-}y} = 3.73$ in, and the HP 14 × 117 has $A = 34.4$ in^2 and $r_{y\text{-}y} = 3.59$ in. Both are similar but the HP is more efficient because it has the greatest area.

Beam substitution is quite common in shoring design because of the use of used beams in existing inventory. For the estimator and preliminary design, beams should be compared on the basis of the properties discussed above. The preliminary calculations should determine area or section modulus required, and at that point beams can be searched for and priced. In the final design all AISC design parameters should be checked.

References

The Aluminum Association, Inc., *Engineering Data for Aluminum Structures, Construction Manual Series, Section 3,* 5th ed., Washington, 1986.

The Aluminum Association, Inc., *Specifications for Aluminum Structures, Construction Manual Series, Section 1,* 5th ed., Washington, 1986.

American Forest & Paper Association, *National Design Specifications for Wood Construction,* ANSI/AF&PA NDS-1997, American National Standard, Madison, Wis., 1997.

American Institute of Steel Construction, Inc., *Allowable Stress Design,* 9th ed., 2d rev., AISC, Chicago, 1995.

Beer, Ferdinand P., and Johnston, E. Russell, Jr., *Vector Mechanics for Engineers: Statics and Dynamics,* McGraw-Hill, New York, 1972.

McCormac, Jack C., *Structural Steel Design,* 2d ed., International Textbook Company, New York, 1971.

Parker, Harry, *Simplified Mechanics and Strength of Materials,* John Wiley & Sons, Inc., New York, 1951.

Sharp, Maurice L., *Behavior and Design of Aluminum Structures,* McGraw-Hill, New York, 1993.

CHAPTER 3

Excavation Work Planning

3.1 Introduction

There is a process that every project must go through in order for the concept to become a reality. In the long term, the cradle-to-grave progression happens whether anyone plans it or not. Left to chance alone, there is little likelihood that the final product will wind up as it was envisioned. The planning process is about controlling the outcome. In the excavation planning process there are two major players, the project design engineer and the contractor, each with a different responsibility in the process and, believe it or not, the same overall objective—to achieve an excavation that ensures three things will result:

- Proper construction of the final product
- Protection of existing buried and aboveground facilities
- Worker protection

Both players are entwined in the process and have a legal and financial stake in the success or failure of the outcome. The relationship is complicated because the actions of one player can affect the success or failure of the other. This chapter examines the responsibilities of the major players and how the excavation planning process can ensure a successful outcome for both.

3.2 Excavation Planning Responsibilities of the Design Engineer

For the project design engineer, the excavation work focus on the project falls into three categories: temporary excavations, structure foundation stability considerations, and stability of final grades including soil retainment structures. To a large extent the means and methods for temporary excavation work are the responsibility of the contractor while the other two categories are tightly controlled by the design engineer; and as a result, planning for temporary excavation work never gets as much attention as the other two. In a perfect world for the design engineer, the contractor would commission his or her own engineering design firm to do the complete investigation and design for the excavation and shoring project. Time, cost, and overlapping requirements from interested parties make this impossible. The contractor usually comes into the picture after the final project design is completed and has no contractual relationship with surrounding property owners, utility owners, government agencies, and the project soils engineer (nor do they have an obligation to the contractor). Thus, the design engineer is forced to develop, and passes on to the contractor, information about existing aboveground and belowground facilities and the soils. The design engineer needs to develop this information for her or his own purposes anyway. The existing facilities information shows up in the plans and specifications, and the soils information is contained in the geotechnical report. The contractor relies on this information and bases the bid and shoring design assumptions on its being complete and accurate. It is the contractor's reliance on this information that is the major source of lawsuits between the contractor and the design engineer.

The basic function of the design engineer in the excavation process is not to excavate but to generate information and a plan that the contractor can use to perform the excavation work. The following are the major steps that the project design engineer goes through to complete the excavation aspect of the project. The discussion of each is for the purpose of identifying problems related to the contractor's reliance on the information.

1. *Commission a soil investigation.* The geotechnical engineer will only deliver the information that he or she is commissioned to develop. The exploration should be extensive enough in depth and location to develop shoring alternatives. Recommendations regarding dewatering and shoring design should be a part of that commission and should include identification of contaminated soils, dewater levels and prospects for success of dewatering, potential for bottom heave and boiling, expected and allowable settlement and ground movement, requirements for underpinning, allowable sloping, allowable shoring systems, and removal of shoring systems.

   Most lawsuits generated around soils reports stem from the contractor's perception that conditions are not as shown in the boring logs and, as discussed above, were the only reliable information at bid time. The concept that the soils report represents a bandwidth of information, and that within it many possible soil conditions can result, should be reinforced in the report and specifications. The owner and contractor should understand that outside that bandwidth changed condition compensation is due to either the owner or the contractor.

One approach to solving the reliance problem has been to require that, prior to developing excavation plans, the contractor obtain the services of his or her own geotechnical engineer for the purpose of developing shoring design parameters. There can be problems with this approach. The original geotechnical investigation still needs to address temporary excavation work because contractors need the information to bid the project. The second report can be very time-consuming and expensive; just look at the time and money that have been spent on the original project geotechnical report. If conflicting information is generated, the project engineer has to rely on her or his expert, thereby rendering the contractor's geotechnical engineer null and void anyway.

Another problem that can arise with geotechnical information, especially on pipeline projects, is that there are most likely other geotechnical reports already generated from the area which may conflict with the report commissioned by the owner. If the design engineer is aware of these reports, it should be made clear in the specifications that they exist so that there cannot be a claim that the owner held back information that would be costly to the project.

2. *Perform existing subsurface utility investigation.* One of the prerequisites for this investigation to take place is that the design (project) engineer (DE) develop a working relationship with every party that has a stake in the outcome of the excavation. At the end of the process the DE must pass on certain aspects of that relationship to the contractor so that he or she can continue it through the construction phase of the project. Again, note that the stakeholder's connection is to the land and the owner of the project, and not to the contractor who comes into the game later in the process. In large part the owner's obligation to protect the interests of these stakeholders is what motivates the design engineer to pay so much attention to the contractor's excavation plan. The contractor and shoring designer should understand that part of the reason for developing the excavation plan, and the review and acceptance process, is more than a power play by the project engineer. The project engineer has a commitment, and if there is a failure, she or he will be liable to the stakeholders.

Another aspect of the subsurface investigation is to prioritize and clarify critical portions of the work. Most of the location verification work can be performed during the construction phase, but some of it has to be done prior to final design. It makes no sense to design a structure or pipeline in a location if it might have to be moved or altered later due to obstructions. Hazardous pipelines, critical water and power supply lines, and fire and emergency vehicle access should be identified. A full assessment of the risks and requirements surrounding these critical items must be conveyed to the contractor so that the gravity of the situation is unmistakeably clear. A decision must be made about which subsurface structures and lines must be exactly located below the surface prior to contract award. In a competitive bid atmosphere the contractor assumes that all subsurface situations are like the ones seen every day on similar projects unless they are highlighted in some way. If they are flagged, the contractor needs information to help decide if it is a 1 or a 10 on the scale of risk and associated cost. Tendencies to overkill on this issue by declaring everything critical just blur the picture. Trying to push this critical aspect of safety onto the contractor invites accusation that the owner is trying to get something for nothing. The project engineer is in the unique position to develop and prioritize the risks and has a duty to convey that information to the contractor.

3. *Structure or utility design.* The critical issue here is that the design structure–soil interaction has to be preserved during the excavation process. The contractor's excavation plan has to provide sufficient dewatering and a solid compact base that lasts through the course of construction. Two other issues are (1) provision of enough working space for proper construction of the structure or utility and (2) removal of shoring so that it does not damage the final product. These are quality control issues for the design engineer, and the DE's acceptance of the excavation plan indicates that she or he believes that they will be met.

4. *Specify excavation methods and alternatives*. Based on the results of the above steps, decisions are made about what type of excavation and shoring systems will fulfill the requirements and will be allowed for use by the contractor. If open cut is possible but the risk of settlement or easement encroachment is too great, it may be eliminated. Risk of bottom heave or boiling may preclude the use of shields, slide rail, and H-pile and lagging. The more excavation systems that are eliminated, the more costly the excavation work will be. The main reason for doing this instead of allowing the contractor to make these decisions is so that they can be rejected without

review and the cost of the more expensive systems is contained in the bid. The contractor's theory on this issue is that if his or her approach to shoring is not precluded at bid time, then the contractor should have the opportunity to submit it and try and make it work even if it defies geotechnical and engineering logic. There is a point to this argument that is hard to refute. The contractor is the expert in excavation shoring and has experience and knowledge that may be better than the theoretical information that the engineers use. The contractor should have full control of means and methods. With a large excavator and a rapid excavation process, the contractor may be able to beat the inevitable collapse of vertical walls and the effects of bottom heave by installing a shoring shield and structure base in a 25-ft-deep excavation when soil mechanics says it is not possible. Contractors are ingenious, and despite extreme failure at the start they can usually adjust their operations to be successful and defeat the logic of the engineers. To the extent that it provides a cost-effective project to the owner and a profitable one to the contractor, this approach should be allowed. Elimination of alternatives stifles innovation and runs up cost. Where extreme unrepairable risk to life and property is concerned, this method should be eliminated prior to bid.

5. *Postaward planning.* This planning involves two major items.
   a. *Following through on the subsurface identification and exploration work.* This is the time for a complete and thorough handover to the contractor of all stakeholder information, contacts, and requirements. Planning for surface location of subsurface utilities and contract-required potholing is generally a separate operation from the production excavation work. A clear distinction of the work that must be performed prior to the start of production excavation and what will be performed in conjunction with the production operation should be made. Exact location of hazardous lines should not be part of the production work because it does not receive the attention and time required to do it safely. If these locations are properly flagged on the plans and in the specifications, the result will be money in the bid for them and time and a focused crew on the job during construction.
   b. *Development and review of the excavation plan.* This is the point where the contractor invariably ends up asking these questions: "Since I am the only one who is financially responsible for the outcome of the excavation work, why does the engineer get to say what I can and cannot do in the plan? Why do I have to prove to him ahead of time that it will work?" For the reasons discussed above, the fact is that the engineer

has a vested interest in success of the plan, and failure of the plan is also failure for the engineer. The engineer is aggressively protecting his or her interests and is providing insurance for success, and the cost to the contractor is definitely not an overriding factor in this pursuit. This is the nature of the relationship, and it sets the tone for the process. Problems and litigation between the contractor and the engineer on this issue can stem from several sources.

The engineer has contracted with an expert in the field of soils and excavation work, the geotechnical engineer. Just as a general practitioner would not override the opinions of a heart specialist, the project engineer cannot ignore the recommendations of the soils engineer. On the issue of interpreting the soils and applicability of shoring systems, the project engineer is just a messenger between the contractor and the soils engineer. The contractor must have a clear understanding at bid time that what is stated in the soils report and required in the contract documents mean what they say. Therefore it is imperative that the project engineer use unambiguous language in the contract, ensuring that the geotechnical report requirements will be met. Should the project engineer fail at this he or she will be caught in between with no leverage on the contractor or soils engineer to get consensus on an excavation plan.

The concept of acceptance or approval of the excavation plan is not clearly or legally defined. The DE is concerned that her or his approval of the excavation plan will be seen as confirmation that everything in the plan is letter-perfect. Because of the DE's professional degree and central position in the project, the DE is often perceived in comparison to the contractor and his engineer as the more responsible party. The objective of the review should be to establish the following:

- The plan was prepared by experienced qualified professionals.
- The geotechnical requirements are met.
- Contract requirements are met.
- Engineering design standards are met.
- The plan is constructible and can be followed.
- Safety issues are addressed.
- Stakeholders are protected.
- Quality control is ensured.

Complete separate and parallel design calculations for the shoring option submitted or line-by-line editing of the submitted calculations is not necessarily a requirement for proper review. However, engineering judgment based on risk and safety has to be exercised in determining the depth of the review. The name and reputation of the person approving the plan are definitely in the mix; however, he or she is not the engineer of record for the plan, the contractor is. The author has seen stonewalling and outright refusal of some reviewing engineers to attach their name to a review of something that was not prepared under their supervision and control. Those with the "what's in it for me to accept this risk" attitude should be left out of the process.

6. *Construction.* The plan must be followed or revised. During construction the excavation work is completely under the control of the contractor. Good faith, consideration for the design engineer, and integrity are the only things that will cause the contractor to follow the plan. To complicate matters, it is nearly impossible to exactly plan every step. The plan is always in transition and should be viewed as such. The progression is this: follow the plan, adjust for what does not work, change the operation, review to make sure it conforms to the original intent, involve the shoring design engineer, and if changes are significant, redraw the plan. The contractor should have the freedom to change as necessary as long as the changes meet the contract requirements. Unlike a structure or a pipeline, the shoring system has no exact form and will not exist at the end of the project. The project engineer's interference with the contractor's ability to hone his or her means and methods can cause a lot of friction on a project, as can a contractor's disregard for the stake that the engineer has in success of the project. Again, this is part of the nature of the relationship between the contractor and the engineer.

3.3 Excavation Planning Responsibilities of the Contractor

On the contractor's side of the equation, the list of objectives is added to and prioritized in a slightly different manner. The overall objective is to achieve an excavation that ensures four things will result:

- Worker protection
- Profit
- Protection of existing buried and aboveground facilities
- An excavation that allows proper construction of the final product

For the contractor, failure on the first two is not an option, as it will quickly put him or her out of business. On the objective of worker protection, everyone involved in the excavation process agrees on the goal—absolute and unequivocal zero tolerance for death and injury in excavation work. The day when the idea that progress and furtherance of the public well-being outweighed loss of life in the process is certainly gone. Technologically we have the means to achieve this goal. On the issue of profit and how it affects this goal, here is a concept that should be nurtured and driven into the heads of the project owner, the project engineer, and the contractor:

> Worker protection cannot be effectively achieved at the expense of profit; worker protection should be viewed as a product and should be profitable to produce.

Everyone receives value from producing worker safety. The project owner should understand the value of the purchase. Loss of life and injury in the process of delivering a project are a blemish of huge proportions and have lasting effects on the end users as well as those involved in the process. Whatever the financial cost, money spent on safety is indisputably a bargain. For the project engineer, planning and setting up the atmosphere for worker protection to succeed should be prominent on the list of deliverable items in the contract with the owner. This includes designing projects that are safe to construct. When negotiating the contract with the owner, the engineering firm should not consider the hours spent on the worker safety deliverable any different from hours spent on any other deliverable; the firm must make a profit on all of them, or it will be out of business. Where the project engineer sees the project as structure or pipeline construction, the contractor may see it as construction of a shoring system that a structure or pipeline can be inserted into. Here the contractor has a clear vision of what he or she is going to produce and sees an opportunity to make a profit. When it comes to worker protection, the vision is not so clearly defined; there is no picture or image, and yet the opportunity to produce it and derive a profit from it is exactly the same. The profit motive is one of the dynamics that drives everyone to efficiently do the very best in business endeavors; without it there is something lacking. In the case of worker protection this profit motive should be harnessed for the purpose of improving the protection.

Partial failure on the second two objectives is often seen by the contractor as tolerable as long as the first two are met. The fact is that protecting a worker from injury is a loftier goal than making sure that the neighbor's foundation does not settle or making sure that a gravity pipeline has an exact 0.02 percent slope. With due respect to the loftiness of safety, it is still possible to achieve perfection on all four goals through planning and diligence in the field. It is the responsibility of the contractor to ensure that this happens.

The function of the contractor in the excavation process is to dig the hole. The excavation planning process for the contractor goes through the following steps, and this discussion is intended to identify pitfalls and clarify the link between the contractor and the engineer.

1. *Estimating.* In addition to the work that the design and geotechnical engineers have done, the contractor has to investigate the site and the soil. The contractor's perspective is different from the engineers' and through that perspective the contractor will see different things. He or she is the one with the knowledge of the equipment to be used and the physical constraints or space within which to operate. Surface obstructions, traffic constraints, the effect of heavy equipment on the streets, environmental impacts from use of this equipment, risk and danger associated with working around high-risk facilities, and the fact that something is impossible to construct are all considered areas of the contractor's expertise and not the design engineers'. The design engineer has general knowledge on this subject; however, the contractor has specific information because she or he ultimately decides what equipment and methods will be used. Failure to notify the engineer about impossibilities and lapses in the plans and specifications at bid time regarding these subjects will get little sympathy and no money from the owner after the award of the contract. Change order approval and litigation will center on what the contractor should have been able to determine at bid time.

   The owner who required that the contractor perform a complete geotechnical investigation as a precondition for bidding a project would end up with very few bidders and a separate and possibly conflicting geotechnical report from each of the bidders. Limiting the quantity of bidders is expensive, and conflicting soils reports are the seeds of disputes and claims as well as calling the original geotechnical report into question. The contractor has to rely to some extent on the project soils report, and exculpatory language forbidding reliance is practically useless. Exploratory excavation work by the contractor at bid time is strictly for the purpose of gaining knowledge to use to eliminate risk and refine the bid. The contractor should have the ability to read and gather from the soils report all the parameters required to determine the ability to dewater, excavate, and design shoring. The contractor relies on this information to prepare the bid.

2. *Preconstruction planning.* At this point the contractor picks up the subsurface investigation where the engineer left off. In addition to the owners' needs and resulting contract requirements to protect existing facilities, the contractor has the OSHA requirement to protect workers in the excavation. As required by OSHA, 1926.651(a) surface installations and (b)(1),(2),(3), and (4) underground installations must be located on the

surface. At this point the key language for OSHA and the courts is "underground installations that reasonably may be encountered." This means that if there is a way to reason that it exists and find it, then it shall be done. Using the information that the project engineer gathered from stakeholders and the information provided by the USA line location system are not enough to fulfill this requirement. Litigation toward the contractor on this issue will center on the fact that the contractor is in his or her realm, the street and into the earth, and therefore the contractor's knowledge and experience have to be used to discover the hazards that may have been missed by others. The excavation contractor knows how closely the markings typically match the actual location; knows the materials used and how the lines were installed; knows where to expect joints, bends, and thrust blocks; knows how deep and fragile the lines are; and knows that if there is no buried electrical line marked that leads to a transformer across the street, it had better be located before digging. If there is a path of reasoning that suggests that a line should exist, the contractor must investigate it. Just as the contractor relies on the preliminary location work that the engineer did at the start of the project, the engineer relies on the contractor to carry it forward and successfully protect the workers and existing facilities. At the end of the day the contractor is the last one looking prior to inserting the excavator bucket into the ground.

The second preconstruction planning function is to develop an excavation plan and take it through the review process. A thorough understanding of the objectives and contract requirements for this plan is critical. The design engineer can only react to what is submitted, and if not everything that is required is submitted, then valuable time is lost. Excessive and generalized specifications on the part of the design engineer and trying to get by with as little as possible by the contractor are defeating to the overall purpose of the plan. The plan should be thorough, clear, and concise.

On small projects sometimes the excavation planning approach is to submit a generalized plan that covers all situations that might arise such as committing to conforming to the OSHA soils classification and tabulated data for sloping and benching, trench jacks, and manufacturers' tabulated data. Submitting literature on these alternatives does not constitute a plan. In order for there to be a plan, the following elements must be there:

- An effort should be made through excavation or soils report review to identify the specific soil types on the project.

- Excavation locations should be identified.
- Hazards associated with those locations should be identified.
- The plan should indicate the method used to safely work in the soil and protect existing facilities.

The purpose of the process is to stimulate thought and planned actions prior to production work at the site where safety is only one in a clutter of issues that the crew will encounter. If the excavation planning process is taken seriously, it will prevent damage, injury, and death.

On large projects there should be a specific plan for each separate structure excavation and pipeline or utility excavation operation. Long-term projects should anticipate preparation and submittal throughout the course of the project as the excavations come up on the critical path. This allows for correction of previous excavation problems to be incorporated into subsequent plans. To ensure the profit objective, the best-case scenario should be planned and pursued first; however, good alternative plans will provide a level of comfort to the reviewing engineer and allow for greater latitude on the best-case plan. On large excavation projects especially long reaches of pipeline and deep excavations, financially there is more at stake and therefore more disputes arise in the plan review and acceptance process. The sources of these problems arise from several situations.

- Interpretation of the specification language
- Interpretation of the soils report
- Risk level that the contractor is willing to accept being higher than that of the engineers
- The contractor believing that precluded shoring alternatives can accomplish the goal and pushing for their acceptance
- The contractor submitting shoring alternatives that are not addressed by the contract or geotechnical report

One of the purposes of the submittal review process is to iron out these differences. This is the time and place where it is important to remember that both the contractor and the engineer have a stake in the outcome, and like it or not, they are in it together.

3. *Construction.* At this point in the process, success or failure of the excavation work is definitely in the hands of the contractor. Always when there is friction between the design engineer

and the contractor, it is about the assertion that the contractor is ignoring the agreed upon plan. The engineer cannot just look the other way because of the liability associated with it; however, the responsibility for failure, if the contractor does things outside the scope of the plan, lies squarely on the shoulders of the contractor. The contractor should also be cognizant of the fact that if failure comes about when the contractor is in conformance with the plan, the engineer could end up sharing some of the liability. At the start the contractor needs latitude and time to make the approved plan work or figure out how it needs to be changed. If good-faith effort by the contractor to adhere to the plan is made, interference by the project engineer will be met with financial damage claims.

During the excavation process the shoring design engineer should visit the site and confirm that assumptions made in the design were correct and that inevitable changes to the plan are acceptable. The shoring design engineer needs to buy into what has been done and decide how much paperwork and planning discussion need to go into the changes. The incentive for the contractor to make this happen is to keep the liability pathway to the engineers open.

3.4 Legal Requirements to Generate an Excavation Plan

The requirement to generate an excavation plan is for the purpose primarily of promoting worker safety and secondarily of protecting existing facilities. In this context the concept of planning centers on the idea that planning will trigger a thought process that will prevent accidents. The level of detail involved in a plan progresses from verbal instruction to written instruction to a drafted plan. The level of detail required varies with the task and situation. OSHA requires commitment by the employer to plan and promote worker safety by performing a hazard assessment of every work site. In excavation work there is a requirement for a competent person to do an initial site review and continually monitor the work site. The excavation plan requirements escalate in the following progression:

- For excavations under 5 ft deep, the competent person determines whether the excavation is safe for workers. If the decision by the competent person is that a vertical walled trench less than 5 ft deep is safe, verbal instruction should be given to the work crew that the trench is safe; and if ground conditions change, the competent person needs to redo the decision.
- For excavations 5 to 20 ft deep, worker protection systems can be selected by first categorizing the soil in accordance

with Appendix A and then selecting a protection system from Appendix B, sloping and benching; Appendix C, aluminum hydraulic shoring; Appendix D, timber shoring; or manufacturers' shoring using their engineer-stamped tabulated data. With the OSHA tabulations the selections can be verbally transmitted to the crew. If manufacturers' information is used, it is required that the tabulated data be kept at the site while the shoring system is in use, and the selection and configurations can be verbally transmitted to the crew. Technically when one is using manufacturers' tabulated data, there is no requirement that a written plan be developed; however, it is important that the crew completely understand the configuration and safe installation and removal of the system. All this information should be available and contained within the tabulated data whether it is OSHA's or a manufacturer's.

- For excavations over 20 ft deep and any combination of open cut and shoring that is not covered in tabulated data, there must be a design by a registered professional engineer with a plan that shows the configuration and the engineer's stamp. OSHA goes no further than this statement except to define a registered professional engineer as one who is registered in the state where the work is being performed and that the engineer's stamp for manufacturers' equipment can be from any state, and to define accepted engineering practice as "those requirements which are compatible with standards of practice required by a registered professional engineer." The standard of care used in developing the excavation plan is not well defined. Article 3.4 attempts to outlines a minimum of what that process and the final product should include.

The requirement that the plan be developed by a registered engineer is the point where everyone—OSHA, the owner of the project, the project engineer, the contractor, and the workers—is relying on the shoring plan engineer to go through the thought process that will prevent accidents and protect workers inside the excavation. Due to this reliance by others, the excavation plan design engineer has a duty to be complete and concise when developing the plan.

Aside from OSHA requirements, most states have civil laws that operate to cause excavation safety planning and adherence to OSHA minimum requirements. The objective of these laws is to get the project owner's commitment and involvement in worker safety as well as to make it a contract requirement to plan for worker safety in excavations. In the case of government contracting here is an example—California Labor Code Section 6705:

> No contract for public works involving an estimated expenditure in excess of twenty-five thousand dollars ($25,000), for the excavation of any

> trench or trenches five feet or more in depth, shall be awarded unless it contains a clause requiring submission by the contractor and acceptance by the awarding body or by a registered civil or structural engineer, employed by the awarding body, to whom authority to accept has been delegated, in advance of excavation, of a detailed plan showing the design of shoring, bracing, sloping, or other provisions to be made for worker protection from the hazard of caving ground during the excavation of such trench or trenches. If such plan varies from the shoring system standards, the plan shall be prepared by a registered civil or structural engineer.

Here the requirement to generate and submit a detailed plan is for any excavation over 5 ft deep. This is in excess of the OSHA requirements in that the plan has to be written in sufficient detail or else drafted and submitted for review and acceptance. The other sticking point revolves on what the word *acceptance* means. The awarding body more than likely does not have excavation safety planning experience, so the statute allows for review and acceptance by a registered civil or structural engineer, most likely because it requires that level of expertise to determine if the plan is adequate. Acceptance should hinge on determining whether the engineer who prepared the plan used a standard of care compatible with engineering practice, and not on whether the plan is correct in every detail. Everyone involved is still relying on the shoring designer to get it right.

In the case of private contracts, building permits that involve excavation work are not issued until it is verified that the excavation contractor possesses an excavation permit. To obtain an excavation permit, a contractor must have a safety commitment and a general injury and illness prevention program in place and must demonstrate knowledge of OSHA excavation safety requirements. The owner's procurement plans and specifications normally state that the contractor has to conform to all permit requirements, and hence the excavation safety planning process is triggered.

Site-Specific Excavation Plan

Often a detailed excavation plan amounts to a written statement by the contractor agreeing to conform to the OSHA excavation safety requirements. The submittal usually contains OSHA sloping and benching tables, tabulated data from OSHA Appendixes C and D, and manufacturer's tabulated data for shoring equipment that the contractor contemplates using on the project. In defense of this procedure, it does establish that on the specific project, any one of the worker protection systems submitted can be used if the soil conditions are appropriate, and it defines how to determine if they are. This plan gives the contractor an array of options to be used as conditions warrant. The real downside is that it requires very little forethought and planning. The information is not specific to the project and can be applied to any project. Ultimately each specific excavation site on the project has to have soil identification and a specific worker protection plan assigned to it. In the field and in the mix of other logistical tasks that are being performed simultaneously, the specific excavation safety planning task does not get the attention that it deserves. One may argue that with preplanning, other operations and availability of equipment at the time may render preplanning useless; however, after preplanning is completed and then the plan is changed, there still has been more thought put into it than if it were put off until the last minute. At a minimum for a shoring plan to be considered site-specific, it should identify the location, soil type, and specific shoring system to be used.

3.5 Elements of an Excavation Plan

An excavation plan should convey enough information to determine the location of the work, the extent of the work, and how to construct the work. The plan should be backed up with preliminary investigation and calculations. The resulting submittal package is drawings and calculations. With an excavation plan the designer of record has to be the one who will implement the plan, the contractor. The plan must be the contractor's plan with his or her input, resources at stake, and commitment to following it. The contractor should develop the plan with her or his own engineering staff and then involve a shoring design engineer to calculate and draft the plan, if there are no in-house personnel qualified and licensed to do it. There are two advantages to going outside to a shoring design engineer. In cases where there are serious safety and risk issues involved, it eliminates the in-house engineer's conflict of interest between profit and safety; and it spreads the risk. Because the excavation plan is prepared by the constructor, there is no need for exact specifications and detail beyond what it takes for the workers to construct it. The result is a plan that is not as formal as the set of plans and specifications that the project engineer prepares. Aside from constructability instruction, the plan has to convey enough information for others to be convinced that it locks in success of the three stated goals. All the following elements should be present in the plan.

Paperwork

1. *Preliminary investigation.* Identify soils report or soils exploration from which assumptions and conclusions are drawn. Specific bore logs that apply to the site should be in the package so that they can be referred to and confirmed during the excavation and redesign process.

2. *Calculations of open cut systems.* If OSHA Appendix A is being used, the soil-type basis of conclusion should be identified. If design by a registered engineer is being used, slope stability calculations including surcharge loading are required.

3. *Calculations of shored systems.* These should include the following:

    - If dewatering is a requirement for success of the shoring system, dewatering methods and discussions indicating likelihood of success, continuity of control, and backup alternatives
    - Soil parameters and loading diagrams
    - Bottom stability calculations
    - If manufactured equipment is used, soil resisting system tabulated data indicating that soil loading can be resisted and that it can fulfill other requirements of the system. Should include installation and removal procedure and instruction for safe operation. Structural design calculations for the manufactured equipment generally not required
    - If an engineered design is used, it should include
        a. Assumptions including soil strength parameters, unit weight of soils, and design location of water table
        b. Soil, surcharge loading, and water loading
        c. Design standards and safety factors
        d. Complete allowable stress design or load factor design calculations
    - If ground movement and settlement are an issue, provide settlement calculations.

Drawings

1. Title information

    - Project title
    - Project engineer
    - Contractor
    - Contractor's consultants—geotechnical engineer and shoring design engineer if they are used
    - Source of soils information
        a. Geotechnical report name and number
        b. On-site pothole reports with identification in accordance with OSHA Appendix A
        c. Previous experience with soils in the area; assumption to be verified at time of excavation

2. Location information

    - On small project, state locations that it applies to or statement that the plan is inclusive of all excavations on the site

- On pipeline project, state station numbers that define the reach or structure locations to which the plan applies.
- On large projects, the site of each plan. Pipelines, yard piping, and structures will each have separate plans
- Location, if applicable, of bore logs or exploratory excavations
- North arrow
- Datum elevation and alignment

3. Extent information—site plan
   - The plan view should show everything affected by the excavation including
     a. All existing aboveground features that are affected by the excavation such as topography, trees, overhead lines, property and easement lines. Specifically identify hazards such as slopes too steep to work equipment on, overhead power lines, and emergency vehicle access requirements.
     b. All subsurface facilities. Specifically identify and locate all subsurface hazards such as high-voltage conduit and high-pressure gas mains.
     c. Site plan details where critical dimensions are necessary such as distance from excavation line to surface or subsurface hazard.
   - The plan view should include and call out new excavation or shoring elements such as these:
     a. On open cuts show surface cut line and grade, bottom cut line and grade, and slope designation including direction and run to rise. On pipeline open cuts show end of the trench configuration as well as side slopes.
     b. On shored excavations show outline and callout of shoring type and plan view elements such as sheet piles, wales, and struts, or shoring shield, spreaders and endplates.
     c. On shored excavations show elevation of shoring elements at the surface and the bottom of the excavation.
     d. Critical dimensions.
   - Include the footprint of production work that is to be constructed inside the excavation.
   - The plan view should show planned spoil piles, material storage, equipment access, and setback requirements.
   - Show dewatering wells and piping.
   - Show fencing and barricading.

4. Structural plan and section
   - The plan view should include
     a. Datum elevation and alignment
     b. Layout and callout of shoring elements
     c. Critical aboveground and belowground hazards
     d. Surface elevation and dredge line elevation
   - Sections should include
     a. Critical elevations
     b. Water table
     c. Soil types
     d. Location of critical surface and buried utilities
     e. Slopes and run-to-rise indication
     f. New shoring elements including dimensions (4-in wall, 8 × 16 shoring shield, W14 × 117 wale, etc.)
     g. Production work to be constructed including bottom of dredge line (maximum depth of excavation, design elevation) and wall clearances for formwork
5. Details
   - Include details of structural connections for engineered and field-constructed shoring.
   - For manufactured shoring equipment, provide enough detail to show how parts are connected.
6. Notes
   - Notes should be specific to the plan, not catch-all phrases.
     a. Include locate-before-you-dig requirement and one call center number.
     b. Indicate procedure to be used for existing line location work and whether the location work is to be performed prior to production work or during production work.
     c. For engineered systems indicate material strength and quality requirements.
     d. For manufactured shoring equipment indicate strength requirements such as psf ratings. Manufacturer's name and model number are not necessary.
7. Installation and removal
   - Provide a step by step installation and removal and backfill procedure with section drawings indicating

the excavation level, shoring condition, and allowable safe working area at the end of each step. *Workers can never be allowed to work in unshored areas in order to construct or remove shoring.*

8. If settlement an issue, settlement monitoring plan.
   - Plan should include
     a. A datum line and elevation shown on the plan
     b. Location and frequency of measurements to be taken
     c. Methods and equipment to be used
     d. Method of cataloging and preserving information
9. Drawing work
   - The first sheet of the drawings should have an index and key to linework and symbols used.
   - There should be a drawing scale clear keyed path between plan, section, and details.
   - All drawings should have page number, date, signature of preparer, and engineer's stamp, if required.

Excavation plan drawings can be hand-drawn or computer-drawn provided they are clear and contain the needed information. The "keep it simple" philosophy used by contractors should be promoted in the drawings. However, it is important to realize that on projects with a lot of repetitiveness such as a pipeline with miles of pipe and several manholes or a microtunnel project with several boring and receiving sites, even though open cuts or shoring structures may be exactly the same at every site, the shoring designer must still investigate the entire reach. The entire pipeline length and each site of structure excavations have to be investigated for soil conditions, existing aboveground and belowground facilities, and conditions that affect the safety of the work. The opposite approach to generating a separate site plan is to draw a single plan and list station numbers where it applies and then list stations and detail where variations from the standard plan are found. Paper is the only thing this approach might save. When crews are doing work at a specific site, information that applies to other sites will only cloud the issues at the specific site, and therefore the site-specific information eventually gets ignored. One advantage of CADD work is that it is easy to copy repeatable information and develop very specific drawings for each location. Figure 3.1(a) to (c) is an example of a shoring shield plan for a bore and jack operation in a city street.

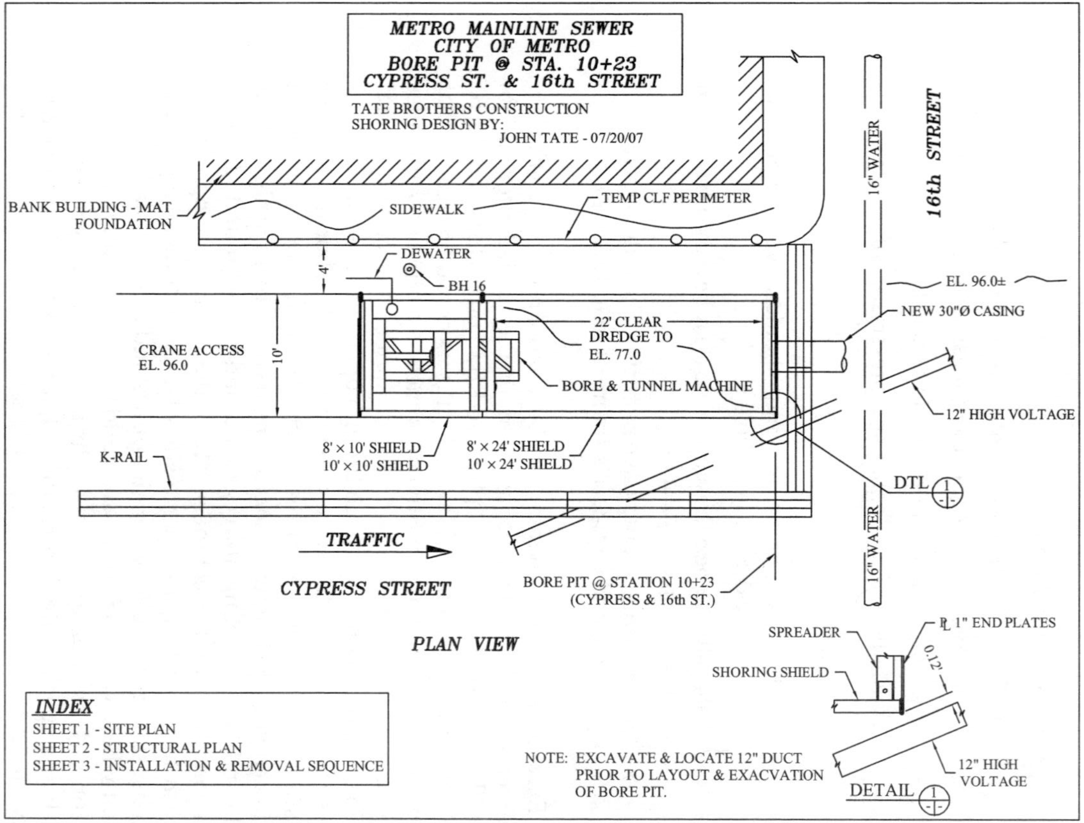

(a)

FIGURE 3.1(a) Site plan for bore and jack excavation.

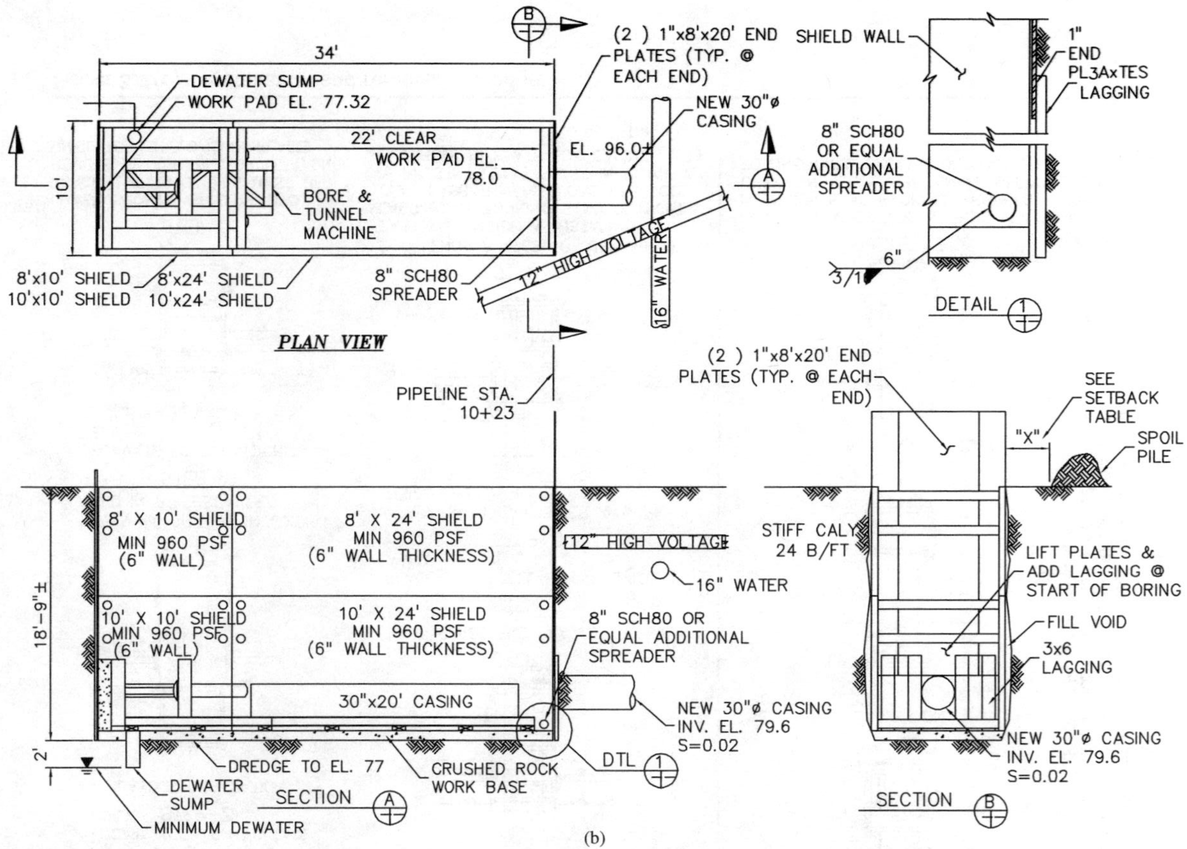

Figure 3.1(b) Structural plan for bore and jack.

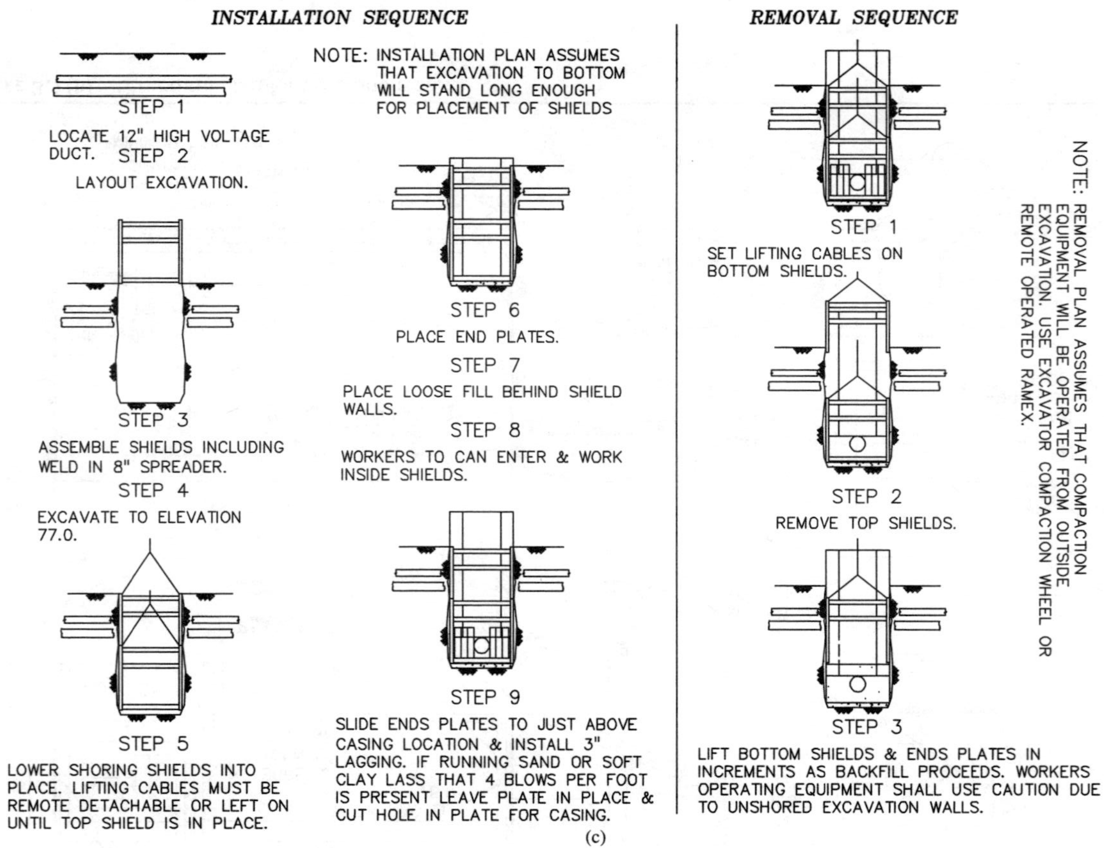

FIGURE 3.1(c) Installation and removal procedure.

3.6 Design Standards for Excavation Plans and Shoring Systems

Design standards are a set of design practices and material specifications that everyone can agree represent good state-of-the-art engineering practices in the specific design task. Even though a person is experienced in the particular area of design, no engineer can know everything about it; so by adhering to design standards the engineer can be sure to produce a state-of-the-art design. The design standard also serves as a guide for design review so that the reviewing engineer is not put in the position of having to question the basis of another engineer's design. Design standards are usually developed and adopted by engineering organizations that represent the particular industry at hand. The shoring industry as it exists today is relatively new compared to other construction-related industries such as construction equipment manufacturing and construction materials production. Consequently in the United States a cohesive shoring industry association does not exist. The National Utilities Contractors Association (NUCA) is the largest and strongest association that is allied with the shoring industry, and it has always been in the forefront of promoting excavation safety; however, excavation safety and shoring is not NUCA's only focus. More recently in 1994 the U.S. manufacturers of shoring equipment formed the Trench Shoring and Sheeting Association (TSSA), primarily to promote the industry and secondarily to develop standard guidelines for the design, engineering, and manufacture of shoring equipment. The standard addressed soil loading diagrams in order to facilitate the development of tabulated data, material specifications, and fabrication. In February 1996, the manufacturers could not come to agreement on the final draft of the specification, and therefore it was never adopted by them.

There are plenty of design specifications that can apply to shoring design; however, they were not developed with shoring design in mind. The problem is that some shoring design practices that shoring designers, contractors, and manufacturers use are different from what is used in the design of permanent structures, such as a one-third short-term shoring use factors, assumptions regarding surcharge loads, and the shape of soil loading diagrams. Project owner–side engineers do not necessarily have a belief that these practices represent substandard design; it is just that there is no way for them to find a basis, such as a design standard, for accepting them. When OSHA came out with the revised excavation safety standard in 1989, they allowed for engineered design of worker protection systems in excavations, but they did not address acceptable design standards. In fact they removed it from the original draft and left it to the engineering community to decide the issue. Debate about what is acceptable continues today. Over the years most of the design practices that shoring designers use today have been accepted. This is evidenced by the fact that they are specified in most shoring design specifications.

If shoring design practices that are different from normal structure design practices are used, there must be a reasonable basis for it. Basic differences between design and construction of permanent structures and shoring structures are as follows:

- The project is temporary for the purpose of facilitating the permanent construction and will be removed after it is no longer needed.
- The project is constructed inside the ground and not on the surface.
- Shoring elements are normally either used materials or manufactured items that are used many times over in short durations.
- One of the primary functions of shoring is safety and elimination of life-threatening hazards for workers and the public.

To the extent that the safety, cost, and practicality of the design are affected by these differences, the design methods should reflect it even though it may be different from permanent design.

The intent of this section of the book is not to develop a design standard for the shoring industry, but to present methods and standards that are in common accepted use today.

3.6.1 Standard Practice for Designing Open Cut Excavations

There are essentially three different alternatives for open cut design.

- *OSHA Option 1—1.5:1 slope.* This is sometimes called the do-nothing option. The unique thing about this option is that it does not require any understanding of the soil or classification of the soil; however, the designer should be aware that in some very loose sands and gravels and soft marine clays a 2:1 or greater slope may be required. When OSHA set this option out, they were only concerned with worker safety and not protection of existing facilities. It is a reasonable assumption that if a 1.5:1 slope moves, the workers will have enough time to remove themselves to a safe area. A design procedure for this option entails slightly more than simply stating that the slope is to be 1.5:1. This option can only be used in excavations less than 20 ft deep. The following steps should always be completed in the use of this option.

  1. Determine the extent of the slope on paper or at the site.
  2. Determine the water table. If the water table is above the bottom of the excavation, another option has to be used. Good construction practice has the water table 2 ft below the bottom of the production product, including bedding material for pipe and compacted base for concrete placement.

3. Determine if surface and subsurface facilities are going to be affected by the cut. If they are, they have to be removed or a different option has to be used.
4. There should be room to set spoil piles and normal surcharge loads from construction equipment and traffic at least 2 ft from the surface excavation line. If there is no room, another option must be used.

- *OSHA design by Appendix A and Appendix B.* OSHA Appendix A is an alternative to the Uniform Soils Classification System. It was developed so that a person who does not have education in geotechnical engineering can classify different soil types in a hierarchy of stability in order to choose worker protection alternatives. Appendix B is a set of tabulated data for sloping and benching of excavations to a maximum depth of 20 ft. These are the steps in developing an open cut plan using this alternative.
  1. Based on the expected soil type, choose the appropriate slope configuration in Appendix B and determine the extent of the slope on paper or at the site.
  2. Determine if surface and subsurface facilities are going to be affected by the cut. If they are, they have to be removed or supported.
  3. Determine equipment and dead load surcharge locations. They should be set back a minimum of 2 ft from the edge of the cut.
  4. Based on the soil as it is encountered in the excavation, use OSHA Appendix A to determine soil type, and confirm original assumptions, or alter slopes in accordance with actual soil type.

  The major advantage to this method of slope design is that the contractor's competent person can determine the slopes in the field as the soil types are encountered. This works well for pipeline excavations where new soils are being encountered with every excavator bucket scoop and the excavation is being backfilled fairly rapidly. The downside for pipeline work is that the basic OSHA soil classifications are conservative and stepped, with no opportunity to make use of the soil stability between the steps. On long pipeline projects the amount of unnecessary excavation work can add up quickly. For structure excavations that will be in place for more than a few days, the method can lead to problems because the soil types and slope requirements can change daily depending on how long the excavation has been open, surcharge loading, and seasonal fluctuations such as irrigation, stream flow releases, heat, wind, and rainfall.
- *Design of open cut excavations by a registered engineer.* This option must always be used for excavations over 20 ft deep and

is highly recommended for long pipeline and large structure excavations because the risk management benefit and cost savings that come with planning and engineering should easily exceed the cost of the design.

Engineered excavations are based on soil mechanics and not on OSHA soil type determinations. Design parameters are based on a soils investigation and have a continual range instead of being stepped. The design process includes evaluation of all excavation-related issues such as water table, soil type, length of time the excavation will be open, existing facilities, environmental factors, surcharge loading, construction accessibility, and worker safety. This design generally results in more slope options and more favorable slopes with less excavation required than those allowed in an OSHA Appendix A design.

Quite often design by a registered engineer has been seen by the reviewing engineers and owners as a method that the contractor uses to sidestep OSHA sloping and benching requirements. They would prefer to see the words "must use whichever is the most stringent" written between the options. It was the express intent of OSHA and the federal government to allow this option so that a cost-efficient, safe alternative to the OSHA method of designing slopes would be available to the contractor. The purpose is to prevent waste and to foster competition and ingenuity. The design by civil engineer option deserves far more respect on both sides of the street than it gets. Far more expertise goes into the design — a geotechnical report, slope stability analysis, the contractor's input regarding means and methods, and follow-up during the excavation process. The major problem for the respect issue is that there is no generally accepted standard as to what constitutes a complete professional design by a registered civil engineer. Even with a standard design process there is still room for disagreement on interpretation of soil design parameters and level of risk management. The submittal and review process is the forum for working out the latter issues. The following is a design standard for open cut excavations that the author developed for use in his office.

Open Cut Design Procedure

Every one of the following steps shall be noted in the calculations and on the final plan.

1. All open cut excavations are to be designed using soil parameters taken from the following hierarchy of sources:
   a. Geotechnical report design recommendations
   b. Closest boring log to excavation site

c. Contractor's exploratory excavations

d. Observation of production excavations at the site

Review all sources available; however, first use the geotechnical design recommendations. If they are not available, use the most reliable of the remaining sources.

2. Determine preliminary slope lines and review entire site within and surrounding those lines for existing surface and subsurface facilities that affect the excavation. Show all the affected facilities on the plan and detail how they are to be dealt with, such as locate prior to excavation, support, remove, separate sloping or shoring at these locations, etc.
3. Determine water table and reliability of dewatering methods. It should be maintained at 2 ft below working base.
4. Determine dredge line and production baseline, start of pipe bedding or concrete base.
5. Evaluate surcharge effects including spoil pile, material storage, equipment size, equipment access and where it is to be located, effect of traffic and structures.
6. Evaluate the time period over which the excavation will be open and environmental exposure during that period. Look at factors such as rainfall, irrigation seasons, cold weather, drying cracking and dust control, surface drainage away from excavation, stability risk over time, and mitigation measures if unforeseen problems arise.
7. Evaluate safety issues such as access and egress routes, safe area for workers during heavy lifting operations, accumulation of hazardous atmosphere inside excavation, barricading, and public safety surrounding the excavation.
8. Perform bottom stability and slope stability analysis. Include sensitivity analysis by varying assumed parameters. There must be a 1.5:1 factor of safety based on reasonable parameter selection.
9. Follow up with a site visit during the initial excavation process to confirm soil design parameter assumptions.
10. Follow up with a site visit shortly after production work has started, and discuss safety issues with the crews working inside the excavation.
11. Follow up with a minimum of a phone conference with project site personnel after every stability-threatening event such as rainstorm, excavation flooding from failed dewatering, broken utility inundation, or earthquake.

3.6.2 Standard Practice for Shored Excavation Design

Shored excavations can use manufactured shoring equipment, site-specific shoring construction, or a combination of the two. Shoring can also be used in conjunction with open cut methods. In design of shored excavations OSHA allows for the contractor to use manufacturer's tabulated data and design by a registered civil engineer (RCE). OSHA requires RCE design where tabulated data are not available for the proposed shoring scheme. There is some confusion about it; however, there is no depth limitation for the use of manufactured equipment as long as the data address the depth and circumstances in which it is allowed to be used. Anything at any depth that is not addressed by tabulated data must be designed by a registered civil engineer. Tabulated data developed for shoring equipment by the manufacturer must also be stamped.

Shoring industry practices and requirements in design procedures and material specifications for both the site design and the manufactured design are not always well defined. Manufacturers are secretive about their procedures for the obvious competitive reasons, and contractors and their shoring designers have a level of comfort between them that allows for a less specific and time-consuming design process. The same level of comfort is not usually enjoyed by the engineers who review the plan. This section discusses shoring industry design-side practices and conventions.

Shoring Design Terminology

The following terms used within the shoring industry apply to design assumptions and types of shoring systems in use.

Flexible shoring system A shoring system that allows enough movement for the soil to mobilize some of its internal strength to resist collapse. For sands and gravels the strength comes from shearing resistance between the grains, and for cohesive soils the strength comes from bonds between the grains. The result of soil movement is reduced load on the shoring system. For the shoring designer the assumption of flexibility allows the use of active soil pressure instead of at rest or passive and the use of a triangular soil loading diagram in some cases. Cantilevered permanent retaining walls and cantilever shoring walls [Fig. 3.2(a)] deflect at the top and are considered flexible.

Restrained shoring system A system in which struts and tiebacks do not deflect into the excavation and therefore add restraint and therefore additional load to the shoring. The primary effect of restraint is to change the soil loading diagram from triangular to trapezoidal. The sheeting and wales in strutted shoring systems [Fig. 3.2(b)] allow enough movement to render the system flexible even though the struts cause it to be restrained. Hydraulic jack shoring is rigid at the jack cylinder; however, the jack blades do

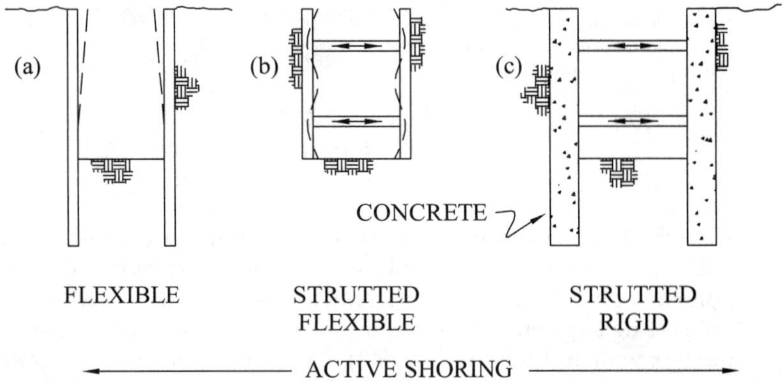

Figure 3.2 Shoring system movement.

not confine the surrounding soil thereby allowing infinitesimal movement of the soil to set up arching and reduce the soil load. In restrained flexible systems, the total soil load is similar to the resultant of a triangular load; however, the distribution of the load is more trapezoidal, and the total resultant acts toward the middle of the wall.

Rigid or fixed shoring system A rigid shoring system [Fig. 3.2(c)] that has very little movement into the excavation and sometimes loads the soil being held back, as in the case of tiebacks or struts that are preloaded beyond the soil load. Examples of rigid systems are strutted concrete caissons, frozen ground walls, stiff sheet piles with preloaded wales, and tieback walls. A rigid system is desirable where prevention of settlement and soil movement toward the excavation is important. Aside from controlling the excavation and shoring installation operations, the best way to control settlement is to increase the stiffness of the sheeting, decrease the strut spacing, and preload the struts. All these measures increase the cost to the shoring materials, increase the time and cost of labor and equipment, and decrease the working space inside the excavation. Also as the fixity of the system increases, the soil loads increase from active to at rest pressure and beyond depending on preload forces.

Active shoring system Most simply stated, a system that takes action to prevent soil movement versus one that does not. In an active system, the shoring and soil touch at the interface and restrict movement toward the excavation. The degree of flexibility and restraint is not addressed with this term. This term should not be confused with *active soil pressure*, where through movement the soil acts internally to prevent collapse and thereby reduces the loading on the shoring.

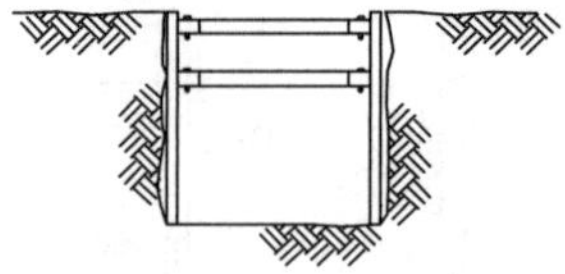

FIGURE 3.3 Shoring shield with void between soil and shore, passive shoring.

Passive shoring Shoring for worker protection only which does nothing toward preventing soil collapse. Shoring shields and manhole boxes are the best example. Most often with this shoring the hole is dug large enough to set the box inside with a void between the box wall and the soil (Fig. 3.3). To some extent shoring boxes can be turned into active shoring by backfilling the void or digging the box in by digging below and pushing it down. Shoring systems consist of different elements (Fig. 3.4) that gather and resist soil loads in different ways.

Sheeting and lagging The material that carries the soil load to support members or prevents soil from falling off the excavation wall. Steel plate set between H-piles is sheeting because it carries the soil load to the piles. The steel plate on the wall of a shoring box is considered sheeting because it carries the soil load to the beams inside the box. Plywood behind trench jacks prevents sloughing of soft clay or raveling of gravel from the wall of the excavation. Due to soil arching to the support members, sheeting does not usually get the full soil load. Lagging serves the same function as sheeting; the difference is that lagging is not as wide as sheeting. The most common lagging is timber boards 6 to 12 in wide and 3 to 6 in thick.

Solid sheeting Term used when the intent is to eliminate the exposure of any ground on the walls of the shored excavation. It also connotates no voids behind the sheeting and no soil running between cracks in the sheeting.

Wale A horizontal shoring member, or beam, that gathers loads from sheeting or soldier piles and carries them to struts.

Soldier pile A vertical member that receives sheeting loads and delivers them to the ground in the case of cantilever shoring or in the case of strutted excavations to wales aboveground and belowground. Soldier piles do not normally carry vertical loads except in the case where they are driven instead of set in drilled holes, as is true with bearing piles, and therefore bending strength is more important than axial strength. Where pile driving is not a consideration, wide flange beams or H-piles are equally adequate for soldier piles. If the vertical members do not extend into the ground, they are called soldier beams.

Sheet pile Piles that gather the soil load similar to sheeting and have a structural shape that is capable of carrying the sheeting

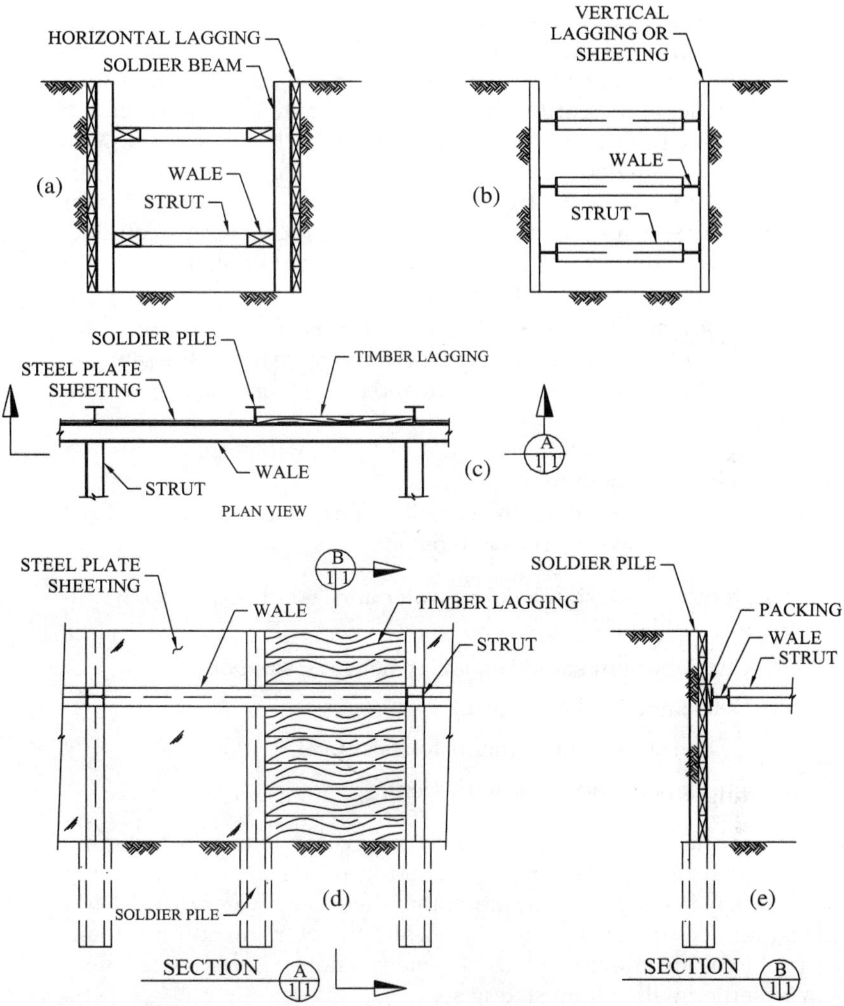

FIGURE 3.4 Shoring elements.

loads into the ground and up to the wales. Their dual nature, sheeting and pile, may be where the name originated. The name may also have originated from the fact that they consist of a thin sheet of steel rolled into a shape that allows it to be driven into the ground like a pile. There are two common pile shapes, U and Z (Fig. 3.2).

Strut Horizontal axial force-carrying members that resist the loads delivered by the wales or soldier piles. Structurally struts are designed the same as columns except that they have additional bending forces due to gravity and possible impact from materials being hoisted into the excavation. In rectangular excavations wales also become struts, carrying both axial and bending loads. Struts react to the entire loading in the system, and failure of one can cause a progressive failure, domino effect, in the entire system. If struts are angled vertically to the ground, they are sometimes referred to as rakers. Trench jack cylinders and shoring shield spreaders are also struts. H-piles, round pipe, square steel tube, and square timber beams make the best struts because they are nearly symmetric in all directions.

Packing material. Usually steel or wood, that is placed between wales and sheeting when there is a gap between the two, for the purpose of preventing sheeting movement and subsequent soil movement.

Shoring Design Procedure

After the size and depth of the excavation are determined, the final shoring design proceeds in five steps:

1. Determine design level of water and likelihood of maintaining that level.
2. Check bottom stability for clay heave or sand boiling.
3. Determine load on shoring system.
4. Design structural elements to resist load.
5. Check deflection and settlement.

Every one of these steps should receive careful thought and notation in the calculations for every shoring design. For the first two items it may be as simple as stating that there is no water present and the bottom is stiff clay or dense sand, and therefore no bottom stability problems are anticipated. If settlement is not an issue, the last item can be settled with a note stating so.

The water level is important because it will determine if the shoring will have to extend below the dredge line of the excavation and it can add significant loads to the shoring system. Because of the fact that it is a slightly different discipline, hydrology and hydraulics, there is considerable debate about whether dewatering design and calculations should be part of the shoring design. At a minimum the design water level and reliability of the dewatering system should be submitted and accepted before the shoring design is approved. If the dewatering system is not reliable, then the shoring system is not either unless it is designed for the worst-case water condition.

Bottom stability calculations will determine if the shoring has to extend below the bottom of the excavation and will affect the required shoring design strength. Commonly used design formulas developed by Terzaghi and NAVFAC The required factor of safety is 1.5.

Shoring Loads

Figure 3.5(a) to (c) shows three basic load conditions encountered in shoring design. The location of the bottom of the shoring and the location of the water table are what separates the cases. In Fig. 3.5(a) there is a basic assumption that all water can drain out at the bottom of the shoring system. The condition shown where the water table is the same on both sides of the shoring wall is a basic assumption in the design of any shoring system where the bottom of the shoring does not go below the bottom of the excavation, and in the tabulation of data for use of manufactured shoring equipment. If this assumption is made, it is important for the contractor to understand that he or she must maintain the water level at the base of the excavation or the shoring system will be underdesigned.

Even though there is no shoring extending below the bottom of the excavation, it is still possible for water to back up behind the shoring Fig. 3.5(b). A layer of sand above a clay layer at the bottom will allow water to flow to the shoring wall and then seep down the wall. Even though the layer of water at the wall is thin, it will still transmit the full hydrostatic force as if there were a whole lake backed up behind it. Even when there is sand that allows flow through it rapidly, pumping down a flooded excavation can draw the water down rapidly inside the excavation while leaving it backed up behind the shoring. Unlike soil forces that can take time to build up, water instantly transmits a force of 62.4 lb/ft^3 times the height of the water on the shoring wall. If there is any doubt about the contractor's being able to control the water level inside and outside the shoring, then a reasonable hydrostatic force should be included. This is one reason why in the use of OSHA Appendix A soil that has water seeping on the sides of the excavation is required to be classified as type C.

Figure 3.5(c) shows the load condition for shoring that extends below the bottom of the shoring. The loads transmitted to the shoring do not care where the bottom of the excavation is. On the resisting side, in addition to the resisting force of the shoring the soil becomes part of the resisting system. The water is neutral and can only serve to counteract some of the water pressure from the load side.

The designer must design for every force that the shoring system is known to have to resist. The three basic loads shown in Fig. 3.5—soil, surcharge, and water—are always considered in shoring design. If water is not present, the assumption and reasoning behind it should be clearly stated. There should always be a minimum surcharge load and a statement of what surface loads are included in it and the minimum setback required. In shoring design, *surcharge* is sometimes used in place of the term *live load* which refers to things that can be moved in and outside the influence of the structure. Surcharge is considered anything that is placed on the ground that creates a load and resulting lateral force on the shoring.

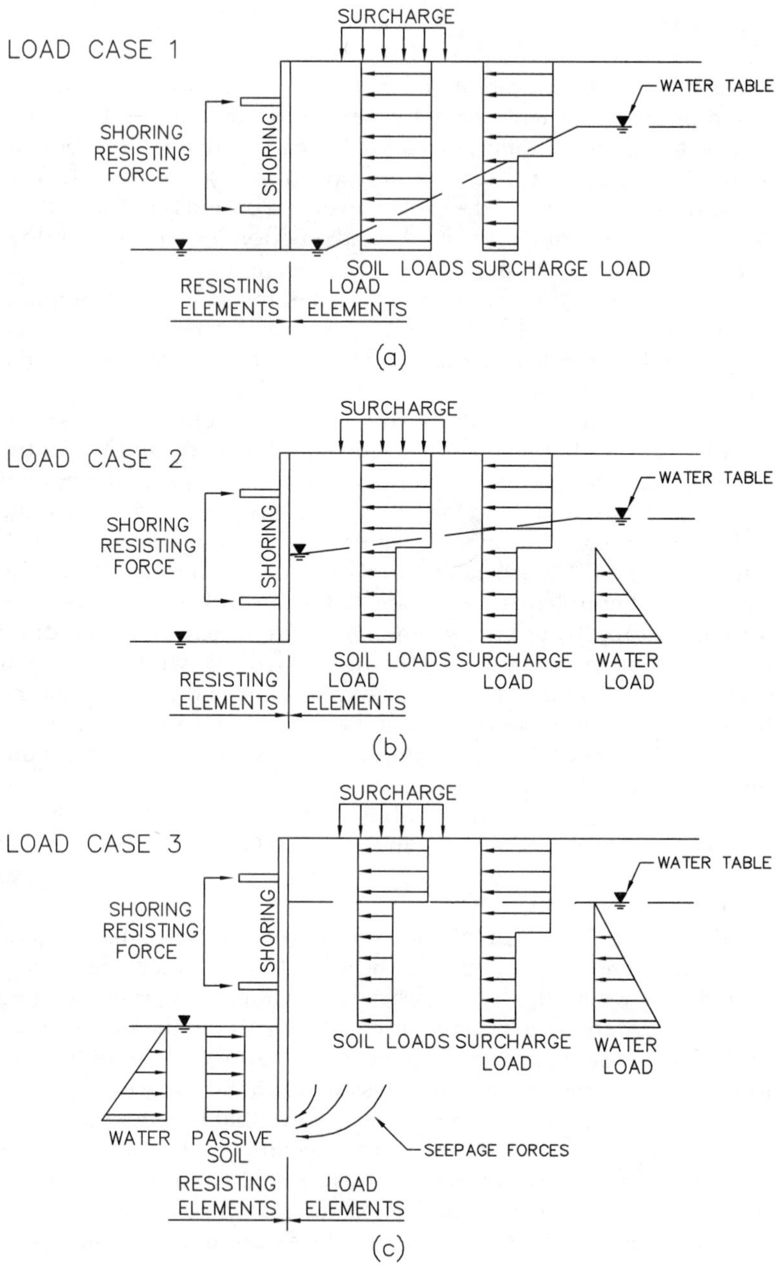

FIGURE 3.5 Shoring design load cases.

Other loads that could be present but are not normally designed for are

- Stream flow, tidal flow, and wave action
- Thermal loads such as ice and steam
- Impact from swinging cabled loads being hoisted into the excavation, equipment inside or outside the excavation, in waterways boats and floating debris
- Lateral forces from jacking and tunneling operations
- Earthquake forces

If any of these forces are anticipated, they should be included in the design.

The analysis should try to predict the loads as accurately as possible so that when allowable strength or safety factors are applied, the results are not clouded by fudge factors brought in through the analysis.

Water weighs 62.4 lb/ft^3; the maximum pressure is 62.4 times the depth and always has a triangular loading diagram. A triangular loading diagram for soil is also referred to as an equivalent fluid pressure diagram.

Surcharge loads are always larger toward the surface and decrease rapidly after about 10 ft of depth into the soil. The calculated surcharge load is P-shaped and is usually rounded off to stepped down rectangular shapes. The important thing about squaring off surcharge loading is to make sure that the load total is equal to or greater than the curved shape. Chapter 7 presents the prominently accepted Boussinesq equation for calculating surcharge loads from soils and equipment and Cooper E80 loads (based on the Boussinesq equation) for railroads surcharge loading. Soil unit weight can vary between 90 and 135 lb/ft^3 (or pcf), and therefore a reasonable approximation of the correct weight of each soil type should be used. In soil mechanics, soils are always first determined to be noncohesive (sands and gravels) or cohesive (clays), because each type behaves differently. Articles 6.2.3 and 6.2.4 present apparent earth pressure diagrams developed by Terzaghi & Peck and NAVFAC for strutted excavations in noncohesive and cohesive soils. These diagrams are the result of measured strut loads in deep excavations in many different locations and soil types. One of the findings of these strut load measurements was that similar struts at a given level could vary as much as 60 percent from average, and the total of the loads in similar vertical sets could vary as much as 30 percent from average. The resulting apparent pressure diagrams are slightly counterintuitive in that normally one would assume that the soil load would increase in a triangular distribution, just as water pressure does. Basic soils analysis calculates the loading in this way, and cantilevered shoring walls and permanent retaining walls are considered flexible and are designed using this equivalent fluid pressure. In the case of shoring construction using restrained shoring systems (strutted excavations), the installation procedure, soil variations, time spans of the installation sequence, and bottom stability all affect the movement and loading of the soil. Soil loads arch (article 6.3) to struts previously installed above, thereby increasing the loads on the upper struts as the excavation proceeds. The total load on the shoring system is roughly equal to the total load calculated from a fluid pressure diagram. However, the pressure diagram is more parabolic with the total force resultant located at about one-half the depth of the excavation, and therefore a rectangular or trapezoidal diagram is used. These apparent pressure diagrams are not just a cranked up superconservative approach to soil load analysis. They are designed to cover a spectrum of predicted strut loads and to prevent progressive failure of the entire system. Progressive failure occurs when one link in the system fails, thereby shifting the load to nearby members that then become overloaded and fail to the point where the entire system fails. There is no built-in factor of safety, and the strut load calculated from these diagrams can be expected to be the same as it would be measured in the field. The factor of safety gets built into the allowable strength of the strut material.

3.6.3 Short-Term Soil Loading

Due to soil arching, a portion of the soil loads goes directly to struts, corners, and the bottom edge of the excavation. These loads bypass sheeting and wales. The loading on these elements is predicted to be less than that on struts. Because of this, coupled with the idea that shoring loads are short-term as opposed to a lifetime of loading that causes fatigue, the shoring industry has adopted the use of a short-term factor, usually a one-third increase, applied to material strength. Project design engineers and agencies that they represent have accepted this to varying degrees, allowing anything from 10 to 33 percent overstress. At the 33 percent level this can be seen as a 75 percent decrease in the apparent soil pressure or an 88 percent use of the ultimate yield of steel material strength in bending, leaving a 12 percent margin for safety. This may seem to be fairly far out there in terms of risk level; however, it is similar to the 33 percent allowable overstress for wind and seismic forces in building construction, which is also too far out there for the comfort of many design engineers in building construction. Given that the life safety risks, in terms of death and injury, in buildings are far greater than in shoring, the use of a short-term factor by the shoring industry should be viewed with the same degree of respect as it is with the building design industry.

The 33 percent short-term design factor was originally contemplated in the original OSHA draft construction safety standards. When the excavation safety standard was adopted in 1989, the short-term factor showed up only in Appendix B, Sloping and Benching. There *short term* was defined to be less than 24 hours, and in type A soil a 0.5:1 slope was allowed to a maximum of 12 ft deep. Shoring manufacturers also picked up on it by allowing additional depth for short-term use. The problem with these uses is that excavation work rarely goes as planned, and slopes and boxes placed in a trench for short-term use end up being used longer than 24 hours. In the case of sloping and benching, the soil can change substantially over a 24-hour period and collapse on workers. Here the 24-hour limit should be adhered to. In the case of active soil support systems, the soil loading builds up over a longer time, and the possibility of imminent collapse of the trench wall onto the worker does not exist. In the case of passive protective systems such as shoring boxes that do not touch the trench walls, after the soil has collapsed onto the box wall, the soil load is less than it was immediately prior to collapse. Soil arching also reduces the loading on these protection systems. The real danger is soil loads building up over long periods, months to years and should be taken into consideration when using the 33 percent factor. For these reasons some shoring manufacturers and shoring system designers have incorporated the 33 percent overstress into their designs. In this book it is referred to as a 33 percent shoring use factor.

The shoring use factor is not based on the same concepts used to develop apparent pressure diagrams and does not create a double count reduction, one on the load side and one on the resisting side. In Chap. 6, Excavation Stability and Shoring Design Loads, this subject is brought up again.

3.6.4 Allowable Strength of Shoring Materials

The most commonly used materials in contractor-constructed shoring and manufactured shoring are steel, aluminum, and wood. Concrete, cement soil mixture, grout in tieback systems, and frozen earth are also used to a large extent in specialized shoring systems that are most often installed by specialty contractors. These contractors take an active part in the design and research of their specific system, and therefore when one is using these types of shoring systems, they should be consulted for design specifics. In terms of experience with contractor-constructed shoring systems and the quality of workmanship and materials that they bring to the task, state highway departments hands down have the most experience. It makes sense that they would have adopted the most practical and safe shoring design specifications. When we look at the production and design of these materials for permanent and manufactured use, there is no question that the respective material industry associations have the largest body of knowledge. As a general rule, design should be in accordance with industry design specifications. Where differences between the recommended allowable stress occur, it is due to the temporary and multiple-use nature of shoring materials. The following industry association specifications have been universally accepted in the United States:

Steel:

Allowable stress design

American Institute of Steel Construction, Inc. (AISC)
1 East Wacker Dr., Suite 3100, Chicago, IL 60601-1920

Wood:

Allowable stress design

National Design Specifications (NDS), ANSI/AF&PA NDS-2005

American Wood Council

PO. Box 5364, Madison, WI 53705-5364

Lumber and grading rules

Western Wood Products Association

522 SW Fifth Ave., Suite 500, Portland, OR 97204-2122

Manufactured sheeting

APA—The Engineered Wood Association

7011 So. 19th, Tacoma, WA 98466

Laminated timber

American Institute of Timber Construction, Inc. (AITC)

333 West Hampton Ave., Englewood, CO 80110

Aluminum:

Allowable stress design

Aluminum Design Manual, 2005

The Aluminum Association, Inc.

1525. Wilson Blvd.,Suite 600, Arlington, VA 22209

Concrete:

Load factor design and allowable stress design

Building Code Requirements for Structural Concrete

American Concrete Institute

P.O. Box 9094, Farmington Hills, MI 48333-9094

The other highly recommended source for material and design specifications is

*Bridge Design Specifications**

The American Association of State and Highway Transportation Officials, Inc. (AASHTO)

444 N. Capitol Street NW, Suite 249, Washington, DC 20001

Short-Form Material Design Standard

Due to the excessive amount of excavation and bridge falsework shoring projects involved in road construction, most highway departments have adopted the following short-form standards for allowable design stress, loadings, and deflections for timber and steel:

Timber:†

These design stresses are based on Douglas fir larch group II or equivalent lumber or timber.

Compression perpendicular to grain	$F_{\perp} = 450$ psi
Compression parallel to grain	$F_c = \dfrac{480{,}000}{(L/d)^2}$ psi ≤ 1600 psi
	where L = unsupported length, in
	d = least dimension, in
Flexural stress	$F_b = 1800$ psi for depth > 8 in
	$F_b = 1500$ psi for depth ≤ 8 in
Horizontal shear	$F_v = 140$ psi
Axial tension	$F_t = 1200$ psi
Modulus of elasticity	$E = 1{,}600{,}000$ psi

Timber connections shall be designed in accordance with *National Design Specifications for Stress Graded Lumber and Its Fastenings* except that reductions in allowable loads required therein for high moisture condition of the lumber and service conditions shall not apply.

*All states have adopted these guidelines as a minimum and have their own published bridge design specifications.

†Timber is considered nonsurfaced lumber that is full nominal dimension. Lumber is considered surfaced on four sides, S4S, and is ½ in less than nominal dimension for boards to 7 in wide and ¾ in less than nominal for boards 8 in wide and greater.

The designer should carefully evaluate the grade and condition of the proposed lumber, especially if it is used material. These short-form design stresses have been in use since before 1997 when NDS came out with adjustment factors for the effects of knots, slope of grain, splits, checks, size, moisture content, duration of load, and repetitive use. When one is buying new lumber, a select structural grade should be requested.

Steel:

Until approximately 1980, the most commonly used steel for plate, I beams, and shapes was A-36 manufactured with a yield strength $F_y =$ 36 ksi. As the mix of recycled steel from autos and other higher-strength items increased, the steel manufactured under the A36 specification was testing as high as 50 ksi. Today most steel I beams and plate are manufactured under a dual A36/A572 specification with a minimum $F_y = 50$ ksi. Most of the beam and plate being used for shoring is 50 ksi material; however, due to reuse, tracking, and multiple owners, even if there are mill certifications for the steel, it is hard to match the steel with the certifications and verify the steel strength. Obtaining and keeping mill certifications with the material they belong to is important because doubt about material strength when steel shoring is designed close to allowable strength can add cost in conservative measures taken and cause designers to lose sleep. For critical items it is fairly inexpensive to send a small piece of the steel, a cookie, to a testing laboratory to certify the yield strength. It is better to have it done before use because it is a given that it will be done later in the event of a failure.

On this subject the highway department prefaced the short-form allowable stresses for steel with the following statement:

> For identified grades of steel, design stresses, except stresses due to flexural compression, shall not exceed those specified in the *Manual of Steel Construction* as published by AISC.

When the grade of steel cannot be positively identified, design stresses, except stresses due to flexural compression, shall not exceed either those specified in said AISC Manual for ASTM Designation: A36 steel or the following:

Tension, axial and flexural: F_t and $F_b = 22{,}000$ psi

Compression, axial:

$$F_a = 16{,}000 - 0.38\left(\frac{L}{r}\right)^2 \quad \text{psi}$$

$$\text{Except } \frac{L}{r} < 120$$

where L = unsupported length

r = least radius of gyration

Shear on gross section of web of rolled shapes: $F_v = 14{,}500$ psi

Web crippling for rolled shapes provide web stiffeners if

$$F_p > 27{,}000 \text{ psi}$$

For all grades of steel, design stresses shall not exceed the following:

Compression, flexural

$$F_{b\,(\text{compressive})} = \frac{12{,}000{,}000}{Ld/bt} \text{ psi}$$

where L = unsupported length, in

d = least dimension of rectangular columns, in

b = beam width, in

t = thickness of compression flange

$$F_{b\,(\text{compressive})} \leq 22{,}000 \text{ psi for A36 and unidentified steel}$$

$$F_{b\,(\text{compressive})} \leq 0.6F_y \text{ psi for other identified steel}$$

Modulus of elasticity E = 30,000,000 psi*

The state of California has developed its own trenching and shoring design manual and a separate falsework design manual. These manuals have been referenced and used all over the world.

Sheet piles:

Sheet pile grades are

ASTM A-328	F_y = 38,500 psi	$F_b = 0.65F_y$
ASTM A-572 grade 50	F_y = 50,000 psi	$F_b = 0.65F_y$
ASTM A-690	F_y = 50,000 psi	$F_b = 0.65F_y$

Steel tube and pipe:

Steel tube and pipe are used within the manufactured shoring industry mostly for manufactured shoring box walls and struts. In the site-constructed shoring they are used almost exclusively for struts. Grades in common use are as follows:

Steel tube:	ASTM A500 grade B†	F_y = 46,000 psi
	ASTM A500 grade C	F_y = 50,000 psi
Pipe:	ASTM A53 type E grade B	F_Y = 35 ksi
	ASTM A53 type S grade B	F_y = 35 ksi

*AISC defines E as 29,000,000 psi; deflection and buckling calculations are most affected by E. A higher E results in less deflection and a higher allowable strength for buckling.

†Mill certification usually runs around F_y = 52,000 to 58,000 psi.

References

American Forest & Paper Association, *National Design Specifications for Wood Construction*, ANSI/AF&PA NDS-1997, American National Standard, Madison, 1997.

American Institute of Steel Construction, Inc., *Allowable Stress Design*, 9th ed., 2d rev., AISC, Chicago, 1995.

State of California Department of Transportation, *Standard Specifications*, Central Publication Distribution Unit, Sacramento, 1984.

The Aluminum Association, Inc., *Engineering Data for Aluminum Structures, Construction Manual Series, Section 3*, 5th ed., Washington, 1986.

The Aluminum Association, Inc., *Specifications for Aluminum Structures, Construction Manual Series, Section 1*, 5th ed., Washington, 1986.

Yokel, Felix Y., and Stanevich, Ronald L., *Development of Draft Construction Safety Standards for Excavations*, vol. 1, National Bureau of Standards, Department of Commerce, Washington, April 1983.

CHAPTER 4

Existing Subsurface Installations and Outside Force Damage Protection

4.1 Introduction

Excavation work is generally thought of in the context of soil stability only. There are really two separate influences that must be dealt with in excavation work—the soil stability and the stability of existing structures within and on the surface of the soil, the infrastructure. The work involved with handling the effects of the soil has always been the same: define the soil and forces associated with it and develop a method of containing them. The quantity of work involved with handling the effects of the existing infrastructure on an excavation has been steadily increasing over the last century. Early on, the solution to the problem was to locate existing infrastructure as much as possible on the plans and then to require that the contractor deal with the effects during construction. As the amount of infrastructure increased, the impact of delays and damages due to this approach skyrocketed, and the results were getting worse not better. In the early 1980s it was becoming obvious that a change needed to be made. There needed to be a new understanding of the problem and an approach to solving it. There was recognition of the fact that the project designer needed more tools to deal with discovery during the planning stage and that more work needed to be done before the projects got to the construction phase. The necessary tools were not just in the form of new technology. Research, development of organizations devoted to the task, education, and funding were necessary. Most importantly a new step had to be introduced into the excavation design and construction process that had been heretofore avoided due to cost and complication. New technology developing at the time—air/vacuum excavation, surface geophysical equipment; new computer mapping, and data management systems—that promised to make it easier to locate subsurface infrastructure and do what previously seemed impossible also prompted the change. Today a new branch of engineering discipline has evolved around this, *subsurface utility engineering* (SUE), and a new segment of the excavation industry has evolved. This segment of the excavation industry most likely has greater importance and impact on safety than the shoring side of the industry because of the fact that it is focused not primarily on worker safety, but on public safety.

For the person new to excavation work, it is important to understand that working with existing facilities is a science of its own and that the impact of failure in this portion of the work has a greater potential for disaster than failing to deal with the soil properly. This chapter looks at excavation planning work that is not directly related to soil stability.

4.2 Major Players in the Process of Existing Facility Depiction

Four entities have a stake in and some control over the outcome of the process of depicting subsurface information:

- *Project owner.* The entity that owns the ground has a major stake in the outcome of the location process. As the one who funds the project, the owner makes primary decisions about the cost and extent of the search to locate utilities. If she or he financially constrains the design engineer in the face of warnings that it is causing risk, the owner can become culpable in the event of an accident. Also the value of the property that the owner holds is seen as a major asset in the recovery of damages in the event of an accident.
- *Project design engineer.* The project design engineer's job is to collect information, depict it on plans, and advise the owner regarding the extent of the search and level of accuracy needed to safely construct the project. Collection is an iterative process that evolves as the information and project design evolve. Engineering firms usually do most of the collection work in-house and sometimes contract out to subsurface utility engineering firms or line location excavating contractors in cases where it is necessary to excavate and determine the exact location of a utility. The project engineer has a limited budget as determined by the owner and limited accuracy in the information that he or she collects. The project design engineer is responsible for

how thoroughly and accurately the quality of the information is depicted in the plans and specifications.

- *Utility owners.* The job of utility owners is to protect and ensure continuous operation of their facilities. They try to maintain and keep accurate records and drawings and will always disclaim responsibility for accuracy of the work. The fact is that they keep records, but the accuracy is not always reliable. They are obligated under federal and state statutes to participate in and share the costs of a regional notification center, maintain records, and mark the location of their known facilities on the ground surface just prior to excavation. There are penalties and fines for not responding within 24 hours of notification. Operators of hazardous high-pressure liquid pipelines usually will not allow excavation without a representative on site to assist in exact location of their lines. Utility owners often contract with a contractor that specializes in locating and marking lines for them.
- *Excavation contractor.* The job of the excavation contractor is to construct the project while safely protecting and working around existing underground facilities. Civil law requires that he or she notify a regional one-call center 24 hours prior to excavating and designate to line locators the limits of their proposed excavation work. If locators do not respond in a timely manner or contractors do not assist, there are financial and civil penalties that in the case of damage can be assessed on the one responsible. Note that law requires that both parties mitigate damages in the case of delay. Damages should be financial in the form of compensation for delay costs and should never be the loss of life due to the ineffectiveness of anyone to get the job done in a timely manner.

  OSHA states that when responders are not there within 24 hour or cannot establish the location, the contractor can proceed, provided she or he does so with caution and safe means of locating. This should be considered only as a last resort.

  Civil law also requires that the contractor who damages existing facilities in any way notify the one-call center which in turn notifies the owner of the facility. This is an extremely important provision because the consequences of minimal damage can show up years later in the form of explosion, fire, and loss of life. This is definitely not a case where it is easier to get forgiveness than it is to get permission to ignore the damage. The civil law only requires that existing utilities be located on the surface. OSHA law requires that the contractor contact one-call notification systems and then locate and determine the exact location of underground facilities by safe and acceptable means. It also requires the contractor to protect utilities as necessary to safeguard employees. Most often this work is performed in conjunction with production excavation work, unless specific contract requirements state that the work is to be done in a separate excavation process prior to production work. The latter is viewed as far more expensive; however, when lined up against risk to life and production delays, it can become cost-effective quite rapidly.

Obviously all four of these players have different and sometimes competing objectives, but they all have a stake in locating and protecting existing subsurface utilities. The public inclusive of the employees on the project also has a stake in the outcome, but no direct control over the process.

In the interest of everyone involved, the U.S. Department of Transportation (DOT) sponsored the Common Ground Alliance (CGA) to study and promote a one-call system for damage prevention. This study identified and validated existing best practices resulting in publication of a CGA *Best Practices Version 3.0*. This sets the minimum standard of care in operating a one-call system and line location work. CGA promotes a sense of shared responsibility for the protection of underground facilities; but make no mistake about it, if there is an incident, all four parties will not share the responsibility. CGA and the one-call system assist the responsible parties in getting their job done properly.

4.3 Existing Buried Infrastructure

There are many different types of buried infrastructure. Each has unique problems and risks associated with it. They are listed here in a hierarchy of safety risk.

Hazardous Liquid Transmission Pipelines

These consist mostly of natural gas, gasoline, and fuel oil. Of the energy commodities consumed in the United States, 64 percent are transported by pipeline. These pipelines are as much a part of the transportation system as highways and waterways. These lines are intrastate and interstate lines. They are classified as collection lines that go from their source to refining plants, transmission lines that move the product to distribution centers, and distribution lines that move the product to the consumer.

The hazardous liquid transmission industry has experienced rapid growth in the last decade, and with the need for energy independence it is posturing for major additions to the existing 2.3 million mi of natural gas and hazardous liquid pipelines. A major natural gas transmission line is being proposed to go from Prudhoe Bay, Alaska, through the Yukon Territories, British Columbia, and the United States. A 1600-mi pipeline from west Texas to western Pennsylvania and a 1663-mi natural gas project from northern Colorado to eastern Ohio.

Between 1989 and 1998 major hazardous liquid transmission pipeline accidents increased by 4 percent per year. DOT statistics indicate that outside force damage is the leading cause of damage to these pipelines, with 123 deaths and 562 injuries between 1990 and 2006, the second major cause being corrosion. Excavation-related property damage exceeded $500 million in that period.

Damage to high-pressure pipes is likely to result in explosion and fire, sometimes instantly and sometimes hours to years after the incident. Extreme environmental damage results from leakage that does not ignite. Excavation work around these lines should be approached with extreme caution.

Outside force damage to high-pressure lines can come from the following causes.

1. Corrosion after scrapes off the exterior lining or damage to the cathotic protection system. Eventually the wall thickness deteriorates to the point that the pipe ultimate yield stress is exceeded owing to internal pressure, and the pipe bursts.
2. An excavator bucket tooth, a dozer ripper, or a dozer blade can scrape the metal off the side of the pipe. Reduced pipe wall thickness results in rupture.
3. A sharp blow from power or hand excavating tools will add bending stress to the hoop tension stress in the pipe. This can add up to exceed the ultimate yield stress. As an example, a 24-in-diameter × 3/8-in wall A53 steel pipe can operate at a pressure of 612 psi and will experience a hoop stress of 29.3 ksi. A 1/8-in-deep wall scrape and a 450-lb impact force can raise the pipe wall stress to over 60 ksi, which is the bursting strength of A53 steel pipe. This could easily be accomplished with a sharp pick.
4. Remote activities such as trenchless pipe installation, pipe pulling and bursting, drilling, and pile driving can cause damage to high-pressure lines.
5. Excessive overburden pressure results from loads at the surface such as spoil piles, tractor and crane tire loads, and crane outrigger pads.

Pressure Water and Sewer Transmission and Distribution Lines

Force mains are usually constructed at least 5 ft below the surface and run on the contour of the surface. These lines generally have locked or welded joints; however, some ductile iron and asbestos cement pipelines have push-on joints with thrust block restraint.

Damage to these lines results from the same sources listed for high-pressure lines, except that an additional source is removal of thrust block support.

Some consequences from damaging or rupturing water lines are

- Disrupted water service
- Disruption of fire services
- Drinking water contamination
- Flooding of the excavation, causing possible worker inundation and excavation wall collapse
- Environmental and health damage when ruptured sewer mains spill into streams
- Impact force damage to workers from moving pipe and thrust blocks

Specific Hazards Asbestos cement pipe (ACP) has been used in North America for water lines since 1931. By the 1970s it was known that asbestos was a hazard, and in 1979 the U.S. Environmental Protection Agency announced their intention to ban all future use of asbestos in new products. ACP, sometimes called *Transite pipe,* is commonly encountered in excavations. It was manufactured in 20-ft lengths with diameters from 4 to 36 in. This pipe is brittle and easy to break or scrape off with a shovel or excavator bucket. Pipe joints are made with a push-on collar that is separate from the pipe. The collars stick out from the pipe so that it is easy to strike them with an excavator bucket or shovel when digging parallel to the pipe. Aside from problems associated with repairing this pipe, it is dangerous when dry asbestos fibers become airborne and are breathed in. Asbestos-containing materials have been classified into *friable* and *nonfriable* categories. Asbestos cement pipe is regarded as nonfriable and not hazardous. Special care needs to be taken in handling, storing, and disposing of it at an approved landfill site. If it is broken, it becomes friable and a hazardous material at which point the OSHA Construction Industry Standard for Occupational Exposure to Asbestos Subpart Z, 29 CFR 1926.1101 Asbestos, controls handling of this material.

Pressure pipe uses welded, constrained joints or thrust blocks to prevent the pipe from moving at bends and tees. Pressure pipe with rubber gasket joints should be buried prior to pressure-testing it because it can jump vertically and horizontally at the joints and blow apart. If the dirt close to a thrust block is removed, the line pressure can blow the block into the excavation. A 30-in waterline under 150 psi pressure will cause a 106,000-lb force at the thrust block. After the block moves into the excavation, the water can fill a trench before workers can get out.

Gravity Sewer Lines

These lines are found in three major categories:

- *Sewer laterals from house to gravity sewer main in the street.* In neighborhoods that have sewer service every house should have a lateral to the main line in the street. These lines intersect parallel lines running up and down the street. Sewer laterals run at minimum ¼ in/ft, usually have 3 ft of cover at the curb, and dive down from there to connect with the sewer main. Lateral sizes are usually 4 to 6 in and constructed from vitrified clay pipe (VCP), polyvinyl chloride (PVC), cast iron, or Orangeburg pipe. Because there is little safety risk it is usually less expensive to expose and protect or replace these lines as the production excavation proceeds than to prelocate them.
- *Gravity sewer main to wastewater treatment plant or sewage lift station.* Gravity sewer mains generally range in size from 8 to 24 in before coming to a lift station. These lines can run almost flat; however, a 1-ft drop per 100 ft is preferable. The depth of these lines varies from 4 to 30 ft and more. Eventually the cost to pump and lift becomes less expensive than to dig deep. These lines are usually constructed from vitrified clay pipe, PVC, Transite, or reinforced concrete pipe (RCP) and are found below other utility lines.
- *Gravity sewer mains carrying sewage from communities to regional sewage plants with large lift stations along the line.* These lines have a 24- to 144-in diameter. They are laid fairly flat due to the large flows and distances the sewage is carried. At the end of the run they can be 40 ft and deeper.

Specific Hazards If gravity sewer lines are ruptured during potholing or final excavation, workers will be exposed to live sewage. Raw sewage carries bacteria, viruses, and parasites. The Center for Disease Control says that exposure to droplets and sewage gases carried in these lines can cause gastroenteritis, an infection in the intestinal tract, and hepatitis, an infection in the liver. Contact with cuts and scratches can cause infection.

If the escaping gas from broken sewer lines accumulates in nearby enclosed areas, it can cause asphyxiation or explosion.

A broken sewer main or tie-in will create the need for plugging and bypass pumping to tie in or repair the line. Gasoline pumps in confined spaces can create exposure to carbon monoxide poisoning. If pipe plugs are not properly blocked, they can come out of a pipe with explosive force. Bypass and repair plans need to be well thought out and not drawn up on the spur of the moment.

Locating and tapping sewer lines usually involves work in excavations over 5 ft deep and therefore requires shoring or open cut worker protection. In streets with limited right-of-way, shoring is the only option. There are hazards associated with installing and removing shoring.

Storm Drains

A storm water drainage system is a combination of drainage swales, street gutters, drain inlets, culverts, underground gravity conduits, and open channels that carry storm water to treatment plants or streams and rivers. An attempt is made to prevent erosion and contaminated materials from entering the system. Storm drains are mostly gravity flow and can be found at all depths. Drainpipes are also used to drain water from behind retaining walls, and in farm applications to collect and carry away irrigation water.

Consequences of damaging storm drain pipes are as follows:

- Broken and plugged up storm drains will cause flooding and subsequent erosion and destruction of surface facilities.
- Broken or plugged retaining wall drains can cause water to back up behind the wall and cause the wall to fail.

Electrical and Communication Conduits

These break down into the following categories:

- Transmission voltage (cross-country) is typically over 13,800 V.
- Distribution voltage (power plant or substation to user) is under 13,800 V.
- Utilization voltage (end-use power) is lower than 600 V. Household circuits are typicaly 110/120 V and 220/240 V. Voltage over 600 V is considered to be high voltage.
- Communication ducts and duct banks, at low voltage, have an operating current below 250 V.

Electrical duct banks (Fig. 4.1) are typically 2- to 6-in PVC, type EB for concrete-encased and DB for direct burial or metallic conduit, EMT. High-voltage lines are normally encased in red concrete. With the advent of phone cable, fiber-optic cable, computer monitoring control systems (SCADA), etc., concrete-encased duct banks are common. Also new electrical and communications systems have been pulled in through existing abandoned conduits and pipelines. Duct bank systems are usually sloped so that water will drain to pull boxes so that it can be pumped out at that location. Most electrical duct and duct banks are a minimum of 2 ft 6 in from the surface and as deep as 10 ft for large duct banks. There are also direct buried wires. Many times high-voltage conduits are red or encased with red concrete. Also watch for red plastic detectable warning tape buried at the surface.

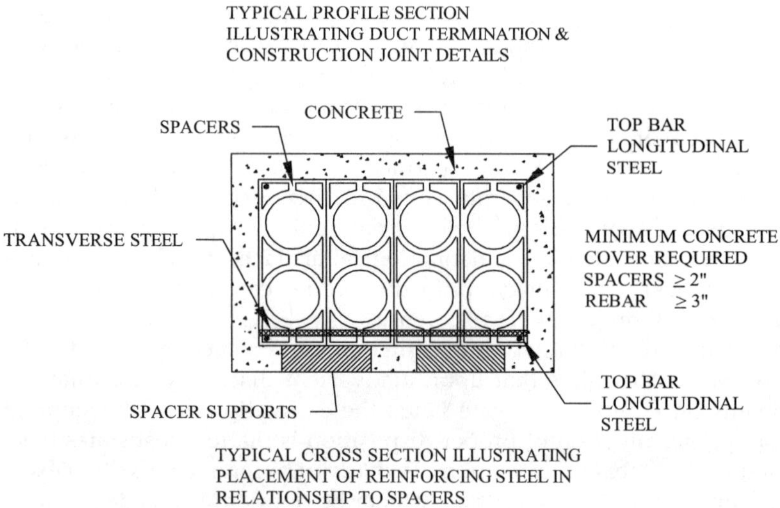

FIGURE 4.1 Typical concrete duct bank.

Consequences of failure to locate and protect include

- Electrocution, electric shock, and burns
- Discontinuance of vital electrical and communication systems

Specific Hazards The fact is that electrical and communications cables are buried, out of sight and out of mind. Codes are not always followed, and line locations are not accurately mapped and many times not reported at all. Always expect to find buried conduits, and do not assume that they are harmless. High voltage and low voltage will cause burns and shock when contacted by the human body and often result in death, even in small voltage doses. A NIOSH study of occupational electrical injuries in the United States found that from 1992 through 1998 there were 37 deaths and 440 nonfatal events related to contact with underground buried power lines. They also found that occupational electrical incidents are disproportionately fatal.

Auger drilling for fence posts and bore and jack operations can encounter and intercept electrical conduits. Because augers are small and some distance from the operator, danger seems less imminent. Operators of posthole diggers are usually fence installers and not familiar with excavation work or tend to forget about subsurface line location requirements.

Discovery of Human Remains, Archaeological Artifacts, and Paleontological Artifacts

There are several federal and state laws aimed at discovering and protecting these resources in the planning process prior to excavation.

The Antiquities Act of 1906 and the Archaeological Protection Act of 1979 made it a crime and a felony to excavate and collect antiquities on federal land and public land without a permit. This also applies to inadvertent discovery during excavation. Laws regarding inadvertent discovery of these items during excavation break down into rules pertaining to public lands and rules pertaining to private land. Furthermore there are separate laws for discovery of human remains versus archaeological artifacts (burial items, weapons, trinkets, bottles and cans, etc.) and paleontological artifacts (fossils and nonhuman bones). These inadvertent discoveries can quickly shut down and delay a project at great expense.

The federal Native American Graves Protection Act of 1990 (NAGPRA) requires that upon inadvertent discovery of human remains on federal or private lands the excavation work be stopped until determination of proper disposition is made. Most states have similar laws that require upon discovery of human remains that excavation stop and the sheriff or coroner be notified. If it is determined that the remains are not the result of a crime, the legal authority is required to notify a Native American Heritage Commission which in turn notifies the most likely descendents of the discovery.

In the case of archaeological and paleontological artifacts, states usually claim ownership of anything found and require that the items not be removed from their setting prior to a determination of their status, proper removal, and distribution.

Contaminated Materials

During the planning phase of government projects, a process of hazardous materials discovery and identification is undertaken. Contaminated soils are reasonably easy to anticipate. A survey of structures and businesses in the area will indicate possible contaminate sources; printed circuit boards (PCBs) from transformers, service stations, dry cleaning, asbestos-related manufacturing, etc., will give an indication of what to expect. In the case of private lands, detection of contaminants is usually avoided by the owner as much as possible. On large private projects, planning and environmental impact reports help to root out potential contamination sites. On small private projects such as farm ponds, irigation and water systems, sewage systems, structure foundations, etc., there are no legal requirements to seek out and find contaminated soil. In these cases the excavation contractor should be extra cautious.

In the case of anticipated encounter of contaminated materials, a hazard analysis and work plan with safety equipment and procedures can be set up prior to excavation. Several tools can be used for anticipated discovery of contaminants in the excavation such as soil testing in the laboratory, metal detection devices, gas detection devices, photoionization detector, soil gas probe, radiation detection devices, visual evidence, and olfactory indications (sense of smell). There is field test equipment available that can determine the level of contamination at the site so that work stoppage is not always necessary.

In the case of inadvertent discovery, the only tools workers have are visual and olfactory. Here identification and quantification of contaminates in the soil are a little bit more complicated. Here are some things to watch for:

- Oil-contaminated soils when freshly excavated have a sheen and a petroleum odor. If a sample of this soil is submerged in water, a rainbow sheen will form on the surface.
- In clay and organic soils contaminants do not migrate far. Due to the negative charge in clay particles, toxic metal ions such as lead, cadmium, chromium, and mercury can be attracted and held. They might show up as a different color than the clay that they are held in.
- In noncohesive soils, water leaches contaminants down to impermeable layers or the water table where concentrations of contaminants will look different from the color of the soil.
- Asbestos is naturally occurring in serpentine deposits (usually bluish green shale).

Once hazardous materials are discovered, they must be identified, managed, and treated or disposed of. Discovery of unanticipated hazardous material will trigger notification requirements. The National Response Center (800-424-8802) is set up to serve as the sole national contact for reporting of all oil, chemical, radiological, biological, and etiological discharges into the environment within the United States.

Consequence of Failure to Locate and Protect Hazardous Materials

Unfortunately the financial incentive to dump and hide hazardous waste is overwhelming. The chance of inadvertently excavating and exposing workers to radiation, PCBs, solvents, asbestos, and a host of other contaminants without ever knowing that it happened is real and probable. Construction workers should at least know that the possibility exists. The result of these exposures can be subsequent contamination of family and friends, illness including cancer and respiratory ailments, and premature death.

4.4 Pipeline Safety Regulations

Most people are aware that there are laws requiring line location prior to digging a hole simply because of the fact that at some time and place they have seen a sign that states there are underground facilities in the area and it is unsafe to dig prior to calling the number on the sign. The "call before you dig" public awareness program has been successful to the point that everyone is aware of it; however, awareness of the requirement is about as far as the knowledge goes.

Due to the severity of accidents associated with hazardous pipelines, federal laws requiring excavation notification and subsurface facilities location were developed through the U.S. Department of Transportation by the Pipeline and Hazardous Materials Safety Administration (PHMSA) through the Office of Pipeline Safety (OPS). The OPS was established in 1968. In the year 2000, OPS published a paper entitled "OPS Research: Past Present and Future," discussing changes to the office in its 32-year history and the events and research that drove those changes. This is one conclusion from the research that is pertinent to excavation work:

> Mechanical damage is the single largest cause of failures on gas transmission pipelines and a leading cause of failures on hazardous liquid pipelines. The mechanical damage usually occurs after a pipeline is constructed and is caused by excavation equipment which deforms the shape of the pipe, scrapes away metal and coating, and changes the mechanical properties of the pipe.

As a source of pipeline failure, corrosion is second to outside force damage. This conclusion has driven laws, research, and funding to the level where they are today.

Even though the regulations were developed with hazardous pipelines in mind, they also protect against damage to any subsurface facilities that might be present. OPS issues and enforces pipeline safety regulations that identify minimum standards for pipeline design, construction, testing, operation, maintenance, and safety. Design and safety of these pipelines are administered in 49 CFR Subchapter D, Pipeline Safety.

4.4.1 Federal Regulations

Federal regulations important to excavation work and protection of existing subsurface facilities are as follows:

> *Title 49,Transportation, Subtitle VIII, Pipelines, Chapter 601, Safety* Sec. 60102(c) Public Safety Program Requirements. (1) The Secretary shall include in the standards prescribed under subsection (a) of this section a requirement that an operator of a gas pipeline facility participate in a public safety program that —
>
> (A) notifies an operator of proposed demolition, excavation, tunneling, or construction near or affecting the facility;
> (B) requires an operator to identify a pipeline facility that may be affected by the proposed demolition, excavation, tunneling, or construction, to prevent damaging the facility; and
> (C) the Secretary decides will protect a facility adequately against a hazard caused by demolition, excavation, tunneling, or construction.

The public safety program referred to here is the *one-call-system*. This addresses the requirement that pipeline operators participate.

The Pipeline Inspection, Protection, Enforcement, and Safety Act of 2006 extends the notification requirement to excavators by amending 49USC60114 with the addition of the following:

Sec. 2 Pipeline Safety and Damage Prevention

(a) One-call civil enforcement-

(1) Prohibitions, Section 60114 is amended by adding at the end the following;

(d) Prohibition Applicable to Excavator—A person who engages in demolition, excavation, tunneling, or construction—

(1) may not engage in demolition, excavation tunneling, or construction activity in a state that has adopted a one-call notification system without first using that system to establish the location of underground facilities in the demolition, excavation, tunneling, or construction area;

(2) may not engage in such demolition, excavation, tunneling, or construction activity in disregard of location information or markings established by pipeline facility operator pursuant to subsection (b); and

(3) who causes damage to a pipeline facility that may endanger life or cause serious bodily harm or damage to property—

(A) may not fail to promptly report the damage to the owner or operator of the facility; and

(B) if the damage results in the escape of any flammable, toxic, or corrosive gas or liquid, may not fail to promptly report to other appropriate authorities by calling the 911 emergency telephone number.

Employers should note that these are public safety issues and not necessarily worker safety issues. The employer is not protected here by the worker compensation laws and can be sued by the public in addition to federal and state criminal penalties. The federal criminal penalties for disregarding these regulations are as follows:

49 CFR Subtitle B, Subchapter—D Pipeline Safety, Part 190

190.229 Criminal penalties generally.

(e) Any person who willfully and knowingly engages in excavation activity without first using an available one-call notification system to establish the location of underground facilities in the excavation area; or without considering location information or markings established by a pipeline facility operator; and

(1) Subsequently damages a pipeline facility resulting in death, serious bodily harm, or property damage exceeding $50,000;

(2) Subsequently damages a pipeline facility and knows or has reason to know of the damage but fails to promptly report the damage to the operator and to the appropriate authorities; or

(3) Subsequently damages a hazardous liquid pipeline facility that results in the release of more than 50 barrels of product; shall,

upon conviction, be subject for each offense to a fine of not more than $5,000, imprisonment for a term not to exceed 5 years, or both.

4.4.2 Worker Protection Regulations

Worker protection requirements come from federal OSHA 29 CFR 1926:

1926.651 Specific Excavation Requirements—Underground Installations.

1926.651(b)(1) The estimated location of utility installations, such as sewer, telephone, fuel, electric, water lines, or any other underground installations that reasonably may be expected to be encountered during excavation work, shall be determined prior to opening an excavation.

1926.651(b)(2) Utility companies or owners shall be contacted within established or customary local response times, advised of the proposed work, and asked to establish the location of the utility underground installations prior to the start of actual excavation. When utility companies or owners cannot respond to a request to locate underground utility installations within 24 hours (unless a longer period is required by state or local law), or cannot establish the exact location of these installations, the employer may proceed, provided the employer does so with caution, and provided detection equipment or other acceptable means to locate utility installations are used.

1926.651(b)(3) When excavation operations approach the estimated location of underground installations, the exact location of the installations shall be determined by safe and acceptable means.

1926.651(b)(4) While the excavation is open, underground installations shall be protected, supported or removed as necessary to safeguard employees.

These requirements are for the purpose of worker protection and not necessarily for the protection of the public.

Some of the language used is notable because it is not very specific. In 1926.651(b)(1), *estimated location* means location on the surface and today is considered to be a width of the pipeline or utility width plus 2 ft centered on where the utility is thought to be. In the same citation the term "reasonably may be expected" refers to the use of reasoning to anticipate unmarked utilities. See Appendix A, Commentary on OSHA Subpart P for more on this.

4.4.3 State Regulations

The Office of Pipeline Safety encourages states to adopt and administer their own pipeline safety programs. States and OPS work together on education, training, and research. The states at a minimum must meet the federal pipeline safety requirements, and in most cases they exceed those requirements. These are typical requirements and provisions of most state programs:

- Excavators must contact regional notification centers prior to excavating.
- All operators of subsurface pipelines must be a member of the one-call center.
- Locator personnel must be qualified by meeting training standards.
- Pipeline operators must maintain plans for their subsurface installations.
- Excavators must immediately notify the pipeline operator if they damage the lines.
- Provisions must be made for consequence of cost to operator or excavator if the law is not followed.
- Provisions must be made for civil penalties for failure to follow the regulations.
- Excavators must dig with hand tools to exactly locate utilities within the approximate location of the line.
- Excavation permits cannot be issued unless applicant has an initial inquiry identification number.

There are some important definitions within these regulations. Some states have gone further than others in defining their language; however, the following or similar language occurs within more than one state regulation.

- "Approximate location of subsurface installations" means a strip of land not more than 24 in on either side of the exterior surface of the subsurface installation. "Approximate location" does not mean depth. See Fig. 4.2.
- "Underground facility" means anything buried or placed belowground for use in connection with the storage or conveyance of water; sewage; electronic, telephonic, or telegraphic communications; cablevision; electronic energy; petroleum products; gas, gaseous vapors, hazardous liquids, or other substances and including but not limited to pipes, sewers, conduits, cables, valves, lines, wires, manholes, attachments, and those parts of poles or anchors belowground. *Qualified person* means a person who completes a training program that meets the minimum training guidelines and practices of Common Ground Alliance current best practices.

The state of California in response to a recent accident has developed a new high-priority category:

- "High-priority subsurface installation" means high-pressure natural gas pipelines with normal operating pressures greater

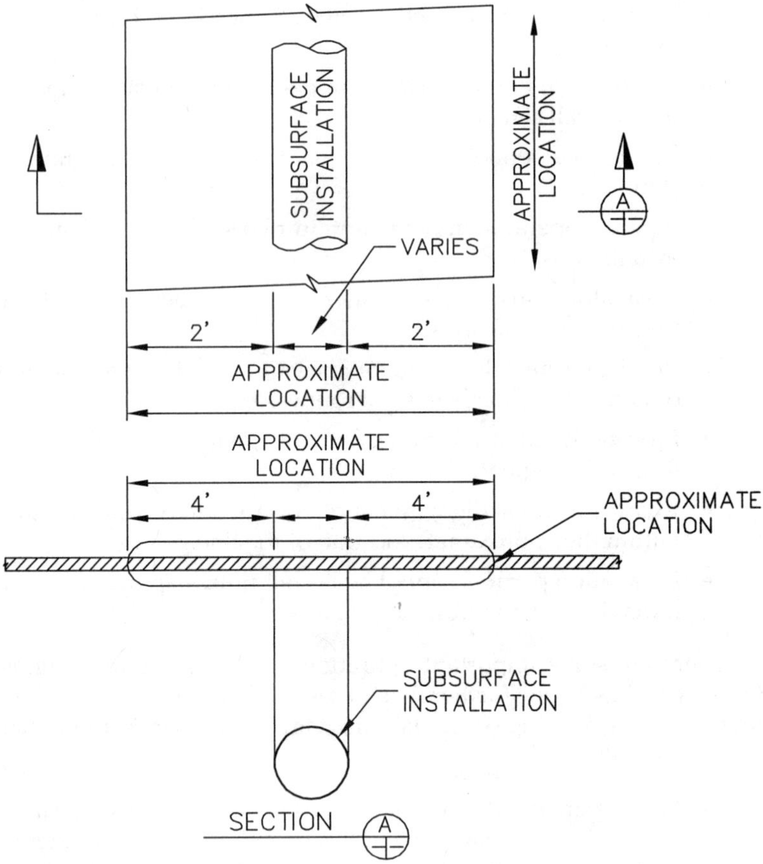

FIGURE 4.2 Approximate location of existing subsurface installation.

> than 60 psig or greater than 6-in nominal pipe diameter, petroleum pipelines, pressurized sewage pipelines, high-voltage electric supply lines, conductors, or cables that have a potential to ground of greater than or equal to 60 kV, or hazardous materials pipelines that are potentially hazardous to workers or the public if damaged.

It is important to note that *approximate location* only refers to the surface and is provided by the owner of the facility, and the *exact location* is determined by the excavator by hand digging unless otherwise agreed on methods are approved. Here is language from a regulation directed toward this:

> The excavator shall determine the exact location of subsurface installations in conflict with the excavation by excavating with hand tools within the area of the approximate location of subsurface installations as

determined by the field marking provided before using power operated equipment, except removal of pavement. Exception:

a. Document notice of intention to use vacuum or excavation equipment and get acceptance from operator of facility.
b. If it cannot be determined by hand digging contact operator for more information.

In some states there are some exclusions from the call-before-you-dig requirement. The following are examples.

- Excludes excavations on real or residential real property when excavation permit is not required
- Excludes persons or entities that rent equipment to excavators as long as rental agreement notifies them that they need to comply with notification laws

4.5 Existing Subsurface Utility Location Standard Practice

There are three phases in the process of locating and protecting subsurface utilities:

1. Project design phase in which existing subsurface utility information is collected and depicted on a set of plans
2. Subsurface utility location-surface marking prior to production excavation work
3. Actual excavation to precisely locate and protect existing facilities

In discussing these phases it is important to make the distinction between two similar terms. The following are the author's definitions with the intent to emphasize the concept that these are two entirely different processes.

Subsurface facility location. Depiction of existing subsurface pipelines, electrical ducts, buried structures, and known buried objects (piles, abandoned shoring, grave site, tunnel, etc.) that are located by lines on a plan or in the field by paint, stakes, driven steel pins, etc. This represents two-dimensional horizontal and lineal markings. The accuracy of these lines to the exact location of the underlying appurtenances is always in question until they are actually dug up and verified. Subsurface facility location work usually does not involve excavation.

Precise location. The true location of existing subsurface pipelines, electrical ducts, buried structures, and known buried objects (piles, abandoned shoring, grave site, tunnel, etc.), that have been exposed, surveyed, and recorded. There are always three dimension coordinates (alignment, length, and depth) associated with this, and if shown on a plan, there is also a benchmark required to tie it into a known survey system. Precise location work always involves excavation.

Subsurface facility location work always precedes exact location work. In facility location work there is risk from traffic and electrocution; however, there is no risk of damaging the line being located. In precise location there are additionally all risks associated with excavation work and a high risk of damaging the line being located.

4.5.1 Project Design Phase Collection and Depiction of Subsurface Facilities

In the project design phase, subsurface facility information is collected and depicted on a set of plans. The best example of standard practice is contained in the American Society of Civil Engineers (ASCE) publication "Standard Guidelines for the Collection and Depiction of Existing Subsurface Data." This document was developed in the year 2000 as a National Consensus Standard and is part of The National Transportation Safety Board's national damage prevention strategy. In this guide they have categorized the quality level of the available information on an existing utility. Based on quality level and project needs, the project owner, design engineer, and constructor can make risk allocation decisions. The quality level is a professional opinion of the quality and reliability of utility information. Quality level is established by different methods of data collection and interpretation. Quality levels are as follows:

- *Utility quality level A.* Precise horizontal and vertical location of utilities is obtained by the actual exposure (or verification of previously exposed and surveyed utilities) and subsequent measurement of subsurface utilities, usually at a specific point. Minimally intrusive excavation equipment is typically used to minimize the potential for utility damage. A precise horizontal and vertical location, as well as other utility attributes, is shown on plan documents. Accuracy is typically set to 15 mm (0.6 in) vertical and to applicable horizontal survey and mapping accuracy as defined or expected by the project owner.

If this quality level is critical to design and necessary prior to contract award, it will require an excavation, including planning, one-call notification, excavation permits from local agencies, traffic safety plans and signage, a surveyor, backfill, compaction, and pavement repair. In the absence of precontract design needs for accuracy, this quality level is usually performed as part of the contract by the contractor prior to production excavation.

For quality level A (QLA) designation, unless an engineer is willing to certify the accuracy of previously exposed and surveyed utilities, they would need to be again excavated and surveyed.

QLA work is usually done at specific locations, and the same as it is with soil boring logs, it does not necessarily depict what is going on in between. Considerable thought should be given to alignment of high-pressure lines that parallel proposed new installations. Alignment for pressure lines is not nearly as critical as for gravity lines. They can be snaked around other utilities, trees, rock outcroppings, poles, etc. Sometimes they are installed by directional drilling and can be off line by several feet. Years later the poles, rocks, and trees disappear, and the reasoning for moving the line out of straight alignment disappears with them.

- *Utility quality level B.* Information is obtained through the application of appropriate surface geophysical methods to determine the existence and approximate horizontal position of subsurface utilities. Quality level B (QLB) data should be reproducible by surface geophysics at any point of their depiction. This information is surveyed to applicable tolerances defined by the project and reduced onto one plan documents.

Surface geophysical methods such as electromagnetic, magnetic, acoustic, and chemical are performed at the surface and do not precisely locate the utility. These methods can determine existence, approximate horizontal alignment, and sometimes approximate the depth of utilities.

QLB depictions on a plan are not extrapolations and should only be shown where positive field location has been accomplished. In the field prior to construction geophysical equipment is often used to mark utility locations and predict or assume the alignment of the utility. This is not the same as QLB depiction on a plan.

QLB geophysical information is checked against existing utility drawings to confirm that they agree. Discrepancies are resolved through additional work in the field.

- *Utility quality level C.* Information is obtained by surveying and plotting visible aboveground utility features and by using professional judgment in correlating this information to quality level D (QLD) information.

Quality level C (QLC) work consists of surveying and depicting on plan the location of surface features of belowground utilities such as sewer water and electrical manholes, pump stations, transformer pads, and storm drain inlets. It also includes confirming the accuracy of existing drawings depicting this information and resolving discrepancies.

- *Utility quality level D.* Utility information is derived from existing records or oral recollections.

QLD work consists of a search for and collection of existing utility records. There is a review of those records for quality, clarity, and ambiguity. There is depiction on a composite drawing including attributes such as pipe material, size, condition, number of conduits, concrete encased, out-of-service, abandoned, approximate depth, and the quality of this information. There is no verification that this information is accurate or that everything that is there is reported.

QLD information is the starting point for making informed risk management decisions about design of new facilities and protection of existing subsurface utilities. Cost associated with obtaining higher utility quality level should be carefully weighed against injury; loss of life; disruption of vital water, fire, and communication systems; and project disruption, redesign, and delay costs. For contractors the more clearly depicted and accurate the subsurface information is, the more closely the job can be bid. The reality is that if the information is not clear, money will be in the bid for the associated risk; and when you get there on the job, the owner will pay again in the form of delay and what it really costs to deal with the problem. Studies have shown that there is rarely a negative return on investment in upgraded utility quality depiction.

4.5.2 Utility Location Surface Marking Prior to Production Excavation Work

In 1998 the Department of Transportation initiated the "Common Ground Study," a study of best practices in preventing outside force damage to existing subsurface facilities and the effectiveness of one-call notification systems. The study was conducted by a joint industry and government team consisting of 160 members. The team investigated and produced a consensus on best practices for locating and protecting existing subsurface facilities. In the year 2000 one result of this work was the formation of the Common Ground Alliance (CGA) and Best Practices 1.0, a collection of best practices in subsurface facility damage prevention.

Today the standard of practice is the Common Ground Alliance document Best Practices Version 4.0. This is a large document, 100 pages, that can be downloaded free from the Internet. It should be on the bookshelf and read by everyone involved in excavation work. The document was last amended in March 2007 and is considered a work in progress to be amended as new technology and better practices evolve. The best practices are divided into eight chapters that include

1. Planning and Design Best Practices
2. One-Call Center Best Practices
3. Location and Marking Best Practices
4. Excavation Best Practices
5. Mapping Best Practices

6. Compliance Best Practices
7. Public Education Best Practices
8. Reporting and Evaluating Best Practices

Prior to any excavation work these Best Practices should be studied and adhered to. In the event of damage to existing facilities, the only defense will be proven adherence to the best practices.

Subsurface Utility Engineering, Standard Guidelines for the Collection and Depiction of Existing Subsurface Data, and CGA's continuing development of Best Practices in subsurface facility damage prevention are the cornerstone of the 21st-century effort to deal effectively with existing subsurface facilities in excavation work. The Department of Transportation through OPS supports CGA as the leading organization in this effort because their membership comprises all the major stakeholders in the process. It is currently estimated that there are over 700,000 line strikes a year and that 40 percent of those strikes do not have a one-call request prior to digging. Reducing this percentage is the primary goal for CGA, and there is no doubt that a reduction in line strikes means an increase in lives saved. CGA is involved in all aspects of developing pipeline safety programs. In 2005 the Federal Communications Commission established a new call-before-you-dig toll-free three-digit number 811 (see Fig. 4.3) that will

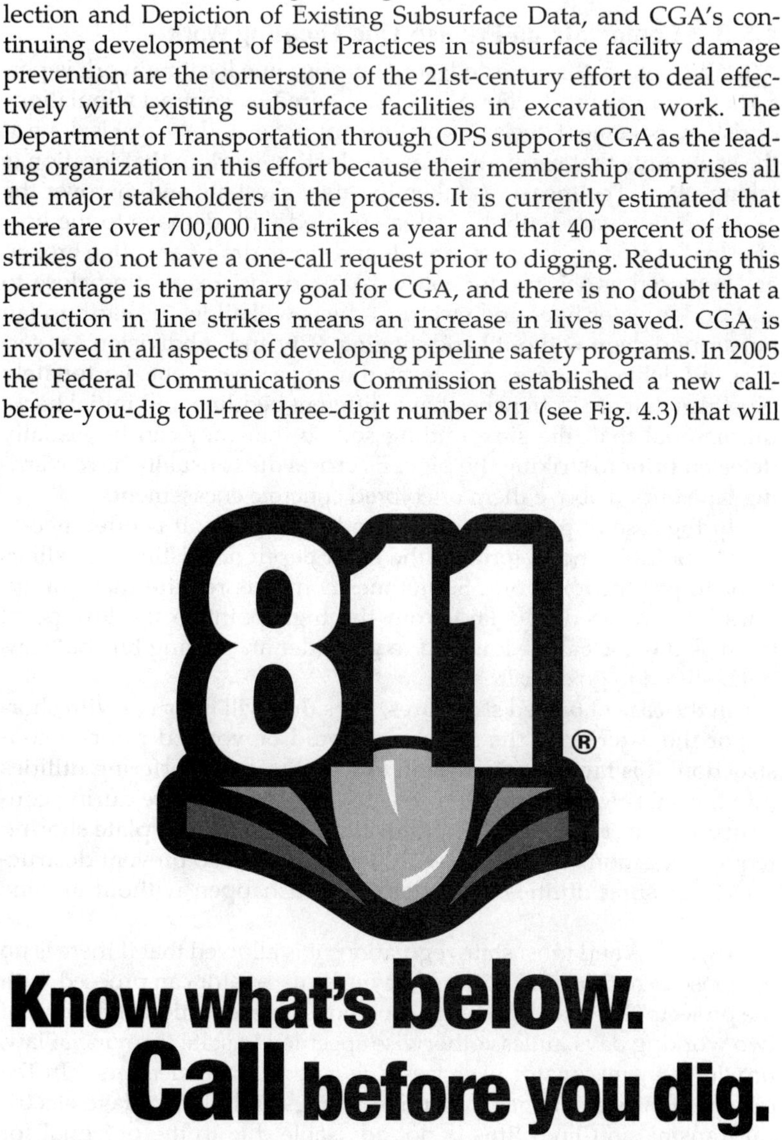

FIGURE 4.3 Call-before-you-dig toll-free number.

connect the caller with the closest one-call center. CGA's Educational Programs Committee has worked with stakeholders to develop an awareness campaign to accompany the national rollout of 811.

Outside force damage prevention is the segment of the excavation industry that will see the most attention, research, and technological advances in the coming years. Everyone involved with excavation work is encouraged to become involved with CGA; they are the key to education and innovation in this field.

4.5.3 Comments on Precise Line Locating Work

After the contract is awarded, some precise line locating work is performed prior to production work as required by contract stipulations or risk management decisions by the contractor, and the rest is done at the same time that excavation for production work in that location is taking place. The measures taken to safely excavate and discover the exact location vary with the risk associated with damage to the line. Due to cost efficiency, sewer laterals and storm drains are often broken and replaced as the production work proceeds. Excavating close to and under water lines and electrical ducts will allow the earth to fall out around them without hand digging. City and subdivision gas service and delivery lines can be located using a steel probe. Fortunately pipelines that are in trenches have different bedding and initial backfill material than the surrounding soil so that they can be visually detected prior to striking the pipe. Electrical ducts usually have warning tape buried above them or colored concrete encasement.

In the case of grade-sensitive production lines, it is often necessary to pothole and determine the exact depth of existing cross lines prior to production work. Sometimes contracts require that gravity lines be excavated and laid from the high point to the low point because the line can be lowered to go under an existing line but cannot be lifted to go over it.

In the case of buried structures, lines that will interfere with shoring or the exterior of the structure should be verified prior to construction. It is far easier to design exactly located interfering utilities into the shoring structure than it is to alter the structure during construction. Sheet pile driving and drilling for beam and plate shoring requires location accuracy to within a few inches to prevent destruction of existing utilities. This damage can happen without anyone knowing it.

By OSHA and most state regulations it is allowed that if there is no response to notification within 2 days, the excavator can proceed with the project: "The excavator may proceed with excavation at the end of two working days, unless otherwise specified in state/provincial law, provided the excavator exercises due care in his endeavors." In the case of high-pressure gas transmission lines and high-voltage electrical transmission lines, this is not advisable due to the potential for explosion and loss of life. In the case of high-pressure gas lines, national pipeline safety standards require that the operator have a representative there at all times during excavation work in the vicinity of the lines. The regulation states

> *49 CFR Part 192, Transportation of Natural and Other Gas by Pipeline: Minimum Federal Safety Standards, Subpart L 192.614 Damage Prevention Program, (c) (6)* Provide as follows for inspection of pipelines that an operator has reason to believe could be damaged by excavation activities: (i) The inspection must be done as frequently as necessary during and after the activities to verify the integrity of the pipeline.

Excavation work without the presence of a representative from the pipeline company precludes the company from fulfilling its legal responsibilities and exposes the excavator to extreme safety and financial risk.

Natural gas companies charge for the cost of repair to their lines as well as for the cost of lost gas. The amount of gas lost in a 4-in line from time of puncture can amount to over $5000, and a 6-in line over $20,000.

High-Risk Pipelines Parallel to New Pipe-Laying Operation

Because utility easements typically carry more than one pipeline, high-pressure gas transmission lines are often parallel and close to new pipeline excavations. It makes no sense to completely excavate and expose the existing line to verify depth and alignment, and then backfill it prior to excavating and installing the new line. Potholing at short intervals does not guarantee location between the potholes. The standard practice is to surface-mark the line and then excavate for the production line with the expectation that the high-risk line does not wander into the limits of the new excavation. This is a recipe for disaster. Anyone with the notion that an excavator operator has the ability to sense and slow down the bucket maneuver when he or she hits an obstruction is badly mistaken. Even at the slowest speed possible when an excavator bucket strikes a pipe, the result will be damage and possible rupture in the case of a high-pressure line. The only safe way to locate high-pressure lines is by hand. The only question is how close the excavator bucket should be allowed before hand digging is started. In the case of parallel high-risk pipelines the following measures should be taken:

- Arrange a job site meeting between the operator and field personnel involved in the excavation work to develop a safe, precise locating plan.
- Develop a team consisting of the operator's locating person, the excavator operator, the excavator operator's trench man, and the competent person. Include alternates in this plan so that if one of the members of the team is not present, someone else can fill in.

- Make sure that everyone is properly trained and the equipment is accurate.
- Develop a safe verification plan. For example,

    1. Make sure that the production trench excavation limits are clearly defined.
    2. Make sure that the expected critical line location is clearly marked, showing the edges of the pipe.
    3. Be sure that the excavator operator knows the expected depth of the line.
    4. When the trench depth or sidewall is within 24 in of the line, use hand-pushed metal probe or hand tools to locate the line. This requires the excavator operator to stop while the trench man does this.
    5. If the line is to be exposed in the excavation, hand dig the last 12 in.
    6. Do not dig without the main players of the team present.
    7. Do not change the plan unless all team members are present and agree that the change is beneficial.

Every high-risk situation is different and site-specific. These situations should be identified in the design stage, they should be emphasized in the bid stage, and the risks associated with failure to protect should be completely understood and planned for in the construction phase. The locating-excavating team members are the bastion between success and failure, and they need to understand the importance of what they are doing.

Both the owner of the project and the contractor constructing the project should understand that the dollar cost of high-risk situations is high and not try to avoid or trim the cost of eliminating the possibility of failure to protect. For pipeline contractors the single most important element of the bid is feet of pipe laid per day. Whether it is 100 or 1000 ft/day, if the bid number is too high, they will not get the job; and if they do not get the production in the field, they will lose money. Everything they do on the project is geared toward making or improving that production number. Contrary to his or her nature the contractor must accept and include the lost productivity resulting from stopping the excavator to wait for the line-locating team to do its job. This is not a parallel activity that can be accomplished while production work continues full throttle, and the bid should reflect that.

If the new excavation is parallel and below the existing line, the open cut or shoring provided for worker protection should also be capable of protecting the adjacent line.

4.6 Surface Damage to Underground Facilities from Wheel Loads

Buried pipelines are designed to support soil loads and traffic loads. During construction of new facilities around existing pipelines, situations can arise where the original design loads may be permanently or temporarily exceeded. These loads can be generated from

- Off-highway earth hauling equipment
- Crane outrigger pad loads
- Horizontal loads from pipe jacking and microtunneling
- Possible increase of existing pipe cover due to spoil piles or permanent fill
- Existing pipe cover that may be decreased and then subjected to surcharge loads from moving equipment
- Requirements of earth removal and recompaction over existing lines
- Change in original trench wall design conditions
- Compromised thrust block soil bearing capacity
- Structures and pipelines that can float if design water level is changed due to burst pipes or there is flooding in new excavations due to heavy rainfall and quick drawdown from pumping

Some of these are permanent effects and others are temporary due to the construction activity. During the subsurface utility locating process, these situations should be identified and planned for. If a new pipeline is installed parallel and close to an existing line, the original trench wall may be compromised. This condition should be planned for and have a solution outlined in the original design. It is a problem for the design engineer to solve. Another common design problem arises where compaction requirements in the vicinity of existing pipelines will exceed the allowable loading on the pipeline. The solution to this problem should be outlined in the plans and specifications. Other problems such as getting off-road dump trucks across an existing high-pressure gas line without damaging it are for the contractor to solve and be responsible for. This type of problem is strictly related to contractor's means and methods and will not exist after construction.

Damage to existing pipelines may not show up for years after construction. Cement-lined and -coated pipes can be cracked, allowing moisture and air to corrode the underlying steel. In welded steel high-pressure lines any additional stress from outside sources is considered added to the internal hoop stress and can quickly exceed safe design limits, causing bursting or buckling. Even if the lining is not compromised, higher stresses always lead to a higher potential for corrosion over time. Old brick sewers, large-diameter concrete, and corrugated metal pipe storm drains are originally designed for cover soil and highway traffic loads in streets. On private property and cross-country the traffic loads may have been left out of the design. The original design assumptions (flexible, rigid, trench wall conditions, etc.) may not even be available at the time of construction. Repairing and replacing damaged infrastructure is extremely expensive, and therefore most pipeline operators are sensitive to any increased loads during construction. They usually require that there be limited or no load increases allowed to their facilities. Many times utility owners will require that loading in excess of soil and highway traffic not be allowed.

Increased loading from construction vehicles and equipment is a common problem for contractors. Typical solutions are to lay steel road plates to spread the load, add soil over shallow pipelines to distribute tire loads, encase the pipelines in concrete, or build temporary bridges entirely over the line. The author has used all these solutions at one time or another.

In the case of steel road plates, the idea is that the plates due to bending strength spread the load and thus decrease the pressure load per square foot at the top of the pipe. A critical assumption that goes with this is that there will be movement at the soil plate contact level for this to work. If a steel plate is laid on top of pavement or concrete that does not flex, there will be no movement and the load will go directly through as if the plate were not there. If there is enough soil movement for the plate to take up the load, during the soil movement process the underlying pipeline most likely received a pressure spike equal to what it would see without the plate being there. In some cases repetitious high-pressure loads are what the pipeline owner is trying to avoid, and the pressure spike that initially deforms the soil is tolerable. In most cases the initial pressure spike is not acceptable, and laying a steel plate directly on the surface should be avoided. If a steel plate is being used, it should be set on top of 6 in of loose soil or sand that can deflect and allow the plate to take up the load.

4.6.1 Determining Traffic and Soil Loading Pressure on Existing Pipes

Surcharge loads at the surface spread laterally into the soil at an angle of 30° to 45°. A conservative approximation of vertical surcharge pressure at depth is to use a 2:1 slope as shown in Fig. 4.4. The pressure at depth is

$$\sigma_z = \frac{P}{(a+z)(b+z)} \tag{4.1}$$

Typical construction situations in which this applies occur where a maximum external pressure is allowed on the pipe and the contractor is not allowed to exceed that pressure during construction.

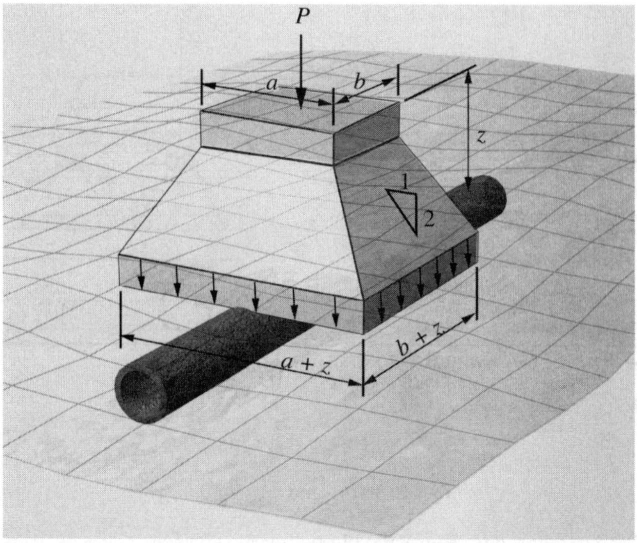

FIGURE 4.4 Pressure at depth.

Design Example 4.1 On an earthmoving project it is necessary to run a 657G scraper fully loaded across a pipeline easement that has a 24-in high-pressure petroleum line buried 48 in below the surface. The pipeline operator says that the line was designed for 4 ft of cover, a standard traffic load, HS20-44. It is assumed that two overlapping wheel loads are possible so the allowable traffic pressure is double. The pipeline operator will allow a 10 psi surcharge pressure loading on the pipe.

The scraper will add a surcharge on the pipe greater than traffic loading. The solution to this problem is to add loose soil and steel plates at the pipe crossing area. As the plate deflects, it bridges and expands the tire footprint area around the tire. The expanded area causes the loaded area at the pipe level to be less. The plate will cantilever beyond the tire to the point where the plate bending strength is exceeded, so the assumed cantilever length has to be checked to make sure the allowable bending strength is not exceeded at that point.

Step 1 To determine the vertical stress at depth from a tire load, use Eq. (4.1). Determine the allowable vertical stress at the top of the pipe from HS20-44 loading, (Fig. 4.5). HS20-44 traffic loading allows a 32,000-lb axle load and a 1.3 impact factor. Highway design specifications require the design tire contact to be a rectangle with an area equal to $0.01P$ in^2, and the length in the direction of traffic is 0.4 times the tire width. Find the tire contact area.

$$\text{Axle load} = 32{,}000 \text{ lb} \times 1.3 = 41.6 \text{ k}$$

$$\text{Tire load } P = \frac{\text{axle load}}{2} = 20.8 \text{ k}$$

$$\text{Tire area} = 0.01 \times 20{,}800 = 208 \text{ in}^2$$

$$\text{Area} = \text{length} \times \text{width} = \ell \times w = 208 \text{ in}^2$$

$$\ell = 0.4w$$

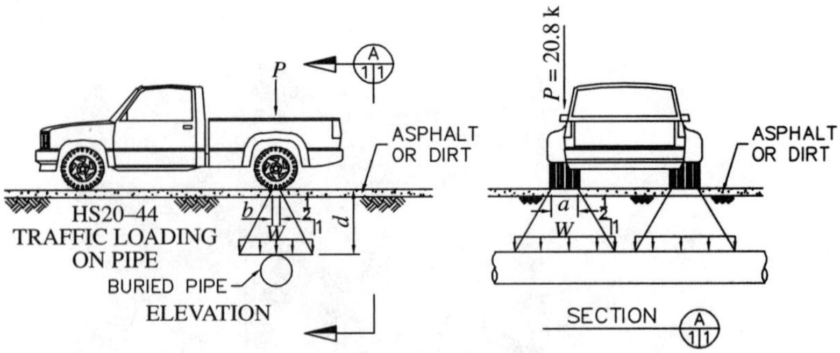

FIGURE 4.5 HS20-44 wheel load on buried pipe.

$$\therefore 0.4w \times w = 208 \text{ in}^2$$

$$w = 23 \text{ in} \qquad \text{and} \qquad \ell = 9 \text{ in}$$

Find the surcharge pressure at 48 in below the surface.

$$\sigma_z = \frac{P}{(w+z)(\ell+z)} = \frac{20{,}800}{(23+48)(9+48)} = 5.1 \text{ psi}$$

Therefore, since a double traffic load is allowable, the allowable surcharge pressure is 10.2 psi.

Step 2 Determine the vertical stress on the pipe from the 657G scraper. These 657G scraper data are from *Caterpillar Performance Handbook*.

647G scraper capacity	44 yd^3
Front axle loaded	128.3 k
Rear axle loaded	133.5 k
Tires	40.5/75 R39
Tire width w	40.5 in

Total rear axle tire load plus impact is

$$P = \frac{133.5 \text{ k} \times 1.3}{2 \text{ tires}} = 86.7 \text{ k}$$

Tire area = 86,700 × 0.01 = 867 in^2

Tire length $\ell = \dfrac{867}{40.5} = 21.4$ in

Find surcharge pressure at 48 in below the surface.

$$\sigma_z = \frac{P}{(w+z)(\ell+z)} = \frac{86{,}700}{(40.5+48)(21.4+48)} = 14.1 \text{ psi} > 10.2 \text{ psi}$$

Step 3 Determine pipe loading from steel plate scraper wheel footprint. The scraper tire load on the pipe is 14.1 psi, and the allowable load is 5.1 psi. To decrease the tire load, propose to add 12 in of dirt over on

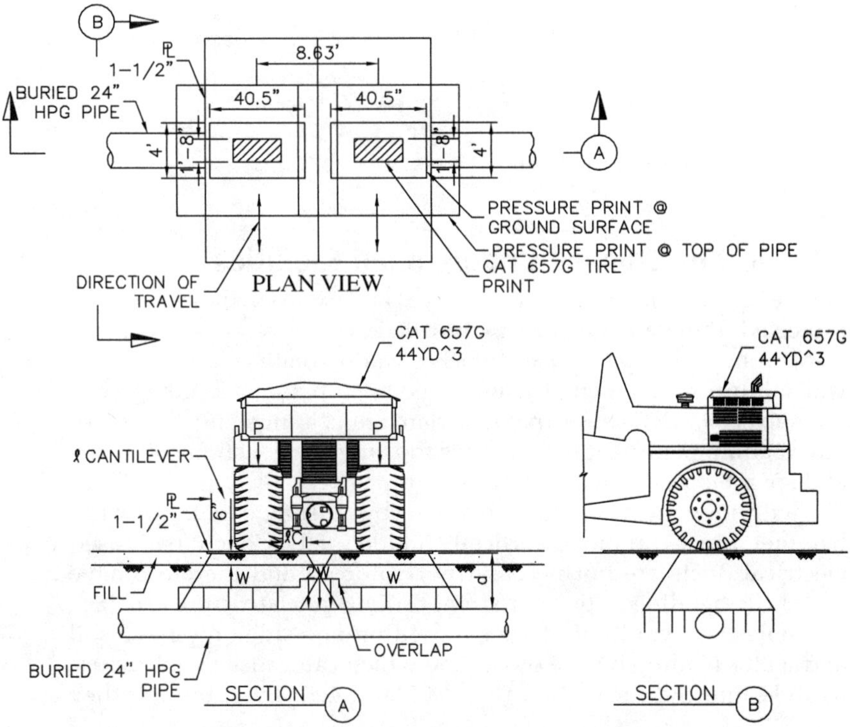

FIGURE 4.6 Layout for 657G scraper plate bridge on two 1.5-in × 8-ft × 12-ft steel plates.

the surface and use 1.5-in × 8-ft × 12-ft steel plates over the pipe (Fig. 4.6).

The fill will settle enough for the plate to deflect and transfer the load to the plate.

The tire width is 40.5 in, and the steel plate is 96 in wide. Assume that the 1.5-in thick plate will span in $\frac{96-40.5}{2} \approx 28$.

The plate footprint is $\ell = (21 + 2 \times 28) = 70$ in and $w = 96$ in.

The pressure at the pipe is $\frac{86{,}700}{(96+60)\times(70+60)} = 4.3$ psi.

Add 12 in of soil at 120 psf $= \frac{120}{144} = 0.83$ psi.

There is a pressure overlap as shown in Fig. 4.6 Section A so the total pressure at the pipe is

$$\text{Surcharge pressure at pipe} = (2\times 4.3) + 0.83 = 9.43 \text{ psi} < 10 \text{ psi} \qquad \text{OK}$$

Step 4 Check the plate cantilever. The plate cantilever is 28 in, and the plate loading is 9.5 psi.

$$\text{Cantilever plate bending is } M = \frac{w\ell^2}{2} = \frac{9.5\times 28^2}{2} = 3724 \text{ in} \cdot \text{lb.}$$

The section modulus for a 1.5-in × 1-in section is $s = \frac{bd^2}{6} = \frac{1 \times 1.5^2}{6} = 0.375\ \text{in}^3$.

$$\text{Bending stress } f_b = \frac{M}{s} = \frac{3724}{0.375} = 9.9 \text{ ksi} < 27 \text{ ksi} \qquad \text{OK}$$

For A36 steel plate $F_b = 0.75F_y = 0.75 \times 36 = 27$ ksi.

4.7 Support of Exposed Underground Facilities

After existing utilities have been located and exposed, they have to be protected from damage during construction.

Lines that are parallel to trenches can be damaged with trench wall failure or settlement. Unsupported lines that intersect the trench are subject to deflection that can damage coatings and linings or have bending failure if their span is too long. Pipe sections with bell and spigot joints such as ductile iron pipe (DIP), RCP, and VCP can fail with movement at the joints. During seismic events or water hammer, pipes can move vertically and laterally. Concrete-encased electrical ducts are not necessarily reinforced and therefore have very little bending and shear strength. Concrete structures such as manholes and vaults that are exposed on one side only have soil and water loading on the other side which can cause the structure to slide into the excavation (Figs. 4.7 and 4.8). Even though they

FIGURE 4.7 Exposed pipes supported across wide open cut.

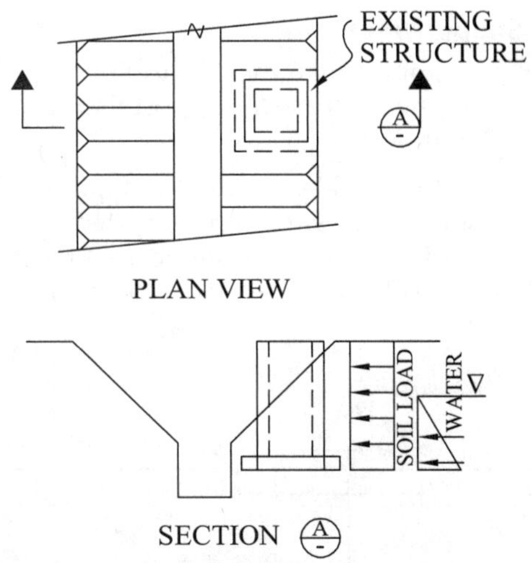

FIGURE 4.8 Structures with no lateral support.

may not be exposed, thrust blocks can have their bearing capacity compromised due to removal of soil. Empty pipes and tanks that have the soil removed above them can float if the water table is high or the trench is flooded for some other reason.

4.7.1 Requirements for Protection of Exposed Existing Facilities

From the standpoint of worker safety, protection of existing facilities is mandated by the following:

> *OSHA 1926.651 (b) Underground installations.* (4) While the excavation is open, underground installations shall be protected, supported, or removed as necessary to safeguard employees

This requirement does not necessarily require prevention of damage to the facility since it only addresses worker protection.

Contractors certainly understand that they cannot damage existing facilities; however, damage is not always evident until long after the project is finished. Sometimes the urgency or danger of broken and disrupted facilities is not clear to the contractor, especially when he or she has limited time and money in the bid for protection. If it is important to the owner, the design engineer should not leave it up to the contractor. The contract should identify each critical line and specify the level of support and protection required. Risk assessment should be based on consequences of failure, similar to the seismic importance factor. The contractor should give serious consideration to risk to workers in the vicinity of the facility.

4.7.2 Pipe Support Plan

Figure 4.9 shows a support plan for a 48-in RCP line that crosses an excavation. In designing this type of support there are several important design considerations:

- Always use turnbuckles or similar support method so that deflection and settlement can be corrected at all times. Preload the system using the turnbuckles so that the beam deflection is equal to the calculated dead load beam deflection.

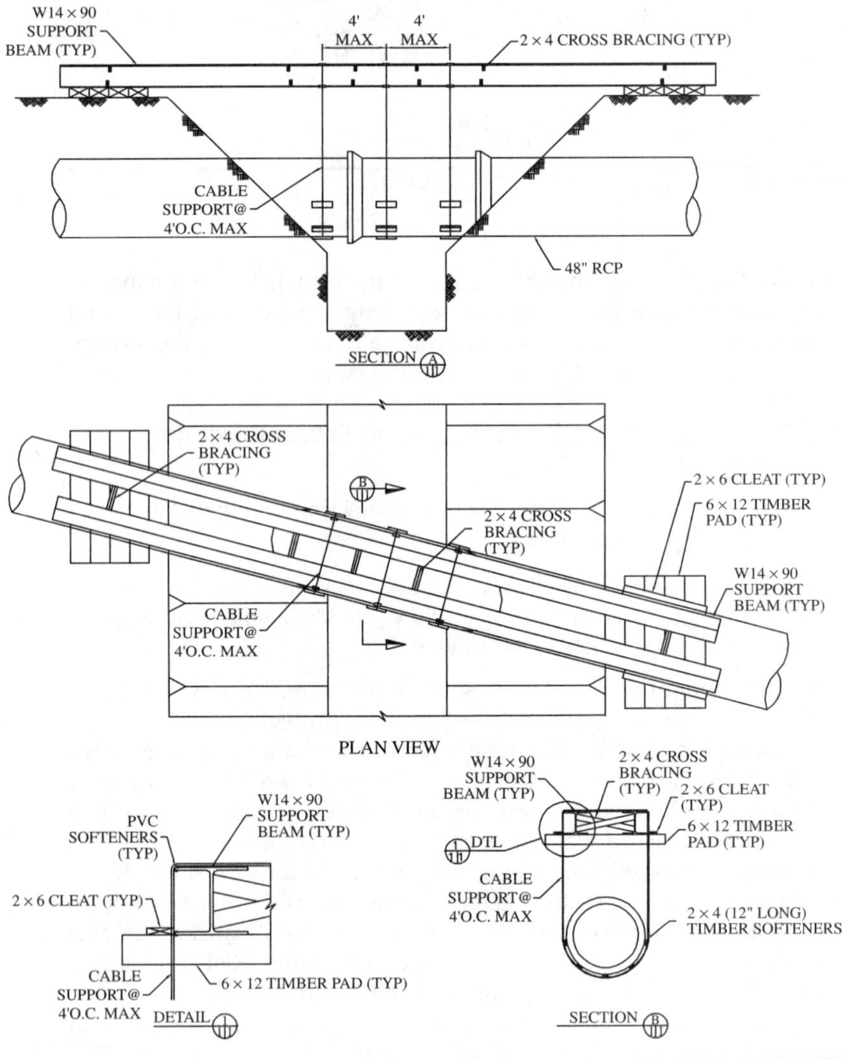

FIGURE 4.9 Pipe crossing support plan for 48-in RCP.

- If there are seismic concerns, the system should be designed for lateral forces, usually 0.3 to 0.5 times the dead load. Uplift should be considered in this situation and can be prevented by blocking between the top of the pipe and the beams.
- Cables should have a 5:1 factor of safety, typical for use of wire rope in construction.
- Cable spacing in the case of flexible joint pipe should be determined by the length of the joints, deflection requirements, and allowable distribution of load at contact points. Provide at least two cables per joint. Avoid line and point loads by using wood blocking between the cable and the pipe. If the pipe is a high-pressure system or has sensitive linings and coatings, the allowable point loads in pounds per square inch on the exterior of the pipe and the allowable deflection should be specified by the manufacturer of the pipe or the design engineer. The cable should have softeners at the beam flange edges.
- The timber pads at the bank can affect slope stability and deliver surcharge loads to the top of the pipe. Generally if the reaction is reduced by use of timber or steel plate padding to less than 500 lb/ft^2 (or psf) and set back 2 ft, it is no different than surcharge loads from spoil piles and construction equipment and will have minimal effect on the stability of the slope. At this loading if the pipe is over 4 ft below the support pad, the load delivered to the pipe is negligible.

Design Example 4.2 An existing 48-in RCP line crosses an excavation (Fig. 4.9). The proposed support system uses two W14 × 109 beams and cable support. The pipe sections are 8 ft long with tongue-in-groove rubber-gasketed joints. Wire rope spacing is 4 ft on center.

Step 1 Determine the loading, bending moment, and deflection.

Beam loading in pounds per linear foot (lb/lf or plf) and span

48 -in RCP with 6-in wall	950 lb/lf
Water, full pipe	785 lb/lf
Support beams, 2 W14 × 109	218 lb/lf
Beam span	40 ft

Due to the long span the support beam's weight will affect the bending and deflection. Calculate the support beam bending and deflection using continuous loading of 218 lb/lf, and then calculate loading from pipe using point loads at 4 ft on center. Use beam tables (Fig. 2.8).

The beam geometric properties for W14 × 109, shown in Fig. 4.10, are as follows:

$$I = 1240 \text{ in}^4$$

$$S = 173 \text{ in}^3$$

$$F_y = 50 \text{ ksi}$$

$$E = 29{,}000 \text{ ksi}$$

Beam flexure and deflection, single beam:

$$M = \frac{wl^2}{8} = \frac{0.109 \times 40^2}{8} = 21.8 \text{ k·ft}$$

$$\Delta = \frac{5w\ell^4}{384EI} = \frac{5 \times 0.109/12 \times (40 \times 12)^4}{384 \times 29{,}000 \times 1240} = 0.175 \text{ in}$$

Bending from pipe load:

$$\text{DL} + \text{LL} = 950 + 785 = 1735 \text{ lb/lf}$$

Point load at 4 ft on center = 4 × 1735 = 6940 lb say 7 k or 3.5 k per beam
From the Fig. 4.10 loading diagram $R_a = R_b = 5.25$ k
Moment = area of shear diagram (Fig. 4.10)
$M_{max} = 5.25 \times 16 + 1.75 \times 4 = 91$ k · ft

Step 2 Check beam for bending and determine prestress loading. Check beam for bending stress:

$$f_b = \frac{M}{s} = \frac{(21.8 + 91) \times 12}{173} = 7.8 \text{ ksi} < 0.6\,F_y = 0.6 \times 50 = 30 \text{ ksi} \qquad \text{OK}$$

Check for allowable unbraced beam length. From AISC beam properties for W14 × 109

$$B_f = 14.605 \text{ in}$$

$$\frac{d}{A_f} = 1.14$$

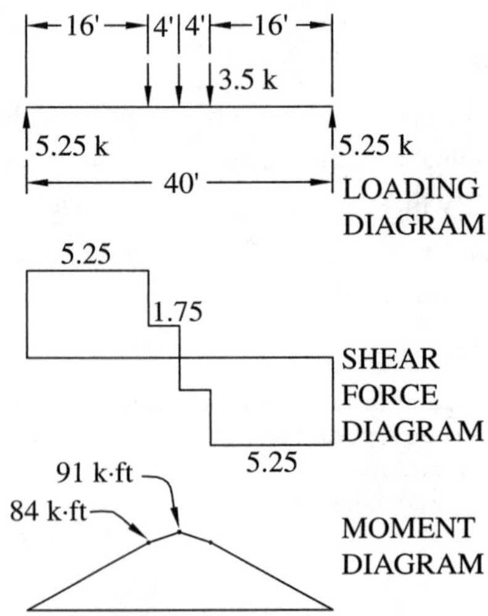

FIGURE 4.10 Loading, shear, and moment diagrams for pipe support.

From AISC (F1-2)

$$L_c < \frac{76b_f}{\sqrt{F_y}} = \frac{76 \times 14.605}{\sqrt{50}} = 13 \text{ ft} \quad \text{or} \quad \frac{20{,}000}{(d/A_f)F_y} = \frac{20{,}000}{1.14 \times 50} = 29 \text{ ft}$$

For block compression flange at 13 ft on center maximum, see Fig. 4.9 Section B.

Preload deflection

$$\text{Deflection, use } \Delta = \frac{P\ell^3}{48EI} = \frac{(3 \times 3.5) \times (40 \times 12)^3}{48 \times 29{,}000 \times 1240} = 0.67 \text{ in}$$

Preload to DL only (including beam weight deflection)

$$\Delta_{preload} = 0.67 \times \frac{950}{1135} + 0.173 = 0.54 \text{ in or } ½ \text{ in}$$

Step 3 Wire rope design. The wire rope load is 7000 lb total and is a two-part system, so the working load is 3500 lb. Use a factor of safety of 5, and find a wire rope with an ultimate strength of 5 × 3500 = 17,500 lb. Use softeners at beam flanges and 2 × 4 × 12 in blocks at three locations to distribute point loads on pipe. Use turnbuckle with a safe working load of 3500 lb (Fig. 4.10).

Step 4 Abutment design. The reaction at the abutment is 5.25 k + 40 × 0.109 k = 9.6 k per beam or total 19.1 k. Use 4-ft × 10-ft pad consisting of 6 × 12 timber matting. Then

$$\text{Soil loading} = \frac{20 \text{ k}}{4 \times 10} = 500 \text{ psf}$$

Check timber for cantilever moment and shear. Beam cantilever is 3 ft.

$$M = \frac{w\ell^2}{2} = \frac{500 \times 3^2}{2} = 2250 \text{ ft} \cdot \text{lb}$$

$$V = 3 \times 500 = 1500 \text{ lb}$$

Allow $F_b = 1200$ psi.

$$s_{required} = \frac{M}{f_b} = \frac{2250 \times 12}{1250} = 21.6 \text{ in}^3 \quad \text{actual } s = \frac{bd^2}{6} = \frac{12 \times 6^6}{6} = 72 \text{ in}^3$$

Check the shear.

$$v = \frac{3v}{2bd} = \frac{3 \times 1500}{2 \times 12 \times 6} = 32 \text{ psi} < 140 \text{ psi} \qquad \text{OK}$$

4.8 Open Trench Traffic Bridges

Steel plate trench covers (Fig. 4.11) have been used since the late 1800s; however, in the 1980s they changed the nature of the pipeline construction operation. At that time 1- and 1.5-in-thick steel plates became a common rental item at shoring rental outlets. At the same

Figure 4.11 Steel plate traffic trench cover.

time excavators who could lift and maneuver these plates were replacing rubber-tired backhoes as the common tractor on the pipeline project. This was the beginning of the end for the open barricaded trench in the street. Prior to this at the end of the day barricades with oil lamps and later battery and solar lamps were put up and taken down again in the morning. Contracts typically were allowed a certain number of feet of trench left open at the end of the day. Accidents from cars and bicycles running into open trenches were considered commonplace and hard to prevent. Another issue was children and animals getting into open unshored trenches.

Today when trenches and open pits in the street are not being worked in, it is standard practice to completely cover them. The change impacted the nature of the excavation operation. Prior to plated excavations the excavator was left at one end of the trench and the loader at the other end, to prevent vehicles from running into the end of the trench. Barricades were strung along the sides of the trench. The next day the crew simply picked up where it left off, excavating and laying pipe. Today the excavator, loader, and barricades are parked away from the excavation, usually in a fenced construction yard. The first operation of the day is to remove and store the street plates until they are reset at the end of the day. The plate pick-up and lay-down operation has taken about 2 hours away from the daily production work. At the end of every day there is a complete cleanup and street-sweeping operation, and then the street is returned to the public to resume normal activities. Accidents due to public incursion into the work zone have been practically eliminated. Additionally plates allow utility potholing to be left open until production work progresses through them. Backfill, paving, and reexcavating around previously potholed lines can be eliminated. Something that was seen at the time as a productivity inhibitor can be seen today as having an extremely high cost/benefit ratio.

4.8.1 Road Plate Installation Considerations

There is a basic tenet about traffic safety that should always be kept in mind when things are constructed that will interact with traffic flow.

> The public has the right to expect the road to be same today as it was yesterday. If the road was smooth and clean yesterday, then the expectation is that it will be that way today. If it is changed during the course of the day, then it should be put back the way it was at the beginning of the day or the public should be fully informed of the changes prior to their getting there.

In the case of road plates the expectation is that the road will be much the same as it was prior to construction, smooth and drivable at normal speeds. Bicyclists and pedestrians have the same expectation. Car tires can easily negotiate a 1-in rise in the street at the edge of a plate while a bicyclist can break a tire when hitting it and a pedestrian can trip on it. This is one of the reasons why a wedge of cold asphalt mix is placed around the edges of the plate. The other reason is to help prevent the plate from moving. Another expectation is that there are no open linear cracks or gaps in the road such as you would find when two plates are separated at the edge. In one case a bicyclist was killed when his tire fell into a crack between two plates. He had ridden across the plates several times; however, one night heavy concrete trucks drove over and separated the plates, and his front tire fell into the crack. Even if he had never ridden over the plates before, the family would have won the lawsuit because his normal expectation was that there is not a longitudinal crack in the road for the tire to fall into. Other installation considerations for trench plates are as follows:

When plates are left for long periods, traffic and weather can deteriorate the plate installation. Plate installations should be checked as often as necessary to ensure that they are adequate. Weekend traffic, special event traffic, construction equipment traffic, etc., can quickly ravel out cold mix at the edges and allow movement of the plates where normal traffic might not.

- In highway traffic at speeds over 45 mi/h plates can be lifted by the airstream. This can be an extremely dangerous situation due to the speed and frequency of traffic on high-speed roadways. In these types of installations it is common to grind the asphalt and wedge anchor bolt the plates down (Fig. 4.12).

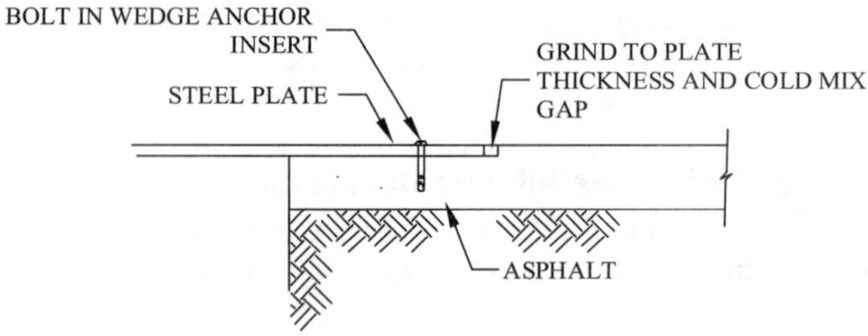

FIGURE 4.12 Trench plate installation with cold mix and concrete baricade.

- The most effective way to prevent uplift and horizontal movement is to tack-weld plates together, thereby rapidly increasing the mass of the plates relative to the mass of the vehicle.
- The coefficient of friction for rubber on asphalt is higher than for rubber on steel, causing a potential tire sliding on the plate surface hazard. Rental outlets supply plates with a skid-resistant wearing surface. Installation and maintenance of the wearing surface are an added expense to the cost of the plate so the rental cost is slightly higher than that for a plain steel surface.
- If a tire engages the plate, there is a potential for the plate to slide on the asphalt. The resisting force is friction from steel on asphalt and the shear capacity of the cold mix on the asphalt. Typical installations require a minimum of an 18-in-wide section of cold mix around the plate.
- If there is an open trench or dropoff at the edge of a plate, proper barricading such as K-rail should be set to prevent pedestrians, bicycles, and vehicles from falling off the edge.
- If plates are set on sloping road surfaces, additional consideration should be given to anchoring them so that they will not slide.

The trench walls must be stable. With an 18-in overlap at the edge of the trench, the load imparted to the road surface when the tire is at the edge of the trench is 20.8 k/(1.5 ft $\times$ 4 ft) = 3.5 ksf. If the pavement section is 4 in thick, the vertical pressure at the top of the base rock is 2.4 ksf. It is questionable whether a base rock course has the unconfined compression strength required to hold up; however, in the field there is little evidence of the base course raveling out. Figure 4.13 shows a graph of a Boussinesq lateral pressure analysis for this loading. The lateral pressure quickly decreases to approximately 0.5 ksf within the first 18 in and drops to 0 at 5 ft down. A very soft clay has a shear strength of 0.5 ksf or

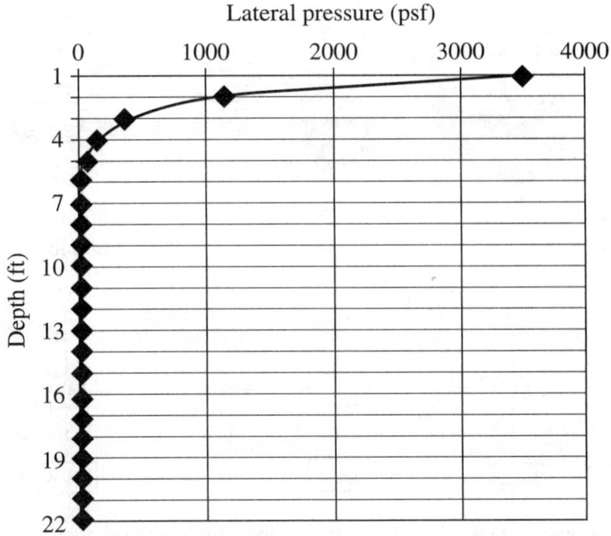

FIGURE 4.13 Boussinesq analysis lateral pressure diagram.

less. If the trench wall is a cohesive material firmer than soft clay, the trench wall does not necessarily need shoring to support traffic over trench plate bridges. If the trench walls in the upper 5 ft are noncohesive sands and gravels that will not normally stand up when excavated, then shoring will be required When the trench is over 5 ft deep, a trench stability analysis should be used to determine if it needs to be shored prior to placing road plates.

4.8.2 Road Plate Engineering

Road plates can be considered a simply supported steel bridge deck. The asphalt or dirt at the edge of the trench can be considered the girder to which the deck delivers the load. The design of highway bridges in the United States is governed by the American Association of State Highway and Transportation Officials (AASHTO). The association sets design standards that are adopted by state highway departments as a minimum; however, states usually modify them to deal with specific issues such as seismic, cold weather, wet weather, and soil conditions. The design standards consider functionality, structural integrity, lifetime durability, and public perception in the case of deflections. AASHTO basic design requires that bridges carry dead loads (structure weight), live loads, and impact loads. Standard highway loadings are HS15-44 and HS20-44 (Fig. 4.14). H loads apply to two-axle trucks and cars, and HS loads apply to a truck and trailer with three axles loaded. There are also permit loads where a single-axle

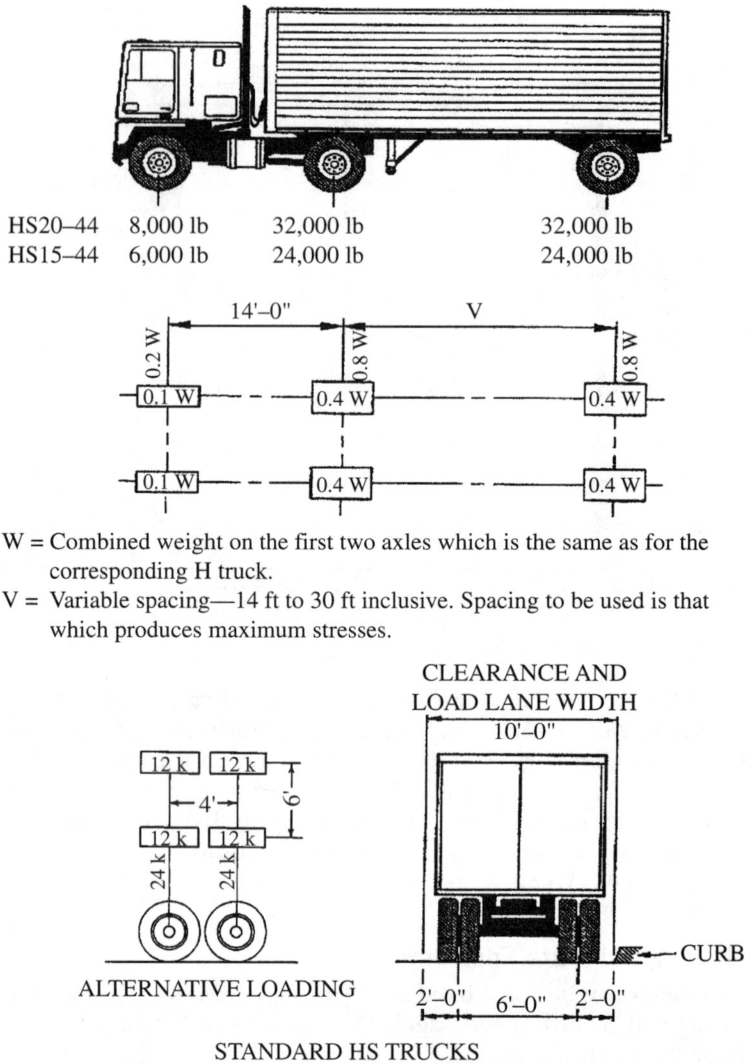

FIGURE 4.14 HS20-44 and HS15-44 traffic loading.

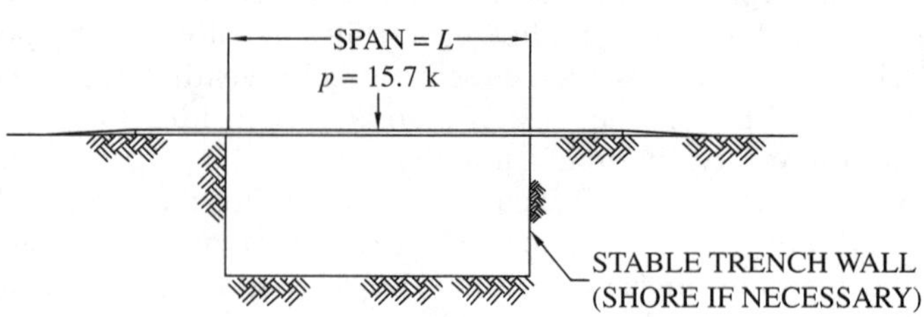

FIGURE 4.15 Road plate span.

load never exceeds 26,000 lb and a dual-axle load never exceeds 48,000 lb. The numbers 15 and 20 following the H refer to the tractor gross vehicle weight in tons. The 44 indicates the year in which the loading standard was adopted, 1944. Figure 4.15 shows the basic highway loading. Bridges and all highways and streets are designed for HS20-44 loading. AASHTO also allows the use of a one-axle load of 24 kips or two-axle loads of 16 kips each spread 4 ft apart in the design of timber and steel grid bridge decks. This provision can be used in the design of temporary steel plate bridges. AASHTO also requires an impact factor

$$I = \frac{50}{L + 125} = 0.30 \qquad \text{for any } L < 40 \text{ ft}$$

Plate Loading

A 24-k single-axle load has a 12-k wheel load. The plate design load is

$$P = \text{LL}(1 + I) = 12 \text{ k} \times 1.3 = 15.6 \text{ k}$$

where LL = 12 k is the live load and $I = 0.3$ is the impact factor.

Plate Physical Properties

In the past steel plate used for road plate was either ASTM A36 min $F_y = 36$ ksi or A572 min $F_y = 50$ ksi. Most of this steel came from remelted scrap steel which contained alloys that added strength to the steel. Consequently mill certifications for ASTM A36 steel always showed $F_y = 50$ ksi or greater. Today all steel shapes and plates are manufactured to a minimum 50 ksi yield strength. AASHTO requires that local bending stress in deck plates caused by wheel load plus 30 percent impact not exceed 30 ksi unless justified by detailed analysis. AISC allows $F_b = 0.75\ F_y$ for solid, round, and rectangular shapes bent about their weaker axis. For $F_y = 50$ ksi steel this results in $F_b = 37.5$ ksi. Even though steel road plates are a temporary structure, it is important to adhere to AASHTO standards. Since deflection is ignored, the support at the trench edge is not completely rigid, and calculated allowable spans are often exceeded it is best to be conservative. The use of $F_b = 30$ ksi seems to be conservative.

Geometric Properties

Given an 8-ft-wide steel plate and a 6-ft wheel base, it is possible to get two wheels on one plate. That would leave a 4-ft-wide section of the plate to support each wheel load. This is the worst possible wheel load case on a single plate, so that when one is calculating section properties for trench plates a 48-in section width is used. The section modulus for plates is

$$s = \frac{bd^2}{6} = 8, 12.5, 18 \text{ in}^3$$

Thickness		Width		Length		Weight	
in	mm	ft	m	ft	m	lb	kg
1	25.4	5	1.5	8	2.4	1832	740
↓	↓	8	2.4	8	2.4	2611	1184
↓	↓	8	2.4	10	3.0	3264	1481
↓	↓	8	2.4	16	4.9	5222	2369
Thickness		**Width**		**Length**		**Weight**	
in	**mm**	**ft**	**m**	**ft**	**m**	**lb**	**kg**
1.25	31.8	8	2.4	10	3	4080	1851
↓	↓	8	2.4	12	3.7	4896	2221
↓	↓	8	2.4	16	4.9	6528	2961
↓	↓	8	2.4	20	6.1	8160	3701
Thickness		**Width**		**Length**		**Weight**	
in	**mm**	**ft**	**m**	**ft**	**m**	**lb**	**kg**
1.5	38.1	8	2.4	10	3	4896	2221
↓	↓	8	2.4	12	3.7	5875	2665
↓	↓	8	2.4	16	4.9	7834	3553
↓	↓	8	2.4	20	6.1	9792	4442

Table 4.1 Road Plate Sizes and Weights

where s = section modulus
b = 48 in
d = 1, 1.25, 1.5 in

The moment of inertia is

$$I_{x\text{-}x} = \frac{bd^3}{12} = 4, 7.8, 13.5\ \text{in}^4$$

where $I_{x\text{-}x}$ = moment of inertia about x axis
b = 48 in
d = 1, 1.25, 1.5 in

Table 4.1 gives typical road plate sizes and weights.

Plate Span

Find the allowable plate span L (ft) (Fig. 4.15). For a simply supported beam

$$M = \frac{PL}{4} = 3.9\ \text{L k} \cdot \text{ft}$$

where $P = 15.6$ k and L = simply supported span (ft). The basic flexure formula is

$$F_b = \frac{M}{s}$$

where $M = 3.9L\ \mathrm{k \cdot ft}$
$S = 8, 12.5, 18\ \mathrm{in}^3$

and $F_b = \dfrac{3.9L}{s}$ and $L = \dfrac{sF_b}{3.9}$

where $F_b = 30$ ksi
$s = 8, 12.5, 18\ \mathrm{in}^3$

$$\text{For 1-in plate } L = \frac{8\ \mathrm{in}^3 \times 30\dfrac{\mathrm{k}}{\mathrm{in}^2}}{3.9\ \mathrm{k}} \times \frac{1\ \mathrm{ft}}{12\ \mathrm{in}} = 5.1\ \mathrm{ft}$$

$$\text{For 1.25-in plate } L = \frac{12.5\ \mathrm{in}^3 \times 30\dfrac{\mathrm{k}}{\mathrm{in}^2}}{3.9\ \mathrm{k}} \times \frac{1\ \mathrm{ft}}{12\ \mathrm{in}} = 8\ \mathrm{ft}$$

$$\text{For 1.5-in plate } L = \frac{18\ \mathrm{in}^3 \times 30\dfrac{\mathrm{k}}{\mathrm{in}^2}}{3.9\ \mathrm{k}} \times \frac{1\ \mathrm{ft}}{12\ \mathrm{in}} = 11.5\ \mathrm{ft}$$

The spans shown in Table 4.2 are typical throughout the United States.

Worst-case shear occurs when the wheel load is at the edge of the trench. A quick calculation for a 1-in plate shows

$$\text{Plate shear } f_v = \frac{P}{bd} = \frac{15.7\ \mathrm{k}}{48\ \mathrm{in} \times 1\ \mathrm{in}} = 0.327\ \mathrm{ksi}$$

which is negligible.

Deflection

In permanent design of bridge decks it is recommended that deflection due to wheel load plus 30 percent impact be less than l/300. For

Plate Thickness		Allowable Span	
in	mm	ft	m
1	25.4	5	1.5
1.25	31.8	8	2.4
1.5	38.1	11.5	3.5

TABLE 4.2 Allowable Spans for Road Plates using HS20-44 Loading

Plate Thickness		Allowable Span		Calculated Deflection		Deflection L/300	
in	mm	ft	m	in	mm	in	mm
1.0	25.4	5.0	1.5	0.6	15.2	0.18	4.6
1.25	31.8	8.0	2.4	1.49	37.8	0.29	7.4
1.5	1.5	11.5	3.5	2.23	56.6	0.42	10.7

TABLE 4.3 Plate Deflections at Allowable Spans

bridges, deflection is a visual and fatigue issue. Deflection is generally ignored in the design of trench plates because they are a temporary installation. Table 4.3 gives deflections for allowable spans for common road plate thicknesses. The calculated deflections seem tolerable and expected on surface streets considering that it is common to encounter manhole covers, drain inlets, and surface irregularities sloped to 1 to 2 in below the surface. At speeds over 45 mi/h these deflections could become a problem for stability of the vehicle and anchorage of the plates. Table 4.4 gives allowable spans for deflection = $L/300$.

Check deflection:

$$\Delta = \frac{PL^3}{48EI}$$

where $P = 15.6$ k
$L = 5, 8, 11.5$ ft simply supported span (ft)
$E = 29{,}000$ ksi
$I = 4, 7.8, 13.5$ in^4

Figure 4.16 is a typical tabulated data sheet for road plates.

Plate Thickness		Allowable Span		Calculated Deflection		Deflection L/300	
in	mm	ft	m	in	mm	in	mm
1.0	25.4	3.0	1.5	0.12	3.0	0.12	3.0
1.25	31.8	4.0	2.4	0.16	4.1	0.16	4.1
1.5	1.5	5.3	3.5	0.21	5.3	0.21	5.3

TABLE 4.4 Allowable Road Plate Spans at Deflection = L/300

Tablulated Data For Steel Trench Plates

ALLOWABLE SPANS FOR TRENCH PLATES
HS20–44 LOADING

THICKNESS		ALLOWABLE SPAN	
(in)	(mm)	(ft)	(m)
1	25.4	4.8	1.5
1.25	31.8	7.5	2.3
1.5	38.1	10	3.0

NOTES TO TABLE
(1) HS20–44 IS A 3200Ib (14515 kg) AXE
LOAD WITH A 0.3 IMPACT FACTOR

(2) PLATES ARE ASTM A36. MIN. COMMON TRENCH PLATE SIZES AND WEIGHTS

THICKNESS		WIDTH		LENGTH		WEIGHT	
(in)	(mm)	(ft)	(m)	(ft)	(m)	(lb)	(kg)
1		2.554		1.85		126.432	740
↓	↓		8	2.84		226.411	1184
			8	12.04		332.064	1481
			8	12.64		542.922	2369

THICKNESS		WIDTH		LENGTH		WEIGHT	
(in)	(mm)	(ft)	(m)	(ft)	(m)	(lb)	(kg)
1.25		0.88		12.04		430.080	1851
↓	↓		8	12.24		438.796	2221
			8	12.64		645.928	2961
			8	2.04		861.160	3701

THICKNESS		WIDTH		LENGTH		WEIGHT	
(in)	(mm)	(ft)	(m)	(ft)	(m)	(lb)	(kg)
1.25		0.18		12.04		438.096	2221
↓	↓		8	12.24		538.775	2665
			8	12.64		748.934	3553
			8	2.04		967.192	4442

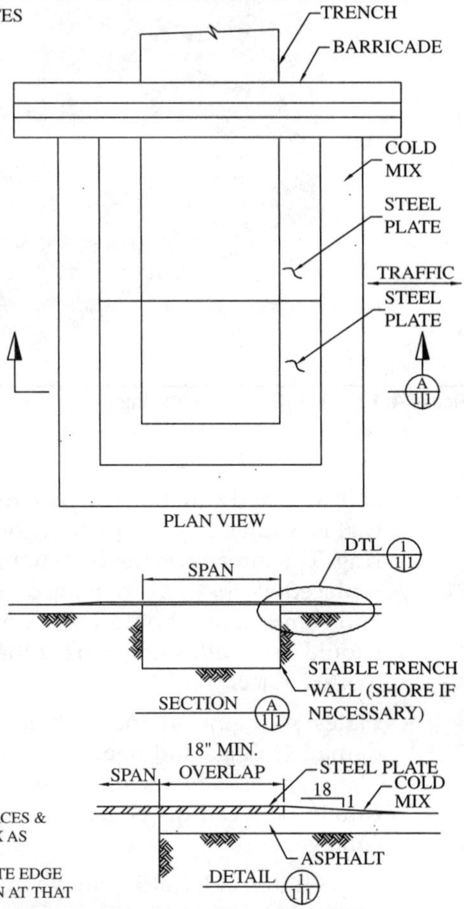

NOTES:
(1) PLATES SHALL BE PLACED ON FLAT SURFACES & BACKED UP AT ALL EDGES WITH COLD MIX AS SHOWN ON DETAIL–1.
(2) IF THERE IS AN OPEN TRENCH AT ANY PLATE EDGE PROVIDE ADEQUATE BARRIER PROTECTION AT THAT LOCATION.
(3) THERE SHALL BE NO GAPS BETWEEN ADJACENT PLATES.
(4) AT SPEEDS OVER 45 MPH THE MIN. PLATE WEIGHT SHALL BE 6500 LB IF PLATES WEIGHT LESS THAN 6500 LB, THEY CAN BE TACK WELDED TOGETHER USING A 1/4" FILLET WELD 4" PER 4 FT OF EDGE.

FIGURE 4.16 Typical road plate data sheet.

4.8.3 Road Plate Handling and Safety Issues

Road plates are normally loaded and hauled out of the supply yard and delivered to the job site with a boom truck. At the job site the plates are handled by hook and chain with an excavator or loader. Worker safety issues are as follows:

- Lift chain and connections fail, resulting in plates dropping on feet and hands. The lift ring shown in Fig. 4.17 swivels

FIGURE 4.17 A 360° swivel lift ring.

360° so that the lift bolt can be screwed into a threaded lift nut that is welded into the plate without having to release the lift ring. The threads on the bolt should be checked regularly and replaced if they are damaged or loose-fitting. Lift chains and connections should have a 5:1 factor of safety. Workers should never allow any part of their bodies to get underneath a lifted object.

- Plates swinging at the end of the chain result in impact damage to legs and hands. The inertial force of a swinging plate can easily break leg bones and shear off limbs if they are caught between the edge of a plate and a hard spot such as concrete curb.
- Plates intercept traffic lanes while being moved. Airborne plates being struck by moving vehicles or construction equipment can cause severe accidents to those in the vehicle or to workers in the vicinity of the plate.

4.8.4 Trench Bridges for Larger Spans

The trend toward giving the streets back to the public at the end of the day combined with the increasing use of trenchless technology with its wider trench openings has created a need to cover larger trench spans. One way to do this is to weld beams to road plates, as shown in Fig. 4.18. These trench bridge sections can be removed and replaced at the beginning and end of the shift with a boom truck. Section properties for the new beam plate shape have increased bending strength at the middle portion of the beam and enough shear strength at the ends to allow the I beam to be held back so that the bridge can

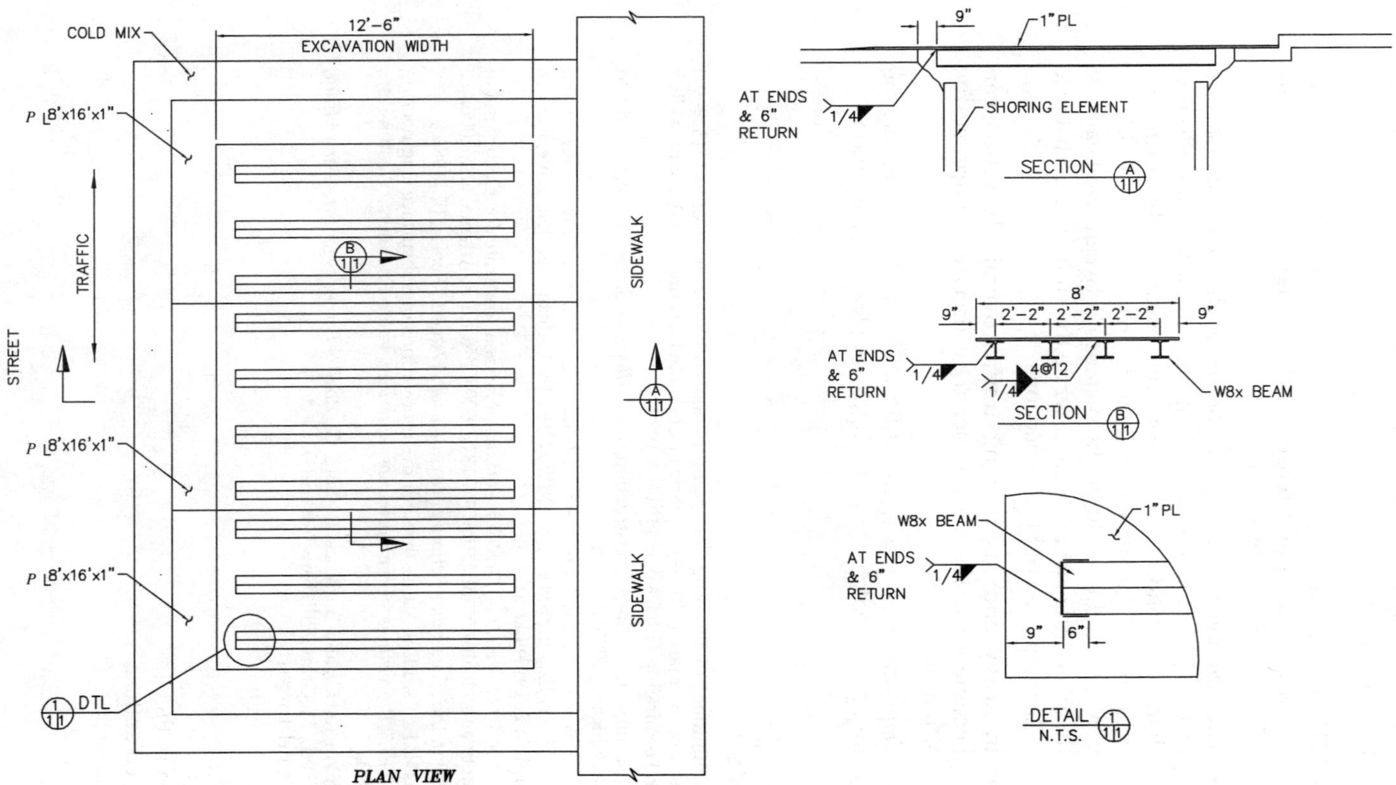

Figure 4.18 Removable trench bridge.

be set flush with the street. Potential problem areas for this type of bridge are as follows:

- There is a stress concentration at the ends of the beams. Quality welding and return welds at the ends of the beam will solve the problem.
- The cantilever loading condition on the plate at the intersection of two plates has a larger bending moment and deflection than the condition between two beams. The increased deflection at the plate intersections creates a bumpy ride for vehicles. In this example the distance between two beams at the plate intersection is 18 in while the simple span distance is 26 in, to even out the deflections.
- In heavy and fast-moving traffic, overall deflection can become an issue. Deeper beam sections will improve the situation.
- Depending on the span additional wheel loads can come into play. Look closely at all possible loading conditions.

References

ACI Committee 318, Building Code Requirements for Structural Concrete (319–99) and Commentary (318–99), American Concrete Institute, 38800 Country Club Dr.,Farmington Hills, MI 48331 USA 1999

American Institute of Steel Construction, Inc., *Allowable Stress Design*, 9th ed., 2d rev., Chicago, 1995.

American Society of Civil Engineers, *Standard Guidelines for the Collection and Depiction of Existing Subsurface Utility Data*, 1801 Alexander Bell Drive, Reston. Virginia 20191–4400 2002.

Casini, Virgil, *Overview of Electrical Hazards*, National Institute for Occupational Safety and Health, Publication 98–131, 1600 Clifton Rd, Atlanta, GA, 2006.

Caterpillar, Inc., *Caterpillar Performance Handbook*, 17th ed., Peoria, Ill., 1986.

Stevens, R.E., "Adding Value through the Innovations of Subsurface Engineering (SUE)," *Proceedings of the Society of American Value Engineers*, Washington, 1993.

Young, Warren C., and Budynas, Richard G., *Roark's Formulas for Stress and Strain*, 7th ed., McGraw-Hill, New York, 2002.

CHAPTER 5

Interpreting Soils Information for Excavation Planning and Shoring Design

5.1 Introduction

Understanding the nature of the soil that you are working in is crucial to successful and profitable excavation work. The soil is the medium that the earthwork contractor works in. The contractor has to shape it, control it, and construct things in it. Rarely is any construction project, whether aboveground or belowground, designed without regard to the soil that it is founded on. Building codes require soils investigations for all governed structures. Soils investigations, better known as geotechnical reports, are primarily developed for the purpose of determining strength and material properties that are used to design foundations which will hold up structures and facilities intended to remain there for many years. Sometimes the geotechnical firms that develop the soils information are also commissioned to develop information and recommendations regarding construction operations within it, such as control of groundwater and excavation stabilization. The purpose for including this information is twofold, to protect the condition of the soil that is being used as the foundation for the structure and to provide information to the contractor that will help him or her bid and plan the excavation in the most successful and economic way.

For the contractor the perspective is always short-term. The contractor focuses on how the soil acts when working with it. Basic knowledge of geology and soils is available in textbooks specializing in those subjects. This chapter focuses on understanding the soil for the purpose of controlling it and working in it during construction. The basic information about the soils comes to the contractor in two forms, the soils report and observations at the site from the surface and excavations. This chapter looks at ways to sort the soil information into categories that help form general conclusions about how the soil will behave when excavated.

5.2 Soil Classification System

The word *soil* is used here as a general term for everything below the earth's surface. This generally includes air, water, ice, solid rock, sand, gravel, silt, and clay. Other things found below the surface such as roots, stumps, wood, concrete, pipes, buried ships and cars, etc. would be referred to as existing facilities or obstructions. For the construction engineer, the task is to sort the soil information so that it gives some indication of how the soil will behave when it is excavated and to obtain some physical properties that can be used to calculate stability for sloping or loading for shoring. Other considerations concern whether the soil will support the equipment to be used, how easy it is to dewater, and whether the bottom of the excavation will remain stable. Determining the location and nature of water and ice is fairly easy, while sorting differences between rock, sand, gravel, and clay is more challenging and definition-sensitive. Air is present in some form in all these; however, knowing the quantity of air is not very helpful.

Sorting systems or classification systems are devised to shake out information needed for a certain purpose. The Unified Soil Classification (USC) System shown in Table 5.1 was devised to provide information needed for use in constructing structures on top of it and for use as a building material such as road base, drainage medium, and concrete. This system is the standard of the industry, and all geotechnical reports for use in construction follow it. The system also works well for the purpose of understanding how the soil will behave during dewatering, excavation, and shoring operations.

The major division in this system is between cohesive and noncohesive soils. There are significant differences between how the two soils behave. Table 5.2 lists differences between the two that are relevant to excavation stability. When one is investigating the soil for excavation stability, the first question to ask is whether the soil is cohesive or noncohesive.

5.3 Attributes of Soils in General

All soils are a mixture of different-size particles and particle types from large rocks to fine clay particles and are derived from many different minerals. The smallest constituents have the

Major Divisions			Group Symbol	Typical Names
Course-Grained Soils More than 50% retained on the No. 200 sieve	**Gravels** 50% or more of coarse fraction retained on the No. 4 sieve	Clean Gravels	GW	Well-graded gravels and gravel-sand mixtures, little or no fines
			GP	Poorly graded gravels and gravel-sand mixtures, little or no fines
		Gravels with Fines	GM	Silty gravels, gravel-sand-silt mixtures
			GC	Clayey gravels, gravel-sand-clay mixtures
	Sands 50% or more of coarse fraction passes the No. 4 sieve	Clean Sands	SW	Well-graded sands and gravelly sands, little or no fines
			SP	Poorly graded sands and gravelly sands, little or no fines
		Sands with Fines	SM	Silty sands, sand-silt mixtures
			SC	Clayey sands, sand-clay mixtures
Fine-Grained Soils More than 50% passes the No. 200 sieve	**Silts and Clays** Liquid Limit 50% or less		ML	Inorganic silts, very fine sands, rock four, silty or clayey fine sands
			CL	Inorganic clays of low to medium plasticity, gravelly/sandy/silty/lean clays
			OL	Organic silts and organic silty clays of low plasticity
	Silts and Clays Liquid Limit greater than 50%		MH	Inorganic silts, micaceous or diatomaceous fine sands or silts, elastic silts
			CH	Inorganic clays or high plasticity, fat clays
			OH	Organic clays of medium to high plasticity
Highly Organic Soils			PT	Peat, muck, and other highly organic soils

Prefix: G = gravel, S = sand, M = silt, C = clay, O = organic
Suffix: W = well graded, P = poorly graded, M = silty, L = clay, LL < 50%, H = clay, LL > 50%

TABLE 5.1 Unified Soil Classification (USC) System. The top half is noncohesive soils, and the bottom half is cohesive soils

Cohesive Soil **Over 50% of particles pass the No. 200 sieve**	**Noncohesive Soil** **Less than 50% of particles pass the No. 200 sieve**
• Plasticity—sticks together and can be molded • Particles less than 0.002 mm affected more by electronic bonding than by gravity • Can have vertical trench walls • Strength that changes with vibration and movement of heavy equipment • Has as primary engineering property cohesion *c* • Consistency that varies from very soft to hard • Can experience bottom heave • Does not give up water easily. Can be dewatered from inside the excavation • Fails by shear cracking and sloughing	• Has no plasticity and breaks apart easily when excavated • Affected more by gravity, particle shape, how closely packed, and particle friction • Vertical trench walls more likely to fail • Subject to raveling with drying and vibration • Has as primary engineering property the angle of internal friction ϕ • Relative density that varies from very loose to very dense • Can experience boiling at base of excavation • Can allow so much flow that dewatering becomes impossible. Should be dewatered from wells outside the excavation • Experiences rapid failure

TABLE 5.2 Excavation-Sensitive Soil Attributes

greatest effect on the strength of the soil. For example, the strength of concrete is determined mostly by the cement, the finest particles in a mixture that is mostly sand and gravel. The fines in noncohesive soils play the decisive role in the shear resistance of the material.

Sand, gravel, and silts are generally subangular to rounded, while clay particles are flakelike and rounded particles are not found in the clay matrix. Some other basic attributes of soil in general are as follows:

- The more densely the soil grains are packed, the stronger the soil.
- The removal of water adds strength to the soil.
- Deep clays are generally stiffer than shallow clays.
- Fissures are more common in deep, stiff clays.

5.4 Cohesive and Noncohesive Soils

Cohesive Soils

Cohesive soils stick together because of a combination water and particle attraction attributable to clay minerals. Depending on the amount of water and clay particles, cohesive soils can be molded or deformed (exhibit plasticity) and still hold together. They will look the same in texture no matter what shape they are molded to. Cohesive soils are referred to as fine-grained soils (Table 5.1, bottom half) and meet the requirement that more than one-half of the material will pass through an almost microscopic No. 200, 0.074-mm, sieve hole. This determination is based only on particle size and not on the quantity of water in the sample. Cohesive soils are further divided into silts and clays (Table 5.1, bottom half middle division).

The force of gravity on particles that are less than 0.002 mm is insignificant compared to the electrostatic forces between the particles and the water surrounding them. Clays are flat, flake-shaped particles. Clay particles in water interact to form an electrostatic bond. This adsorption process changes the characteristics of both the clay mineral and the water to form the clay mass. As clay particles are squeezed close together due to the natural force of gravity from soil laid on top of it, or artificial compaction effort, the bond becomes stronger and excess water is squeezed out. Generally the deeper the clay particle is from the surface, the stronger the cohesive force.

At a certain point the clay structure becomes watertight. The adsorbed water that is in the clay cannot be drained out; however, it is important to note that the water can be dried out with exposure to the atmosphere. In the dry state it acts more as a rock although it is usually easier to break and forms cracks more readily.

Clay can be remolded by vibrations and heavy equipment working on top of it. When clays are remolded, they become weaker. Not all particles in a cohesive material are clay, although the clay particles cause most of the cohesion. Silt, sand, gravel, and rock can be present. The more clay particles are present, the more the soil exhibits cohesive soil characteristics. The Unified Soil Classification System (USCS) divides fine-grained soils into those with a liquid limit (LL) first below 50 and then above 50. The liquid limit is an indication of the compressibility of the soil and the amount of water that can be held in the soil through absorption and adsorption.

The more nonclay silt particles there are in the sample, the less compressible it is and the easier it is to drain water out. The USGS symbols for these nonclay silts and clays are ML, CL, and OL (Table 5.1, bottom half top). The M stands for silt, C for clay, O for organic, and L for low liquid limit. Low liquid limit soils are less cohesive and plastic, tending more toward the behavior of sand and gravel.

Soils with a liquid limit above 50 have more clay particles, are more compressible, and hold more water. The USCS symbols for these silts and clays are MH, CH, and OH. The H stands for high liquid limit. The USCS has one more division called *highly organic soils* with the symbol PT. This is peat and swamp soils. It is highly compressible and close to the consistency of very soft clays found in "young bay muds." They are very plastic but low in strength, cause high loads on shoring systems, and will not hold a very steep slope. Bay muds are also referred to as marine clays. The consistency of silts and clays which varies from very soft to hard and the fact that water does not flow through them are major differences between them and noncohesive soils.

Cohesionless Soils

When the soil cannot be molded without breaking apart, it is noncohesive. Cohesionless soils are referred to as coarse-grained soils (Table 5.1, top half) and divided further into gravels and sands. The defining test is that over one-half of the materials will not pass through the No. 200 sieve. There may be silts and clays in this soil but not enough to exhibit plasticity. Even though it is considered cohesionless, there are certain factors that cause cohesionless soils to stay together and stand up vertical to some height when they are excavated. Further divisions are as follows:

- *Clean sands and gravels*. The shape of the particles and how densely they are packed are by far the most important criteria for clean sands and gravels. *Clean* means that there is very little clay and fine particles. Particle shape varies from sharp, pointed edges with angular returns to round (Fig. 5.1). On a large scale think of rocks in a jetty or riprap, to rounded like a marble. Shale, and rock that is excavatable or blasted is also angular. Sands and gravels are the result of the breakdown of larger rocks through mechanical weathering. Air, temperature changes,

Figure 5.1 Angular and rounded particles.

wind, and water are the primary drivers of mechanical weathering. The weathering process breaks down, picks up, and then moves the soil to the location where it is found prior to excavation. As a general rule, the farther away from the original source the sand and gravel are found, the more rounded the particles will be. Ocean beach sand would be more rounded than lake beach sand, gravel found in a riverbed would be more rounded than gravel in an alluvial fan. Wind-blown particles are more rounded than washed particles. How these particles are laid down and what has happened to them since they were deposited determine how densely they are packed. The gravity weight of soil and ice will compact the grains. Seismic activity will further consolidate the grains.

Streambed sands and gravels cause a lot of problems for excavation contractors. In the streambeds, sands and gravels are constantly being picked up and laid down. As the flow location meanders, the particles are left to some degree rounded, segregated, and loose. Ancient and recent streambeds can wander laterally for miles and then disappear over time as lighter silts and clays are deposited on top of them to form valley floors. Pipeline excavation production in these soils can easily go from 300 to 20 ft/day or less as old streambeds are encountered. Because of the lack of uniformity in the soil, boring logs taken at 100 to 500 ft apart have a slim chance of fully defining the expected excavation characteristics of streambed soils.

Well-graded sands and gravels have an even distribution of particle sizes while poorly graded materials are more segregated. Beach sand would be poorly graded while a sample from an alluvial fan or a riverbank would be tending toward well graded. Water flow segregates sands and gravels, and therefore gap-graded loose material would be an indicator of ancient streambeds. Well-graded materials stand at steeper slopes than graded materials. On a grain size chart (Fig. 5.2), the vertical curves indicate poorly graded material, and long, smooth curves are well graded. In the Unified Soil Classification System, clean gravels are identified as GW and GP, and clean sands as SW and SP. The G stands for gravel, S for sand, W for well graded, and P for poorly graded. Capillary action from moisture and friction between the particles is primarily what causes clean sand and gravel to hold itself together and stand up during excavation. Capillary action is the same principle that causes water to bead on a flat surface and allows water skeeters to walk on water. As the moisture is exposed to the atmosphere, it evaporates and the capillary force is lost. Friction is derived from the angularity of the particles. In moist soils subangular grains would be expected to stand

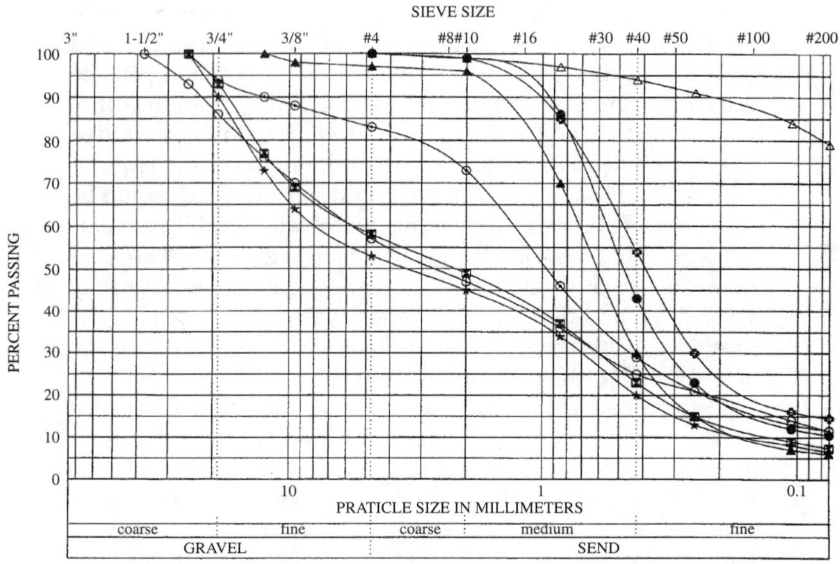

Figure 5.2 Grain size chart.

up vertical for a short time and then rest at approximately a 1:1 slope. As the grains become more rounded, the stand-up time decreases and the slope degenerates to 1.5:1 or worse. When these soils are submerged, they could at best be expected to stand at a 1.5:1 slope. Clean sands and gravels allow water to flow through them easily and can be hard to dewater.

- *Dirty sands and gravels.* If there are over 12 percent fines (fines are silts and clays), the material is no longer considered clean. The silts and clays change the nature of the material. Fine sand and clay particles are washed into the voids of the sands and gravels and form a weak bond. The most common binding agents are clay particles and calcium carbonate. The bonding of the clay particles is present; however, since there is less clay than rock, the cohesion is much weaker. Chemical weathering of certain types of rock dissolves calcium carbonate in water that is then leached into sands and gravels, causing the sand and gravel to bond together as the water table subsides. The addition of water will weaken both clay and calcium carbonate binding agents. Other less common binding agents tend to cement the sands and gravels together and are less susceptible to degeneration from water.

The density and shape of the gravel and sand particles also play a part in the strength or stand-up ability of these soils. They tend to stand up vertically longer than clean sands and gravels and depending on depth of excavation can be stable at a ¾ : 1 slope. When the strength of the bonding agents in these soils is exceeded, a vertical wall or slope failure is rapid. Dirty sands and gravels are harder to dewater than clean sands and gravels, and the fine particles can flow through them, causing settlement and plugging up pumps. The USCS symbols for dirty gravels are GM and GC and for sands SM and SC. The G stands for gravel, S for sand, M for silts and silty clays, and C for clays.

5.5 Fundamental Design Properties of Soils

The major property that we are looking for is the soil's ability to resist failure by shear. Shear failure leads to slope movement, loads on shoring, and bottom-of-excavation instability. Shear failure starts at some point in a mass of soil as a result of a combination of external loads on it such as construction equipment and spoil piles and internal loading created by removal of lateral support during excavation. If the shearing resistance of a soil and the stress that caused the rupture are plotted on a graph (Fig. 5.3), an approximate straight rupture line emerges and is defined by Coulomb's failure equation

$$s = c + \sigma' \tan \phi' \tag{5.1}$$

where s = shearing resistance or shearing strength
c = cohesion
σ' = effective stress
ϕ' = effective angle of internal friction

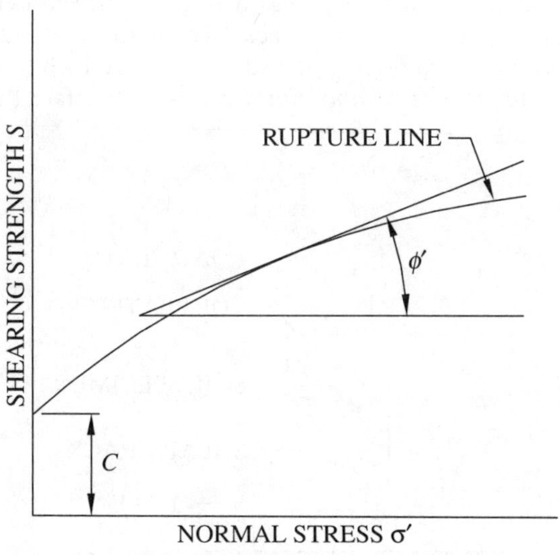

FIGURE 5.3 Plot of direct shear test showing cohesion intercept c and angle of internal friction ϕ'.

The cohesion, effective stress, and angle of internal friction are defined and discussed below. The values for c and ϕ must be determined before any type of stability or soil loading calculations can be performed. For categorizing purposes the term *cohesion* is most often associated with cohesive soils, and the term *angle of internal friction* is most often associated with noncohesive soils.

Cohesion *c*

Cohesive soils have an internal strength that holds them together and causes them to support loads placed on top of them. Cohesion causes the soil to stand up and not deform when lateral support is removed, as happens when it is excavated. The failure mechanism for cohesive soil is a breaking of the bond between clay particles due to excessive loading. This results in sloughing and creep in soft clays and rapid bank collapse in stiffer clays. The cohesion can be thought of as a measure of the strength of bonds at the clay particle level and is reported in units of pounds per square foot (psf or lb/ft^2), tons per square foot (tsf), and thousand pounds per square foot (psf).

Since the electronic bond is the controlling force in clays, the angle of internal friction ϕ will not be mobilized until near failure so a $\phi = 0$ condition is assumed. In this case tan 0 = 0 and Eq. (5.1) becomes

$$s = c \tag{5.2}$$

The most prominent laboratory method for determining cohesion is the unconfined compression (UC) test (Fig. 5.4). This is the same as a concrete cylinder test that places a round concrete cylinder in a press and applies a load until it cracks. The load divided by the area of the cylinder top provides a pounds per square inch (psi) strength of the concrete. The term *unconfined* refers to the fact that there is

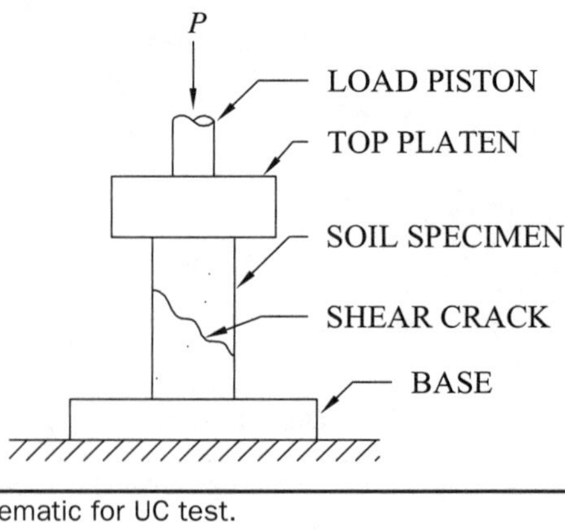

FIGURE 5.4 Schematic for UC test.

FIGURE 5.5 Pocket penetrometer.

nothing confining the sides of the cylinder. In the field the soil would be confined by soil around it. The unconfined compression strength is called q_u, and the cohesion is approximately

$$c = \frac{q_u}{2} \tag{5.3}$$

The results of this test are considered to be conservative, reporting shear strength less than actual, and are increasingly conservative as softer clays are tested.

During the soils investigation in the field the unconfined compression strength q_u is estimated by blow counts taken in the ground during the drilling operation and can also be tested with a pocket penetrometer (Fig. 5.5) on a sample brought up from the drilling operation. This instrument is based on a force per unit area and reads directly in tons per square foot (tsf). The unconfined compression strength is usually reported on a soil boring log under a heading titled q_u or UC. It is important to note that although there is some cohesion in dirty noncohesive soils, the test is only performed on cohesive soils. The test is taken on samples that are as close to the in situ condition of the soil as possible. The test provides the "undrained" strength because it is taken when the soil is saturated or in a natural moisture condition.

Angle of Internal Friction ϕ

Also known as the angle of repose, this is the angle between the horizontal and the plane of contact between two bodies when the upper body is about to slide over the lower [Fig. 5.6(a)]. In the case of cohesionless soils there is a rupture line where the lower plane is stable soil and the upper plane is not stable [Fig. 5.6(b)]. In cohesionless

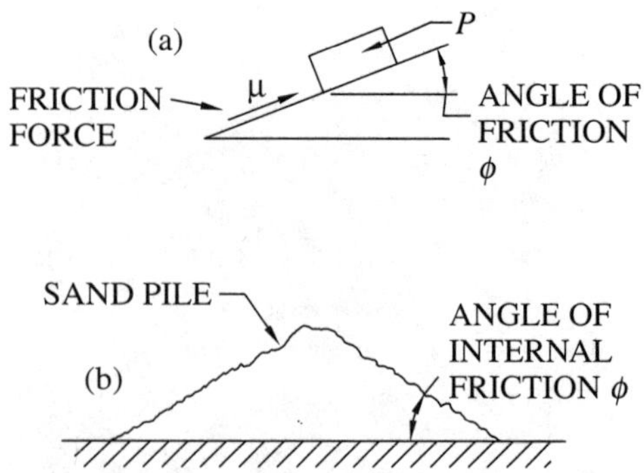

FIGURE 5.6 Angle of internal friction.

soils with no silt and clay present the cohesion $c = 0$, and Eq. (5.1) becomes

$$s = \sigma' \tan \phi' \tag{5.4}$$

In these soils the shear resistance is developed by the interaction of solid grains of sand and gravel. Under pressure the grains can distort (change shape slightly but not break), crush, roll, slide, and interlock, causing shear resistance. At the point where shear resistance is exceeded, there is a massive shear failure, and the final slope will be less than or equal to the angle of internal friction.

During excavation in cohesionless soil, the walls of the excavation will stand up vertical due to capillary action from moisture in the soil and shear resistance developed between the grains. The shearing or failure-causing force is the weight of the soil in the excavation wall and any additional loads placed at the surface. Eventually the loads create a greater shearing force than the resisting force, and the vertical wall fails. If the face of the wall has time to dry out before failure, the capillary forces will evaporate and the wall face will start to ravel, thereby removing some shear-resisting soil grains and causing further potential for failure.

In the laboratory a typical test to determine ϕ is the direct shear test, the results of which report the friction angle and cohesion (Fig. 5.7).

In the field angle ϕ can be approximated from blow counts taken during exploratory drilling.

Water Table and Effective Stress σ'

A basic concept, *Archimedes' principle*, relating to objects immersed in water, either partially such as a boat or totally such as a rock, is that the object is buoyed up with a force equal to the weight of the water

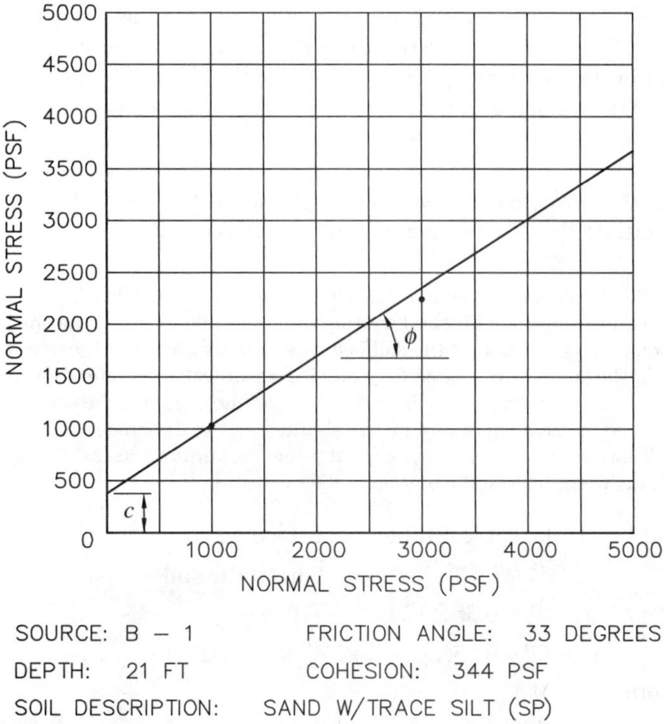

Figure 5.7 Direct shear test results.

displaced. Consequently if a block of concrete that measures 1 ft × 1 ft × 1 ft and weighs 150 lb/ft³ (or pcf) is placed in water, it will displace 1 ft³ of water and will have a buoyant force equal to the weight of water displaced, in this case 1 ft³ of water, or 62.4 lb. On dry land the block will weigh 150 lb, and in water it will weigh 150 lb – 62.4 lb = 87.6 lb. By the same principle soil grains that are below the water table weigh less than soil grains above the water table.

Stress σ is defined as a force applied to a unit area. Stress deforms the shape of the object that the force is being applied to by compressing, stretching, or shearing.

Pressure is also defined as a force per unit area and is often used synonymously with the word *stress*. Pressure that deforms an object is said to be *effective pressure* and is given the symbol σ'.

Pressures that act simultaneously in all directions, such as atmospheric pressure and water pressure, do not necessarily deform an object and are referred to as *neutral pressure*. *Water pressure* is defined as

$$u = \gamma_w h_w \tag{5.5}$$

where γ_w = unit weight of water, 62.4 pcf, and h_w = height of water above where the pressure is being calculated.

In soils the *neutral pressure u* is also referred as the *pore water pressure*. The neutral pressure does not have a measurable effect on the void ratio or shearing strength of the soil. The total pressure on an element of soil at a given depth is

$$\sigma = \sigma' + \mu \tag{5.6}$$

where σ' = effective pressure and μ = neutral pressure from water. To calculate the effective pressure use $\sigma' = \sigma - \mu$.

Example 5.1 A simple example will help to illustrate how the effective and neutral stress affect slope stability and shoring design problems. In soil mechanics the lateral stress condition of the soil is always some factor, 0 to 1, of the vertical stress. The lateral stress factor for water is 1, assume that the lateral stress factor k_a for the sand and gravel in Fig. 5.8 is 0.35 and the weight of the soil is 120 pcf. Calculate the active pressure on the shoring wall at the dredge line in case 1, Fig. 5.8(a), where the water table is 5 ft below the surface and case 2, Fig. 5.8(b), with the water table at the bottom of the excavation.

Case 1 — water table 5 ft below surface

$$\sigma' = P_a = k_a(\gamma h_1 + \gamma' h_2) \qquad \text{from the soil}$$

$$\sigma'_a = P_a = 0.35 \times 120 \times 5 = 210 \text{ psf}$$

$$\sigma'_b = P'_b = 0.35(120 \times 5 + 57.7 \times 7) = 352 \text{ psf}$$

where $k_a = 0.35$

$\gamma = 120$ pcf (unit weight of in situ soil)

$\gamma' = 120 - 62.4 = 57.6$ pcf (unit weight of wet soil)

$h_1 = 5$-ft depth to water table

$h_2 = 7$-ft depth of water to dredge line

$\mu = \gamma_{\text{water}} h_2 = 62.4 \times 7 = 440$ psf

$\sigma = P_b + \mu = 352 + 440 = 792$ psf

Case 2 — water at dredge line

$$\sigma' = P_a = 0.35(k_a \gamma h_1 + \gamma' h_2) \quad \text{from the soil}$$

$$\sigma'_b = P'_b = 0.35(120 \times 12 + 57.7 \times 0) = 504 \text{ psf}$$

where $k_a = 0.35$

$\gamma = 120$ pcf (unit weight of in situ soil)

$\gamma' = 120 - 62.4 = 57.6$ pcf (unit weight of wet soil)

$h_1 = 12$-ft depth to water table

$h_2 = 0$-ft depth of water to dredge line

$\mu = \gamma_{\text{water}} h_2 = 62.4 \times 0 = 0$ psf

$\sigma = P_b + \mu = 504 + 0 = 504$ psf

The vertical stress in the wet condition is less than that in the dry condition because of the buoyant weight of the soil below the water table;

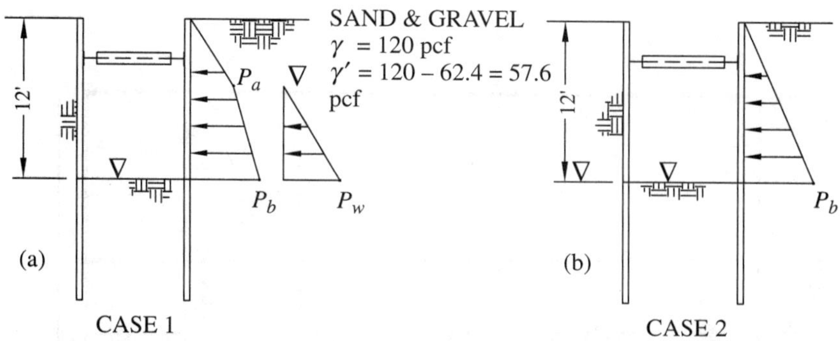

Figure 5.8 Example 5.1: case 1, water table 5 ft below surface; case 2, water table at dredge line.

however, the horizontal stress on the shoring is higher because the weight of the water is also being held back. When water is backed up behind shoring systems the pressure increases significantly. Whenever there is water present in an excavation, it will influence the stress condition of the soil and the stability of the excavation.

5.6 Reading Soils Reports and Bore Logs for Shoring Design

On practically every project involving construction in or on the soil, there is a geotechnical investigation. The report is financed and commissioned by the owner and project design engineer for the purpose of obtaining soil properties relative to structural stability and longevity of the project. The basic exploratory program most often consists of borehole drilling and sample recovery, in situ testing in the borehole, visual and manual testing to classify the soil in accordance with the Uniform Soil Classification System (Table 5.1), and further laboratory testing of samples taken from the drilling program. From the testing program and research about geological setting, current history of the site, seismicity, hydrology, and chemical analysis from potential contaminated sources, a geotechnical report including design recommendations is developed.

From the excavation planning and shoring design perspective, the three most important pieces of information in the soils report are contained in the boring logs:

- The language used in the description of the soil layers
- The field-performed *standard penetration test*
- The results of unconfined compression tests in the laboratory

The bore log lists information obtained from the drilling operation and lab testing. In the description (Fig. 5.9) every layer of the soil is classified and described in detail using terminology defined

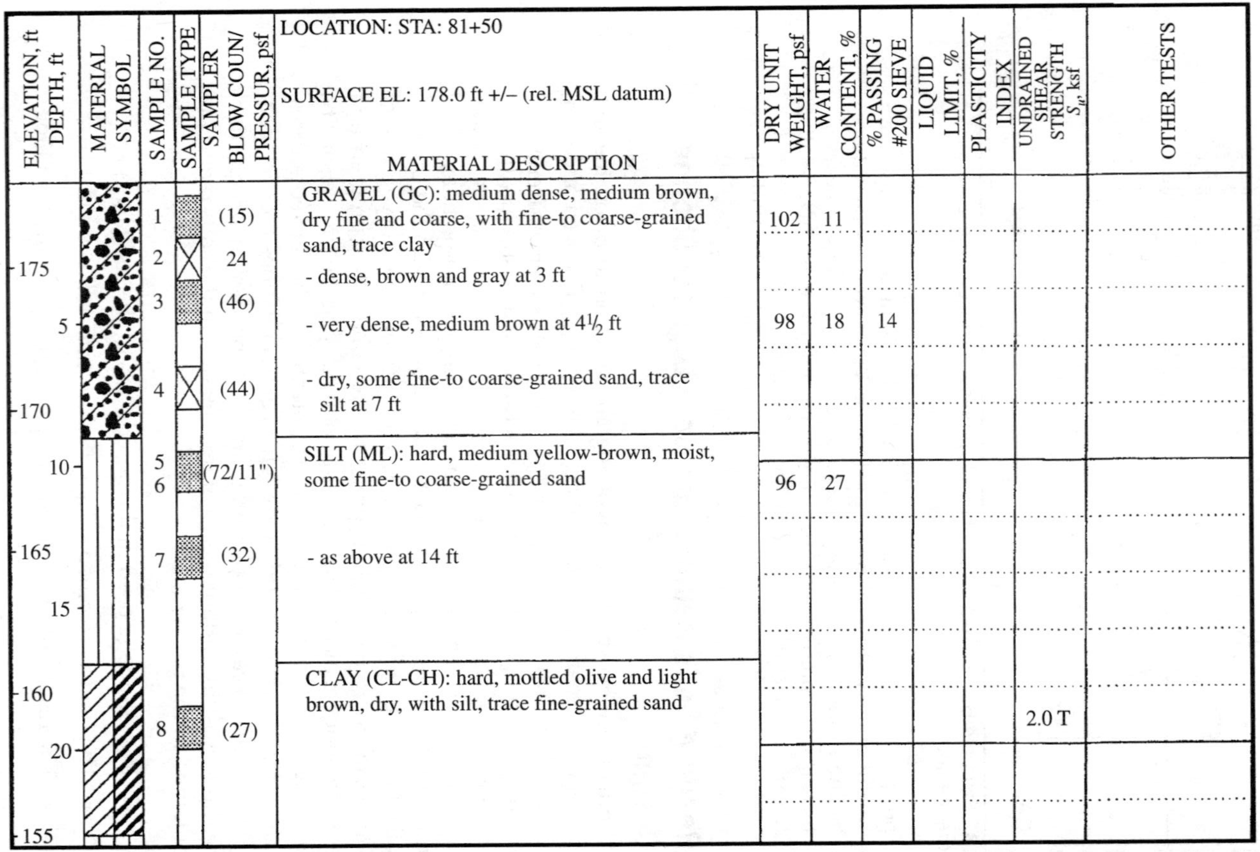

FIGURE 5.9 Soil boring log.

Relative Density		Consistency		
Sands and Gravels	**SPT *N***	**Silts and Clays**	**SPT *N***	**Unconfined Compression Strength (tsf)**
Very loose	0–4	Very Soft	0–2	0–0.25
Loose	5–10	Soft	3–4	0.25–0.50
Medium dense	11–30	Medium stiff	5–8	0.50–1.00
Dense	31–50	Stiff	9–16	1.00–2.00
Very dense	50+	Very stiff	17–32	2.00–4.00
		Hard	32+	>4.00

Table 5.3 SPT *N* Description Correlation

in the USCS. Terms that apply to density—*loose* and *dense*—are used when describing sands and gravels and related terms that apply to consistency—*soft, stiff,* and *hard*—are used with silts and clays. The use of these terms immediately tells the reader whether the sample is cohesive or noncohesive. These terms are defined by blow counts resulting from standard penetration tests (Table 5.3). If a cohesive layer has a blow count of 19, the term *very stiff* is used to describe it, and if the term *very stiff* is used in the description, one knows that a blow count of 17 to 32 and an unconfined compressive strength between 2 and 4 tsf are expected in that material. If the results of an unconfined compression strength test in the laboratory are 2.5 tsf, the term *very stiff* is used in the description. The group symbol also tells exactly which classification the material falls into on the USCS chart. When the term *sandy lean clay* (CL) is encountered, reference to the USCS chart (Table 5.1) indicates that the soil is cohesive and has clay particles with medium plasticity, an indication that a sample will deform without breaking apart. The fact that the sample has sand in it causes it to be termed *lean clay*. This sample would be more likely to give up more water than fat clay.

The standard penetration test (SPT) is taken during the drilling operation using a standard 1.4-in-ID × 2-in-OD × 30-in-long sampler (Fig. 5.10). The sampler is driven by a 140-lb (63.5-kg) weight falling from a height of 30 in. The number of blows is counted for each three successive 6-in drives. The first 6-in drive count is discarded, and the final two 6-in drive counts are added to obtain the number of blows per foot, called the *standard penetration resistance N*. Various terms are used for this on the boring log such as SPT *N*, blow count, penetration resistance, and blows per foot. In sands

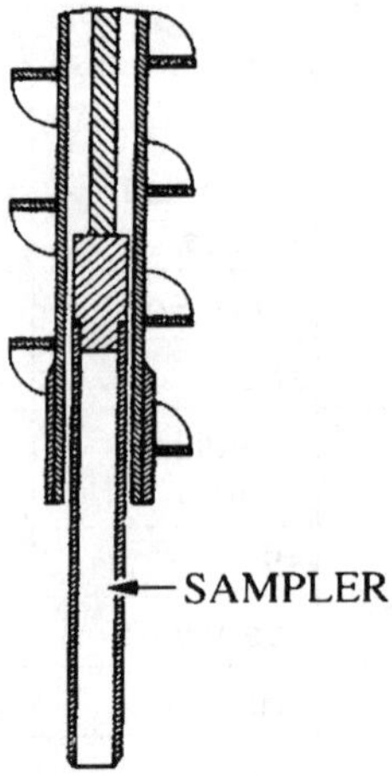

FIGURE 5.10 SPT test sampler.

and gravels the results from this test range from 0 to defined refusal at 50 blows/ft and in silts and clays from 0 to 32 blows/ft. Major factors among many that affect the results of this test are the drive rod length, the borehole diameter, and the type of sampler used. There are approximate correction factors for these variations (Table 5.4); however, they are not universally agreed upon. Two other samplers used for this test are a 2 3/8-in-ID × 3-in-OD modified California sampler (MCS) and a 2-in-ID × 2.5-in-OD split spoon sampler (SSS). The different type of sampler and the depth location where it was used are usually shown on the boring log with a symbol (Fig. 5.11). If the blow count has been corrected for these differences, it should be reported in the bore log; otherwise correction factors should be applied before the results are used.

The results of the SPT test can be used to directly correlate the unconfined compressive strength q_u for cohesive soils (Table 5.3)

Variation			Correction Factor
Rod length (approximate depth of sample)	>30 ft	>10 m	1
	20–30 ft	6–10 m	0.95
	13–20 ft	4–6 m	0.85
	10–13 ft	3–4 m	0.75
Borehole diameter	3–5 in	65–115 mm	1
	6 in	150 mm	1.05
	8 in	200 mm	1.15
Standard sampler (SPT)	1.3/8-in ID/2-in OD		1
Modified California sampler (MCS)	2.3/8-in ID/3-in OD		1.2
Split spoon sampler (SSS)	2 in-ID/2.5-in OD		1.1

TABLE 5.4 Approximate Corrections to Measured *N* Values

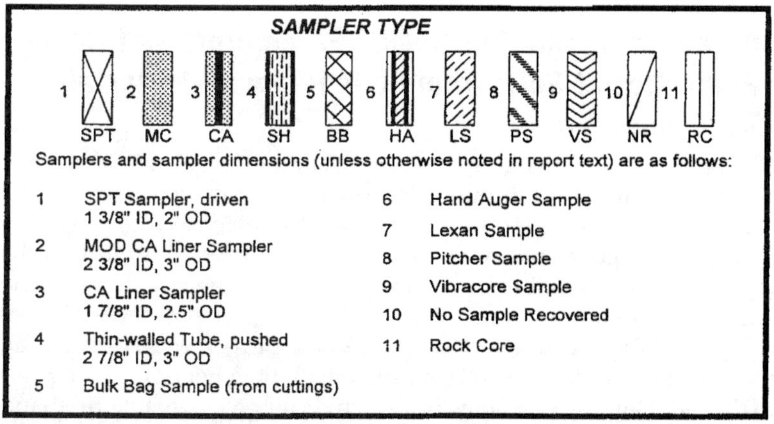

FIGURE 5.11 Example of sample type key on bore log.

and the density and angle of internal friction ϕ for noncohesive soils (Table 5.5).

Blow counts reported on the boring log and correlations are used extensively throughout the engineering community to perform calculations. There is also widespread opinion about the accuracy of the blow counts and reliability of the correlations. In the field, pocket penetrometer tests can be performed on samples of cohesive soils brought up from the borehole, and in the laboratory the unconfined compression test provides an accuracy check on the in situ SPT test. On cohesive soils, a direct shear test in the laboratory will provide a comparison to the SPT results. Customary practice for design of excavation support is to use SPT *N* values and test results to determine an assumed expected *c* and ϕ value for design and then check to see how variations in these parameters will affect the safety of the design. The final dig is the ultimate exploratory soils examination because all the soil within the extent of the excavation becomes visible. At that time there should be follow-up to confirm the design assumptions.

Blows/ft SPT *N*	Angle of Internal Friction ϕ, (deg)	Relative Density
<4	<30	Very loose
5–10	20–35	Loose
11–30	35–40	Medium dense
31–50	40–45	Dense
>50	>45	Very dense

Source: Meyerhoff (1956).

TABLE 5.5 Noncohesive Soils Approximate Correlation SPT *N* to Angle of Internal Friction ϕ

5.7 Determining Shoring Design Parameters and Basic Properties of Soils from Soils Reports—Watertable ∇, ϕ, and c

For the construction engineer the easiest and most reliable method of determining the soil conditions in a planned excavation is to read the soils report. The soils report is the result of an extensive investigation by experts in geotechnical engineering. If we keep in mind that the report was developed for the purpose of designing structures inside and on the soil, the information can be interpreted fairly easily for constructability purposes. The cost and time involved in developing the report preclude a contractor from developing her or his own prior to the bid. Digging test holes prior to bid can give valuable firsthand information about how the soil behaves during excavation operations; however, it does not give design values for stability and shoring design. If shoring design value interpretations were made from prebid test holes, they would still be seen as inferior to the results of a soils investigation. Geotechnical reports will usually state that they are for the use of the client, the owner, and the project design engineer, and not for use by third parties. Given that the geotechnical report is funded for the purpose of developing information that will add quality and cost-effective construction, it seems prudent for the contractor to use, rely on, and base the bid on the information presented. Also any trenching and shoring design recommendations made in the report will be forced on the contractor as a minimum shoring requirement. Changes to the design recommendations are always resisted by the project engineer and require negotiating with a third party that has little interest in doing a double take on the engineer's original decision. The problem is that the geotechnical engineer bases recommendations on industry standard practice and conservative risk management policies that assume that an inexperienced contractor will get the job. This leaves no room for innovation based on a risk management theory of safety first followed by maximum productivity. In shoring design this is the classic battle. Both viewpoints need to be represented, and usually reasonable minds come together to achieve a solution that works.

5.7.1 Geotechnical Report Recommendations

In many instances, especially publicly funded excavation and shoring-intensive projects, the geotechnical engineer is commissioned to develop and include in the report information and recommendations for shoring design. This information is always presented with the caveat that further investigation and follow-up will be necessary; however, it is extremely valuable as it is the findings of a geotechnical expert. Follow-up in the field to confirm the assumptions from the exploratory program is also a crucial element in an excavation shoring plan. Some of the following excavation and shoring requirements are usually contained in the soils report:

- *Selection and successful design and installation shall be made the sole responsibility of the contractor.* This always presents the question, Then how can the engineer have a say-so in how I shore it? The answer is that the shoring has to be designed within the contract specifications. Also the owner and the contractor can both be held responsible if these requirements are not met. The project engineer is protecting the owner with this requirement.
- *All excavations must comply with applicable local, state, and federal safety regulations including the current OSHA Excavation and Trench Safety Standards.* Contained in these OSHA standards is Section 1541.1 Requirements for Protective Systems, (b) Design of Sloping and Benching Systems, (4) Option (4)—Design by a Registered Professional Engineer, and Section 1541.1 Requirements for Protective Systems, (c) Design of Support Systems, Shield Systems, and Other Protective Systems , (4) Option (4)—Design by a Registered Professional Engineer.

  These two options allow the contractor and the shoring design engineer to develop an innovative, cost-effective shoring system. Many times the engineered option appears to be less conservative than the other OSHA-defined options. Project design engineers would like to see the words *different but equivalent or more conservative than the first three options* attached to the requirements for the engineered systems; however, the language is not there. In actuality the engineered option is more conservative because it utilizes specific information from the soils report and is more reliable than generalized information developed for all soil conditions.
- *Dewatering discussion and requirements.* Shoring and dewatering are inextricably connected, and therefore it has become common practice to link the two together as one design. Even though one involves structural design and the other involves hydrology, they are both controlled by the soil conditions.
- *Guidelines for trench slope configurations.* This usually defers to OSHA requirements but serves to fit the major site soil conditions into the OSHA type A, B, and C categories and provide more information about how the specific site soils will behave during excavation. These guidelines are often interpreted as an absolute minimum by the project engineer.
- *Generalized or minimum shoring requirements.* This further defines the purpose of the shoring system, stating that the system shall be designed to:
  - Protect workers entering the excavation
  - Protect facilities affected by the excavation

- Protect against settlement
- Stabilize the bottom of the excavation
- Allow for proper construction of the final product
- Prevent damage to production work during removal of the shoring
- Include in the design the effect of transitory loads from water pressure (failure of dewatering system) and surcharge

- *Specific shoring requirements for each pipeline and structure location on the project.* This information further pinpoints the type of soils to be encountered, types of unacceptable and acceptable shoring systems, and problems anticipated during their construction.
- *Shoring loading pressure diagrams for different types of soil and shoring systems.* These diagrams are usually different for cohesive and noncohesive soils and address water pressure loading, surcharge loading, cantilever shoring, and braced shoring. The shape of the loading diagram (triangular, rectangular, trapezoidal) and values in terms of excavation depth to be used are defined. These diagrams greatly simplify the job of the shoring design engineer who does not have to extrapolate parameters c and ϕ from the soils information and adopt a suitable pressure theory to design from. These diagrams are generally easy to use, safe, and specific to the soils at the project site. They are derived from accepted earth pressure theories backed by experience from previous projects with similar soil conditions.

  Providing minimum shoring pressure diagrams in the soils report and making the use of them a contract requirement add involvement in the shoring design and the associated risk to the geotechnical and design engineer. The advantage is that it establishes minimum shoring design parameters that all bidding contractors can use to establish their bids. Underbidding or overbidding the shoring work is minimized. Later when the contract is awarded and the shoring design is submitted for review, there is a baseline for the reviewer to work from.

5.7.2 Developing Shoring Design Parameters from Boring Logs and Soil Test Data

Developing excavation, dewatering, and shoring design parameters is not the primary purpose of a geotechnical investigation. The information is there, it just needs to be extracted. By reading the boring logs, using the Unified Soil Classification System chart (Table 5.1), the relative density and consistency table (Table 5.3), and SPT N/ϕ (Table 5.5), the soil parameters for design are readily available.

At this point it will be helpful to list what information is needed from the soils report and for what purpose it is needed. After that one can develop the most accurate and efficient process for extracting it from the report. Information needed includes the following:

Parameter	Purpose
• Determine cohesive/ noncohesive	1. Earth pressure theories are different for each type. 2. Dewatering potential and technique are different for each type. 3. OSHA type A soil is only possible in cohesive soil.
• Noncohesive soil fines content	1. It indicates stability and ability to stand up. The smallest particles in the soil mass have the largest effect on stability. 2. It affects dewatering.
• Angle of internal friction ϕ	1. It is required for noncohesive earth pressure calculations. 2. It is required for slope stability calculations. 3. It is required for base stability calculations in noncohesive soil. 4. It indicates angularity of particles, needed for determination between OSHA type B and type C soils.
• Cohesion c	1. It is required for cohesive earth pressure calculations. 2. It is required for base stability calculations in cohesive soil.
• Unit weight of in situ soil γ_{sat}	1. It is required for earth pressure and stability calculations.
• Water table ∇	1. It is required for water pressure calculation. 2. It is required to calculate effective weight of soil γ'. 3. It is required for determination of OSHA type C soil.

Where to Look in the Soils Report for Design Parameters

Cohesive/noncohesive: The group symbols used in the description on the boring log (Fig. 5.9) will tell exactly where the described soil falls on the USCS chart.

Noncohesive fines content: Look at group symbol, clean gravels GW and GP and clean sands SW and SP with less than 5 percent fines. Dirty gravels GM (contains silt) and GC (contains clay) and dirty sands SM (contains silt) and SC (contains clay) have fines greater than 12 percent. Gravels and sands with fines between 5 and 12 percent require dual symbols such as SW-SC, etc. In general the more fines, the less permeable the soil is; and the more clay fines, the more likely the soil will stand up for some period of time.

Angle of internal friction ϕ: After reading blow counts in noncohesive layers on the bore log, use Table 5.5, SPT *N* to friction angle. Follow up by checking direct shear tests in test reports. Note that the direct shear test also reports a cohesion intercept value; however, in calculations a $c = 0$ assumption is made because the cohesion can be temporary, and ultimately when it dissipates, it becomes zero.

Cohesion c: After reading blow counts in cohesive layers use Table 5.3, SPT *N* to description correlation. Under the consistency portion of the table, determine the unconfined compressive strength using interpolation from Eq. (5.7).

$$q_u = \frac{N_{\text{blow count}} - N_{\text{min}}}{N_{\text{max}} - N_{\text{min}}} \times (q_{u\ \text{max}} - q_{u\ \text{min}}) + q_{u\ \text{min}} \tag{5.7}$$

where $N_{\text{blow count}}$ = blow count on bore log
N_{min} = lowest blow count in SPT *N* category
N_{max} = highest blow count in SPT *N* category
q_u = unconfined compression strength
$q_{u\ \text{max}}$ = highest UC in category
$q_{u\ \text{min}}$ = lowest UC in category

Example: The bore log describes a medium stiff clay with a SPT $N = 7$ at 15 ft below grade. What are the unconfined compression strength q_u and cohesion *c* at that level?

From Table 5.3 under medium stiff,

$N_{\text{blow count}} = 7$
$N_{\text{min}} = 5$
$N_{\text{max}} = 8$
$q_{u\ \text{max}} = 1.00$ tsf
$q_{u\ \text{min}} = 0.5$ tsf

$$q_u = \frac{7-5}{8-5} \times (1.0 - 0.5) + 0.5 = 0.833 \text{ tsf}$$

Then use the equation $c = \frac{q_u}{2}$ to determine the cohesion

$$c = \frac{0.83 \text{ tsf}}{2} = 0.416 \text{ tsf} \qquad \text{or } c = 833 \text{ psf}$$

It is important to check the units on q_u on the boring log. Some reports show q_u in kilopounds per square foot (ksf). The conversion is 1.0 tsf = 2 ksf = 2000 lb/ft^2. The pocket penetrometer reports q_u in tons per square foot, and laboratory UC and direct shear tests report cohesion directly, usually in pounds per square foot (psf). The results of any of these tests may not give the same cohesion value; however, the results should be similar.

Unit weight of in situ soil γ_{sat}: In design the soil is assumed to be saturated. Saturation is the point where water will not drain out of something; e.g., a sponge dipped in water and then held in the air until no more water drips out of it is said to be fully saturated. The saturated weight of a soil sample γ_{sat} is the dry weight of the soil plus the weight of the water held in it.

The degree of saturation and the weight of the soil vary from no water to totally saturated. On the bore log there is usually a column for dry density and moisture content. The UC test and the direct shear test usually report the dry density of the sample and the moisture content, sometimes called water content w. From these two parameters the unit weight γ_{sat} of the in situ sample can be calculated from

$$\gamma_{\text{sat in situ water content}} = w \times \gamma_{\text{dry}} + \gamma_{\text{dry}} \tag{5.8}$$

where w = water content and γ_{dry} = dry density (pcf). To calculate the effective weight of submerged soil γ' use

$$\gamma' = \gamma_{\text{sat}} - \gamma_{\text{water}} \tag{5.9}$$

There are several problems with determining the exact degree of saturation and using it for design. The water content reported is not necessarily the water content at full saturation. The water content can increase due to heavy rain or flooding and can decrease with dry spells. Also in every different layer and level of that layer there can be different water contents. Finally the dry weight and moisture content of each layer are rarely determined and reported.

For the purpose of excavation stability and shoring design, it is reasonable to use an average saturated and the submerged weight for the soil. Table 5.6 has typical unit weights for different types of soils. Using these values for each layer and then weighted averaging them over the total depth seem reasonable.

In preliminary design and many times in final design, reasonable unit weight values are assumed. Later the sensitivity of these assumptions can be checked by plugging into the equation higher and lower values. Table 5.7 contains typical unit weight design values for in situ soils.

Water table ∇: In bore logs and engineering drawings, the level of the water table is typically indicated by an inverted triangle. Two groundwater levels are of significance on the bore log—the level where seepage is encountered and the final level at the end of drilling. The final level indicates the water table level in the vicinity that is affecting the water level in the boring and can be above or below the seepage level. The initial seepage can indicate the layer

Description	Water Content w (%)	Dry Unit Weight (lb/ft^3) γ_d	Wet Unit Weight (lb/ft^3) γ_{sat}	Submerged Unit Weight (lb/ft^3) γ'
Uniform sand loose	32	89	118	56
Uniform sand dense	19	109	130	68
Mixed-grain sand loose	25	99	124	62
Mixed-grain sand dense	16	115	135	73
Well-graded gravel, sand, silt and clay mixture, glacial till, loose	—	—	125	63
Well-graded gravel, sand, silt and clay mixture, glacial till, dense	—	—	156	94
Soft glacial clay	45	76	111	49
Stiff glacial clay	22	106	130	68
Soft slightly organic clay	70	58	100	38
Soft very organic clay	110	43	92	30
Marine clays, bay mud, soft	87	50	94	32
Marine clays, bay mud, medium stiff	48	71	105	43
Concrete		150	150	88
Water		0	—	62

Source: After Terzaghi, K and Peck,R; U.S. Army Corps of Engineers.

TABLE 5.6 Unit Weight of Typical Soils in Natural State

that the water is seeping through. It is possible to drill through an impermeable layer into a permeable layer that will allow the seepage water to drain completely out of the drill hole, indicating that there is "perched water" in lenses within impermeable layers. The date on which the soil boring was performed is important because it is the only day that the water table is actually known to be at that level. On any other given day, month, season, etc., the water table can be at a different level or nonexistent within the depth of the boring. River and lake levels, recent storms, seasons, and irrigation activity have the greatest effect on water table levels. The soils encountered in soils exploration programs change insignificantly over large periods of time, sometimes millions of years, while the moisture condition can change daily. In the conclusions and recommendations portion of the soils report there is usually a discussion about water table and dewatering.

Description		Wet Unit Weight (lb/ft³) γ_{sat}	Submerged Unit Weight (lb/ft³) γ'
Sands and gravels	Loose	115	53
Sands and gravels	Medium dense	125	63
Sands and gravels	Dense	135	73
Silts and clays	Soft	120	58
Silts and clays	Stiff	130	68
Marine clays	Soft	95	33
Marine clays	Stiff	105	43

TABLE 5.7 Reasonable Design Values for in situ Unit Weight of Soils

5.8 OSHA Appendix A Soil Classification System and Type A, B, and C Soil

The original OSHA regulations governing excavations, trenching, and shoring practices for the construction industry were promulgated in 1971. In 1976 these standards were reviewed against the following:

- Actual construction practice
- The state of the knowledge in geotechnical and structural engineering

The study, draft recommendations, and changes were completed in 1980. In the summer of 1981 five regional industry workshops were held in Atlanta, Ga.; Boston, Mass.; Dallas, Tex.; Milwaukee, Wis.; and San Francisco, Calif. Representatives from unions, engineering associations, contractors, shoring manufacturers, government, and interested individuals attended, providing commentary and recommendations to the proposed changes. One of the objectives of the process was to develop a "voluntary consensus standard" for the future development of OSHA trenching and shoring regulations. The results of those meetings and the final draft were the beginning of the standard that we work with today. In 1987 federal OSHA requirements became law, and in 1990 the trenching and shoring requirements were adopted.

There were some interesting results from the 1981 workshops as regards soil classification. In developing standard practice in the field, it was evident that a method of determining soil conditions in the field as they are encountered had to be adopted. The basis of slope selection and the use of manufactured shoring equipment were at the heart of the soil type selection system. In order to provide as much versatility as possible to the person deciding on the shoring system in the field, it was suggested that there be as many soil type divisions as possible, while others suggested that a finely nuanced system such as the Uniform Soil Classification System would be too complicated. All the classification work had to be performed on the spot in the field. The basis for classification had to be clear and easy to achieve and not bring confusion to the court room during litigation. Amid recommendations for five soil types it was finally proposed that four with a further division for long- and short-term excavations and a 20-ft-deep limitation would be sufficient and simple to use. It was strongly recommended that an industry standard field soil classification system be developed, although at that time OSHA declined to fund it.

In 1989 OSHA promulgated the current 29 CFR 1926 Subpart P Appendix A, Soil Classification that we use today. The system is based on the soil attributes and environmental conditions, primarily water content, in the excavation as it is encountered and again as conditions change. The system relies on visual and manual analysis of the excavated soil at the site and then classifying it into a descending category of stable rock, types A, B, and C. The soil classification system is a separate appendix from appendixes that enable selection of standard practice worker protection systems, such as sloping and benching, timber shoring, and manufactured worker protective equipment. To select any standard practice protective system, the soil must first be classified in accordance with Appendix A.

In this book in App. 2, the OSHA Appendix A with commentary is presented. It is important to remember when reading this that the commentary was developed by the author and is in no way, other than opinion, a part of the OSHA standard. The intent of the commentary is to clarify the language and bring forward information that will help the reader understand the material. Readers should formulate their own understanding and opinions because they are the ones responsible for making the decision regarding soil type and appropriate worker protection systems. Reading and interpreting requirements of the standard should be done with a focus on

- What they are intended to accomplish
- Their interpretation from a practical point of view
- Their interpretation from a legal point of view
- What exactly the standard requires a competent person to do and whether it is possible to do so with the information given

OSHA Appendix A is intended to stand alone and not require an education in geotechnical engineering, knowledge of other soil identification systems, and laboratory testing. Figure AP2.9, soil type identification tree, shows the decision process and conclusions resulting from use of Appendix A.

5.8.1 Comment on Appendix A and Standard Practice Shoring Application Today

The mandate of the original 1976 review of current practices in shoring applications was to look at them in light of actual construction practice and the state of the knowledge in geotechnical and structural engineering at the time.

It is worthwhile to look today through the same filter and see how things have changed. Table 5.8a shows text submitted for workshop discussion, Table 5.8b shows resulting recommendations from workshop input, Table 5.8c contains the notes to OSHA Table 1, Table 5.9 shows a table, similar to the OSHA, that the author developed showing 2007 standard practice in shoring applications. The trend was from a conservative approach by OSHA to a less conservative approach by industry and then back to an industry standard that evolved over 27 years, was tried and true, and works for everyone involved.

In addition to Table 1 some important current practices (1971 to 1980) were also discussed at the workshops:

- The definition of *short term* was debated to be between 24 hours, 3 days, and 7 days.
- The depth of standard practice excavations was debated to be between 16 and 24 ft. Today it is 20 ft.

TEXT SUBMITTED TO AGENCY*

Table 1. Soil Classification System for the Standard Practice

			Steepest Allowable Slope hor/vert[b]	
Soil Type	**Description**	**w_e lb/ft³**	**Depth <= 12 ft**	**Depth > 12 ft**
A	Intact hard	20[a]	3/4:1	1:1
B	Medium	40	3/4:1[c]	$1\frac{1}{2}$:1
C	Saturated, submerged, or soft	80	$1\frac{1}{2}$:1	2:1

*This table was proposed in 1980. It is here for commentary only.

TABLE 5.8a 1980 Table 1, Developed by NBS for Draft Construction Safety Standards for Excavation

RECOMENDATIONS/SUGGESTIONS* Table 1. Soil Classification System for the Standard Practice				
			Steepest Allowable Slope hor/vert[b]	
Soil Type	Description	w_e lb/ft³	Depth <= 12 ft	Depth > 12 ft
A	Intact hard	20[a]	$\frac{1}{2}$:1	$\frac{1}{2}$:1
B	Medium	40	$\frac{1}{2}$:1[c]	$\frac{3}{4}$:1
C	Saturated, submerged, or soft	80	1:1	$1\frac{1}{2}$:1

*This table was proposed in 1980. It is here for commentary only.

Table 5.8b 1983 Revised Table 1, Developed by NIOSH from Industry Recommendations

- The bank height of vertical walls in the bottom of sloped trenches was 3 ft maximum and was proposed to be changed to 4 ft (primarily needed to meet pipe bedding specifications).
- The exposed bank below sheeted and shielded excavations was proposed to be 3 ft in short-term excavations. Installing lateral pipes under 2-ft-high shoring elements was considered problematic by contractors. The contractors supported the 3-ft option while contractors and the American Society of Foundation Engineers, both in San Francisco, opposed it. The authors guess would be that this was related to the preponderance of soft bay mud in the San Francisco Bay area. Today it is strictly limited to 2 ft.
- In the design of manufactured shoring equipment a 33 percent material working stress increase was allowed for short term. Input suggested that this be eliminated; however, final commentary suggested that "the track record of these systems does not seem to justify such a step." Today it is recognized but not approved by OSHA

In 2002, OSHA began a periodic review of the standard as required by the 1998 Regulatory Flexibility Act, with the purpose of determining if the standard should be continued without change, rescinded, or amended. At that time public comment was again solicited.

In review of actual construction practice and technology several items stand out here:

1. Type A: Intact Hard Soils include stiff clays and clayey (cohesive) sands and gravels* (hardpan, till) above the ground water table which have no fissures, weak layers, or inclined layers that dip toward the bank of the excavation as stipulated in note 3. Stiff clays included have unconfined compressive strength† q_u = 1.5 tsf or more. Intact hard soils subject to vibrations by heavy traffic, pile driving, or similar effects are Type B.
2. Type B: Medium Soils are all soils that are not Type A or C.
3. Type C: a. Soft Soils include cohesive soils* with an unconfined compressive strength‡ of 0.5 tsf or less and soils that cannot stand on a slope of 3 hor to 1 vert without slumping (muck).
   b. Saturated or submerged soils are assumed whenever water seeps into the excavation from soil forming the bank; or water is retained by tight sheeting; or there is a possibility that the excavation may be entered by workers within 1 day after more than half of it's depth was flooded and pumped out.
4. Layered Systems: (two or more distinctly different soil or rock types or micaciouse seams in rock) which dip toward the bank of the excavation with a slope of 4 hor to 1 vert or steeper are considered type C. Layered soils are classified in accordance with the weakest layer.
5. Rock: Unstable rock shall be considered Type B when it is dry and Type C when it is submerged. Stable rock is exempt from shoring requirements.

---

*Cohesive soils are clays (fine grained) or soils with a high clay content which have cohesive strength. They do not crumble, can be excavated with vertical sideslopes, are plastic, (can be molded into various shapes and rolled into threads) when moist and are hard to break up when dry.

†In long-term excavations "Intact Hard" soil is Type B soil.

‡Unconfined compressive strength can be determined by undrained laboratory tests, field tests, or the following thumb penetration tests: Stiff clays with an unconfined compressive strength of 1.5 tsf can readily be indented by the thumb nail. They can be indented by the thumb only with very great difficulty. Cohesive soils with an unconfined compression strength of less than 0.5 tsf can be easily penetrated by the thumb and can be molded by light finger pressure. Tsf = tons per square foot.

Notes:

a. The steepest allowable slope is not necessarily safe in all conditions. A competent person shall determine the safe slope, and if there is any indication of general or local instability, slopes shall be cut back to a slope which is at least 1/4 hor to 1 vert flatter than the specified slope.

b. In long-term excavations steeper allowable slopes 3/4 :1 for depths 12 ft or less and 1:1 for depths greater than 12 ft.

TABLE 5.8c Notes to Table 1

- Today the OSHA soils classification is a separate entity, Appendix A, and its use is a precondition for the use of Appendix B, Sloping and Benching; Appendix C, Timber Shoring for Trenches; and Appendix D, Aluminum Hydraulic Shoring for use in Trenches. OSHA did not mandate that the use of

OVERVIEW OF STANDARD PRACTICE 2007* Soil Classification System in Use for Standard Practice			
Soil Type	**Description**	**w_e lb/ft³**	**Steepest Allowable Slope horizontal/vertical Maximum depth 20 ft**
A-25	Cohesive, q_u > 1.5 tsf	25	$\frac{3}{4}$:1
B-45	Cohesive 1.5 tsf >q_u>0.5 tsf, noncohesive, subangular, no water on sides of trench	45	1:1
C-60	Cohesive q_u<0.5 tsf noncohesive, round grain, no water on sides of trench, stands up long enough to insert shoring	60	$1\frac{1}{2}$:1
C-80	Cohesive q_u<0.5 tsf, noncohesive, round grain, water on sides of trench, does not stand up long enough to insert shoring	80	$1\frac{1}{2}$:1

*This table represents industry practice and not necessarily OSHA policy.

TABLE 5.9 2007 Industry Standard Practice in Shoring Applications

alternative tabulated and manufactured protective systems be predicated on first using Appendix A to classify soils; however, OSHA did require that the data for those systems supply "identification of the parameters that affect the selection of a protective system drawn from such data," in other words a soil classification basis for selection of the system. Manufacturers and their engineers chose to use Appendix A, and today virtually all tabulated data for standard shoring systems that are selected in the field require the use of Appendix A as a basis for selection.

- Although there are other field soil classification systems out there (agricultural, geologist, tunneling, etc.), Appendix A has become the standard for the construction industry. This is so primarily because it does not require an education in geotechnical engineering, is easy to learn, can be performed on the spot, and does not necessarily require laboratory testing. It is self-contained in the sense that it provides definitions, requirements, and acceptable methods of field testing to meet the requirements. Most people today with experience in the industry are familiar with the basis of the system, classification of soil into types A, B, and C, even though they may not fully understand the methods for determining them. In the 2002 review it was determined and backed up by commentary that the method should not be changed because it was successful and everyone was familiar with the language used. The author has taught and used this system in the field since 1990 and believes that it is excellent; it places the responsibility for the soil classification and shoring selection with the person in the field at the time and place where it is critical to the safety of the workers.

It is important to point out here, not as a criticism but in the interest of understanding, what Appendix A does and does not do:

- The system places soil into three categories and does not allow for interpolation between the categories. In comparison the Uniform Soil Classification System has 15 group symbols with further descriptive wording within those categories.
- Appendix A does not attach any effective lateral soil weight to the categories. This aspect of soil loading has been eliminated except to state the soil types are in a hierarchy from good to bad. The 1980 OSHA recommendation was to attach an effective lateral force component W_e to the soil type. At that time it was type A-20, type B-40, and type C-80. Today the construction industry uses type A-25, type B-45, type C-60, and type C-80; however, OSHA does not endorse it. The only place in the OSHA regulation that makes any reference to lateral soil pressure values is in Appendix C, Timber Shoring, where it specifies "For soil type A, $P_a = 25 \times H + 72$ (2 ft surcharge)" and similarly 45 and 80 for types B and C. As this is a prescriptive shoring design that OSHA puts forth, it makes sense that OSHA takes a stand on soil loading for this type of system. This is most likely where the current 25, 45, and 60 came from.
- Today tabulated data for manufactured shoring systems stipulate that first the soil must be classified in accordance with Appendix A and then an allowable depth chart and maximum psf rating for the shoring equipment are provided. The psf rating and the allowable depths are not always directly related. Tabulated data should state the psf loading associated with each OSHA soil type because Appendix A does not provide it.

- Soil loading and soil types are definitely related; however, soil loading can only be determined after classifying soil type and is also related to time, environmental factors, and the interaction between the shoring and the soil. For this reason OSHA correctly chooses to leave it out of Appendix A.
- The system does not address a fourth soil category commonly in use that falls between types B and C soil. This classification, commonly called C-60, was developed primarily by shoring manufacturers to address the fact that a type C soil is seen as being 2 times worse than a type B soil and 4 times worse than a type A soil. The presence of water in noncohesive soils or very soft clay is what causes the soil to load at 80 lb/ft of depth. C-60 soils allow for the case where C soils are present without water and allow loading roughly halfway between C-80 and B-45.
- The classification of type A soil lacks some clarity. Due to the five-point list of exclusion the author hears often that it is "impossible to have type A soil." The impression is that OSHA gave them something and in typical government fashion took it away at the same time. Exclusion "(5) The material is subject to other factors that would require it to be classified as a less stable material" is particularly disenchanting; it would be nice if there were some examples given to spark the imagination in this area. In actuality the exclusions are extremely important and do not occur that often. The advantages to working in type A soil include seven different options for open cut systems versus three options for type B soil and two options for type C soil; and almost one-half the lateral soil loading on shoring equipment that type B soil brings to the shoring system. Appendix 2, OSHA Appendix A Soil Classification and Commentary, presents discussion and examples of the type A soil exclusions.
- OSHA did not address the concept of short-term use when they formed Appendix A. The only reference to it is in Appendix B, Sloping and Benching, where they did carry the use of short-term forward with one open cut option in type A soil that allows a ½:1 slope in excavations less than 12 ft deep and open for less than 24 hours. Here the definition and allowable use of *short term* is found with the type of worker protection system being tabulated.

  In practice shoring equipment is used either short-term as is the case with pipeline and utility installation or long-term in the case of bore pits for trenchless work or shoring for belowground structures. Innovations in pipeline construction such as stronger and faster hydraulic excavators and backfill equipment operated from outside the trench, compaction wheels, and remote control compactors have significantly reduced the amount of time that a worker is inside a shored trench and that the trench is open. In the past, shoring manufacturers tried to take advantage of this by issuing a higher psf rating for short-term than long-term use based on the concept that soil loading on the equipment increases over time. Soil loading does increase over time; however, how much the soil load increases from start to finish and how long it takes to get there are anyone's guess. Measurements taken on strut loading in shored clays show that loads increase with additional excavation below them, surcharge loading during and after excavation, and weeks and months into the time the excavation is open; however, sorting out and quantifying the effects of each influence have not been done. The other problem is that construction rarely goes as planned and usually takes longer. If short-term loading is used and the use goes over the time limit, in borderline cases the shoring would have to be removed before the job could be finished. Consequently in tabulated data for manufactured shoring equipment, the use of short-term has declined significantly even though the psf ratings that were put forth for short-term are still used.

- In 1981 the ½:1 slopes for type A and B soils were desired and proposed while in 2007 flatter slopes were being used. The fact is that in very short term, less than 24 hours, steep slopes will stand up; however, as time wears on, they start to fail. Gravity over time, heat, rain, and the water table degrade them. The cost of repairing or flattening slopes after a structure has been started inside the excavation and work has progressed on the surface around the excavation is extremely expensive, if not impossible. Determining ¾:1 slopes simply because there is no room for 1:1 and the shoring alternative is too expensive can be a costly decision. The amount of time the excavation is going to be open should have as much impact as the soil type on the sloping decision.
- One of the concepts that has changed significantly since the 1981 recommendations were published in the report is that at the time they tried to develop a set of rules for standard practice that everyone—competent person, engineer, manufacturers of shoring equipment, and OSHA inspection—could work by. There was a section on strength requirements for predesigned shoring systems and manufactured equipment, acceptable engineering practices for shoring design, and soil loading diagrams. The problem with this stuff is that it promotes the mind-set that if one follows the published specification and there is an accident, then someone

else is responsible. In matters of cost and quality this may be acceptable; however, in matters of life and death, greater diligence is required. Today OSHA requirements are more performance-based, stating requirements to be met and providing information for the responsible person at the site to use to make a decision. The concepts of "state of the art" and "engineering judgment" have left shoring equipment design standards up to the manufacturers as they are the ones responsible for the proper design and use of their equipment.

C-60 Soil Classification

Manufacturers of shoring equipment have promoted and used in their tabulated data an intermediate type C-60 soil classification that in the hierarchy falls between type B and type C soils (see Table. 5.7). This category is generally accepted within the shoring industry by contractors and those reviewing worker protection plans. Federal OSHA and state OSHA programs do not appear to be opposed to this; however, Appendix A and the excavation safety standards do not recognize or address the type C-60 soil classification.

Prior to establishment of Appendix A in 1989, the original OSHA soil identification system also attached a soil lateral loading factor to the soil designations. They were often referred to as OSHA type A-20, type B-40, and type C-80 soils and indicated that, e.g., an excavation 10 ft deep in type B soil would deliver a 20 × 40 = 800 psf lateral load to the shoring. For shoring equipment such as shoring shields at 20 ft deep in C-80 soil, the shield needed to be rated 20 × 80 = 1600 psf, twice as much as for type B soil and generally out of the range of normal shield construction at the time. In the case of trench jacks some argued that cohesive C soil was too soft to stand long enough to get trench jacks installed, and that the soil would slough in between even if one could get them installed. In noncohesive C soil they argued that excavation walls would not stand up long enough to get the shores in. Consequently tabulated data developed by OSHA for hydraulic aluminum trench jacks only addressed soil types A and B.

In response to the fact that there is plenty of soil that does not meet the type B soil requirements but that stands up and acts more as B than C soil, the shoring manufacturers defined type C-60 soil and developed tabulated data for use of their equipment under this category. The very simple and easy-to-use defining test for C-60 soil is the following:

Manufacturers' definitive test for type C-60 soil: If the soil stands up long enough to install the shoring, it can be considered C-60.

In the design and tabulation of operating data of shoring equipment in C-60 soil, a 60 times excavation depth rectangular load is anticipated on the shoring. In Fig. AP2.9, soil type identification tree, the author has included the C-60 soil classification.

5.8.2 Determining OSHA Appendix A Soil Types Using Bore Logs

The OSHA Appendix A soil classification system is focused on strength characteristics of the soil for the purpose of excavation wall stability, while the Unified Soil Classification System sorts out many more properties and characteristics for several purposes. Fortunately USCS also does a good job of sorting strength characteristics. OSHA Appendix A uses the basic cohesive/noncohesive division and uses many of the same terms as USCS. In soils reports and specifically bore logs, most of the testing and descriptive work required for soil type classifications in OSHA Appendix A has already been carried out. Those persons doing the classification work are trained and have years of experience, and therefore their conclusions can be relied on. The major thing that is not addressed in the bore logs and testing done with the USCS is the effect of environmental conditions on an open excavation. The competent person in the field must do this daily in the field. Although the soils bore logs are indicative of the area, they only apply to the soil in the boring while an excavation can be seen as a continuous soil exploration and needs to be classified as it is encountered.

The required OSHA Appendix A classification information can be determined from the soils report and bore logs contained in it. The following is a discussion of the required information and where it can be found in a typical soils report.

Cohesive or noncohesive: The group symbol used in the description of every layer of soil will pinpoint the soil type.

For noncohesive soils one of or a combination of the following terms must be used: GW, GP, GM, GC, SW, SP, SM, SC.

For cohesive soils one of or a combination of the following terms must be used: CL, ML, OL, CH, MH, OH, PT.

Cementation or hardpan: If cementation is suspected, it is usually tested for and reported in the description. Cementitious particles give off fumes when placed in acid.

Unconfined compressive strength: This can be found in the SPT *N* blow count or directly from test results. Table 5.10 has a direct correlation for this.

Fissures: If fissures are evident, they will be reported in the description. The geological history and compressive strength testing will give the soils engineer an indication of fissures.

OSHA Appendix A Soil Type Based on SPT *N* Blow Counts*						
Relative Density			Consistency			
Sands and Gravels	**SPT *N***	**OSHA Soil Type**	**Silts and Clays**	**SPT *N***	**Unconfined Compressive Strength**	**OSHA Soil Type**
Very loose	0–4	Type C	Very soft	0–2	0–0.25	Type C
Loose	5–10	Type C-60†	Soft	3–4	0.25–0.50	Type C-60†
Medium dense	11–30	Type B‡ ↓	Medium stiff	5–7	0.50–1.00	Type B
Dense	31–50		Stiff	8–11	1.00–1.5	Type A ↓
			Stiff	11–16	1.5–2.00	
Very dense	50+		Very stiff	16–32	2.00–4.00	
			Hard	32+	>4.00	

*This table only addresses consistency and density; all other classification requirements must be met. See Appendix A.

†To be type C-60 soil, the excavation walls must stand up long enough to install shoring.

‡To be considered type B sands and gravels, the grains must be angular.

TABLE 5.10 Blow Count and Appendix A Soil Type

Previously disturbed: All natural soil deposits that recently, say, in the last 100,000 years, have been interrupted by natural forces or more recently by humans are considered previously disturbed. The geotechnical engineer researches the geological history as well as recent history of a site. In that report landslides, site fills, and previous major excavation work is identified. The existing facilities site investigation should turn up all known pipeline excavations. The seismicity investigation will point out any areas on the site where an earthquake fault is expected to be found. The description on the bore log will indicate suspected fill.

Sloped and layered: Layers are well defined on the bore logs; however, sloping cannot be determined from a small round drill hole. Plots between boreholes and the original contour of the surface can give an indication of layer slopes.

Noncohesive soil grain configuration: Flat, round, or angular soil grains are reported on the bore log if they are visible. The terms used would be *broken, crushed, angular*, or *abrasive*. The term *coarse* refers to only gradation of particles and not angularity of the particles.

Water: A discussion about water tables and dewaterability of excavations is usually found in the geotechnical design summary. The water table and moisture content of soils are found in the bore logs.

Table 5.10 is the relative density/consistency to blow count correlation table, Table 5.3, with added columns for OSHA soil type. This is an approximation developed by the author and does not represent conclusions from OSHA or the excavation industry.

References

Bowels, Joseph E., *Engineering Properties of Soils and Their Measurement*, McGraw-Hill, Inc., 1986

Bowels, Joseph E., *Foundation Analysis And Design*, McGraw-Hill, Inc., 1997

Occupational Safety and Health Administration, "Regulatory Review of 29 CFR 1926, Subpart P: Excavations," *Federal Register*, March 2007.

Meyerhof G. G., *Penetration Tests and Bearing Capacity of Cohesionless Soils*, Journal of Soil Mechanics and Foundation Division, Proceedings, ASCE, 82, SMI, January 1956.

Peck, R. B., Hanson, W. E., and Thornburn, T.H. *Foundation Engineering*, John Wiley & Sons, New York, 1974.

Sowers, George B., and Sowers, George F., *Introductory Soil Mechanics and Foundations*, Macmillan, New York, 1970.

Terzaghi, K., and Peck, R., *Soil Mechanics in Engineering Practice*, John Wiley & Sons, New York, 1967.

Terzaghi, K., Peck, R., and Mesri G., *Soil Mechanics in Engineering Practice*, 3d ed., John Wiley & Sons, New York, 1996.

Yokel, Felix Y., and Stanevich, Ronald L., *Development of Draft Construction Safety Standards for Excavations*, vol. 1, National Bureau of Standards, Department of Commerce, Washington, April 1983.

CHAPTER 6

Excavation Stability and Shoring Design Loads

6.1 Soil Loading on Excavation Shoring Systems

There are forces of gravity within a mass of soil that tend to produce failure in a bank of soil that has been excavated. The cohesion and angle of internal friction of the soil tend to counteract those forces. If the cohesion and friction are exceeded, to maintain stability, the gravity mass has to be redistributed by sloping or supported in the form of a shoring system.

The first step in designing a shoring system is to determine the forces and nature of the soil that the shoring will need to resist. In physics, Newton's third law of motion tells us that for every action there is an equal and opposite reaction. In fact, due to this third law of physics, every action taken on the soil—removing it, compacting it, adding surcharge loads at the surface, and changing the moisture content through dewatering—will have a reactive effect on the remaining soil. Another way of looking at this is that there is no possible way to do excavation work without changing to some extent the surrounding soil. Controlling the degree of change within acceptable limits is the challenge. Preventing shear failure and settlement through the application of shoring is the solution.

In shoring design the approach to developing a lateral earth pressure assumption that is safe and predictive of the actual pressures that will be encountered is to use lateral earth pressure coefficients developed from classical earth pressure theory and then use them to develop apparent—what one actually experiences in the field—pressure diagrams. These apparent diagrams are developed from the cohesion c and angle of internal friction f, values derived from the soils report and applied to typical apparent earth pressure formulations.

6.1.1 Earth Pressure Theories, Active, at Rest, and Passive Soil Pressure

At any level in the ground, the soil exerts a vertical and horizontal pressure on the soil below it and next to it. Since soil is not fluid like water, the horizontal pressure is not the same as the vertical pressure. The vertical pressure on an element of soil at a given depth is the weight of soil above it plus the weight of surcharge

$$P_v = \gamma Z + \text{surcharge} \tag{6.1}$$

where P_v = vertical pressure, psf
γ = unit weight of soil below the ground surface, lb/ft³
Z = depth below the ground surface, ft
Surcharge = any weight added at surface, lb/ft³

It would be nice if the horizontal pressure could be thought of as the vertical pressure times a horizontal factor; however, the presence of cohesive and noncohesive soils, the moisture condition of the soil, and three different states of stress that the soil can be placed in complicate things slightly.

In excavation design it is reasonable to assume that the soil is in a saturated undrained condition. For noncohesive soils the saturated undrained condition occurs when the soil is above the water table. For cohesive soils the saturated undrained condition occurs when the absorbed water is not squeezed or dried out, and cohesion is what is providing the strength for it to stand up when it is cut into. Over a long period of time, months to years, a braced clay becomes drained, causing the shear strength to dissipate and go to zero. At that point the internal friction has been mobilized and provides the strength to stand up. The horizontal pressure also depends on the stress condition of the soil at the time it is being looked at. In soil mechanics there are three conditions of the soil that are of interest—active, at rest, and passive (Fig. 6.1). A good analogy is forward, neutral, and reverse. After digging, if the soil moves toward the hole, it is said to be in the active condition. Movement is activated by the vertical weight of the soil and surcharge. Active horizontal pressure P_a is some factor less than 1 times the vertical pressure.

In the natural state, prior to excavation the soil is said to be in natural equilibrium, the at rest condition. The at rest horizontal pressure P_0 is also a factor less than 1 times the vertical pressure. The coefficient of earth pressure at rest is given by

$$K_0 = \frac{\sigma_h'}{\sigma_v'} \tag{6.2}$$

where K_0 = coefficient of earth pressure at rest
σ_h' = effective horizontal pressure
σ_v' = effective vertical pressure

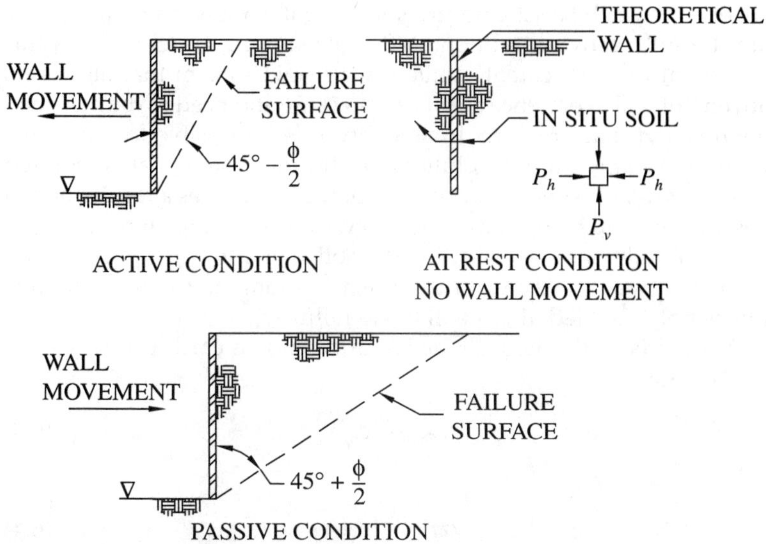

FIGURE 6.1 Active, at rest, and passive conditions.

Due to the range of conditions an in situ soil can be exposed to, such as earthquake, overconsolidation from previous overburden, or glaciations, etc., it is best to use typical predicted values for K_0 as presented in Table 6.1.

In the passive state an outside force such as hydraulic jack or the toe of a pile pushes on the soil, compressing the soil in front of the push. In this case the soil sort of backs up and gathers strength to resist the force being pushed at it. The passive resistance P_p is always greater than the at rest pressure.

In the fully active condition all the shearing resistance has been activated, and the soil is on the verge of failure. The ideal shoring system is one that allows the soil to move enough to activate all the shearing resistance and not go beyond that. Over time the shoring system

Soil	K_0 total, undrained
Soft clay	1.0
Hard clay	0.8
Loose sand, gravel	0.6
Dense sand, gravel	0.4

Source: Sowers, George B., and Sowers, George F., Introductory Soil Mechanics and Foundations, Macmillan, New York, 1970.

TABLE 6.1 Typical Values of at Rest Soil Coefficient K_0

will experience the least amount of loading if it allows the soil to move into the fully active condition. A flexible shoring system allows this movement to some extent, while a very rigid system that allows no movement will experience the at rest pressure and needs to be designed to withstand that pressure. In actuality it is impossible to accurately compute the amount of flexibility in a shoring system that is required to mobilize all the shearing resistance of the soil, and as a result, the soil pressure ends up being somewhere between active and at rest.

In the fully passive condition the soil is on the verge of failure by heaving upward away from the push. As long as the fully passive state is not exceeded, there will be no failure.

Mathematically the fully active and passive earth pressures can be calculated by

$$p_a = \gamma z K_a - 2c\sqrt{K_a} \tag{6.3}$$

and

$$p_p = \gamma z K_p + 2c\sqrt{K_p} \tag{6.4}$$

where p_a = active pressure at depth z
P_p = passive pressure at depth z
γ = unit weight of soil
c = cohesive strength of soil
K_a = coefficient of active pressure
K_p = coefficient of passive pressure

For the purpose of shoring design it is convenient to remember that in noncohesive soils internal friction contributes most to the soil strength, and the small amount of cohesive strength contributes little to strength at the time of excavation and dissipates over time. In this case it is reasonable to assume that $c = 0$. Equations (6.3) and (6.4) reduce to, respectively,

$$p_a = \gamma z K_a \qquad \text{(noncohesive, } c = 0) \tag{6.5}$$

and

$$p_p = \gamma z K_p \qquad \text{(noncohesive, } c = 0) \tag{6.6}$$

In cohesive soils the strength is derived from the cohesion, and the angle of internal friction is only activated over a long time, months to years. In this case it is reasonable to assume $\phi = 0$, resulting in K_a and $K_p = 1$ [see Eqs. (6.9) and (6.10)]. Equations (6.3) and (6.4) become, respectively,

$$p_a = \gamma z - 2c \qquad \text{(cohesive, } \phi = 0) \tag{6.7}$$

and

$$p_p = \gamma z + 2c \qquad \text{(cohesive, } \phi = 0) \tag{6.8}$$

There are several theories used for determining K_a and K_p; Rankine, Coulomb, and log spiral are most often used. The values given will predict the minimum active earth pressure prior to failure and the maximum passive pressure prior to failure. In order for these pressures to be activated, the shoring wall must move. For example, a 10-ft-high cantilever wall in dense cohesionless soil would have to move $\frac{1}{16}$ in at the top to mobilize active earth pressure. In stiff clays behind the same wall the top would have to move 1.2 in due to dissipation of cohesion and not due to shear cracks. It can take several months to years for clay to lose enough cohesive strength to mobilize shearing strength from the angle of internal friction, which is why for short-term shoring design ϕ is assumed to be zero. It is a reasonable assumption that given steel and wood shoring materials used in the excavation and shoring installation process, there is adequate movement to mobilize the shearing resistance in noncohesive soils. Stiff shoring elements such as concrete walls or steel struts that are preloaded prior to excavating below them will not necessarily move enough.

Details of the Rankine and Coulomb theories are presented here, as they are the ones most often used in shoring design. Each of the three theories—Rankine, Coulomb, and log spiral—provides progressively lower active and higher passive pressures. Some would see this as progressing from conservative to less conservative while others would see it as moving from inaccurate to more accurate. If the assumptions are met and the soil parameters are accurate, then any one of these theories provides a safe design. It makes sense to use a more accurate design procedure in the design of permanent retaining walls because there is greater control of the excavation, construction, and backfill process coupled with the fact that the design is for the long term. Due to the nature of shoring design, starting with the fact that the soils report was not developed with shoring as the primary objective of the investigation and extending to variable construction methods, short-term perspective, and uniqueness of every design, it makes sense to take a more conservative tack.

6.1.2 Rankine Earth Pressure Theory

This method of calculation was first used by a famous Scottish engineer Rankine (1857), and the approach has since been termed the *Rankine method*. This method is based on three assumptions:

1. The shoring wall is smooth and has no friction with the soil.
2. The failure surface of the soil is a plane, not curved.
3. There must be movement of the shoring wall.

For a level slope at the top

$$K_a = \tan^2\left(45 - \frac{\phi}{2}\right) \tag{6.9}$$

$$K_p = \tan^2\left(45 + \frac{\phi}{2}\right) \tag{6.10}$$

where ϕ = angle of internal friction (degrees). And for a sloping surfaces less than ϕ

$$K_a = \cos\beta \frac{\cos\beta - \sqrt{\cos^2\beta - \cos^2\phi}}{\cos\beta + \sqrt{\cos^2\beta - \cos^2\phi}} \tag{6.11}$$

and

$$K_p = \cos\beta \frac{\cos\beta + \sqrt{\cos^2\beta - \cos^2\phi}}{\cos\beta - \sqrt{\cos^2\beta - \cos^2\phi}} \tag{6.12}$$

where β = slope at surface less than ϕ. See Tables 6.2 and 6.3 for the calculated values of K_a and K_p.

Pressure diagrams for Rankine horizontal pressures are shown in Fig. 6.2(a) to (d). The resultants, or total pressure, exerted on the wall

Angle of Internal Friction ϕ (deg)	β slope (deg)								
	0	5	10	15	20	25	30	35	40
	K_a level	K_a sloped							
0	1								
10	0.7	0.74	0.98						
12.5	0.64								
15	0.59	0.60	0.66	0.97					
17.5	0.54								
20	0.49	0.50	0.53	0.60	0.94				
22.5	0.45								
25	0.41	0.41	0.43	0.47	0.55	0.91			
27.5	0.37								
30	0.33	0.34	0.35	0.37	0.41	0.49	0.87		
32.5	0.30								
35	0.27	0.27	0.28	0.30	0.32	0.36	0.44	0.82	
37.5	0.24								
40	0.22	0.22	0.22	0.23	0.25	0.28	0.32	0.39	0.77

TABLE 6.2 Rankine Active Earth Pressure Coefficients

Angle of Internal Friction ϕ (deg)	β slope (deg)								
	0	5	10	15	20	25	30	35	40
	K_p level	K_p sloped							
0	1.00								
10	1.42	1.35	0.98						
12.5	1.55								
15	1.70	1.64	1.46	0.97					
17.5	1.86								
20	2.04	1.99	1.83	1.55	0.94				
22.5	2.24								
25	2.46	2.41	2.25	1.99	1.61	0.91			
27.5	2.71								
30	2.99	2.94	2.77	2.50	2.13	1.66	0.87		
32.5	3.31								
35	3.68	3.62	3.44	3.14	2.74	2.26	1.70	0.82	
37.5	4.10								
40	4.59	4.52	4.31	3.97	3.52	2.98	2.38	1.72	0.77

Table 6.3 Rankine Passive Earth Pressure Coefficients

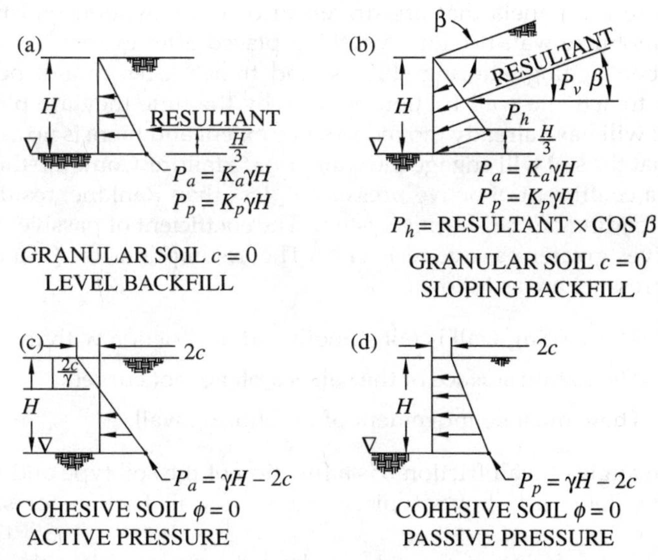

Figure 6.2 Rankine earth pressure diagrams.

are found by calculating the area of the pressure diagram, and the point of action is found at the center of balance of the diagram. For noncohesive soil with a vertical wall and level backfill, the pressure at the base of the wall is

$$p_a = K_a \gamma H \tag{6.13}$$

and total active pressure is

$$P_a = K_a \gamma \frac{H^2}{2} \tag{6.14}$$

located at $\frac{H}{3}$ from the bottom. For a sloping backfill the pressure resultant is parallel to the slope, and the horizontal pressure is

$$P_{a\text{ Horizontal}} = P_a \cos \beta \tag{6.15}$$

6.1.3 Coulomb Earth Pressure Theory

The soil on the back of a retaining or shoring wall has some degree of frictional interaction with the soil that prevents it from sliding downward in the active case and upward in the passive case. This effect adds to the shearing resistance of the soil and reduces the loading on the wall. There has to be movement of the soil and wall for the friction to be activated. Retaining walls are generally constructed from masonry, sheet pile, or wood and are designed to be in place for several years. In these cases it is a reasonable assumption that the wall friction will be activated. In the case of shoring walls, activation of the wall friction is not necessarily a reasonable assumption. Sheet piles and slide rail panels that are driven prior to excavation will most likely mobilize wall friction. Anything placed after excavation such as timber lagging, shoring shields, and trench jacks cannot be assumed to activate wall friction because by the time they are placed the soil will have already moved to some extent and there is no assurance that the soil will engage the wall prior to failure. Coulomb theory yields a coefficient of active pressure K_a less than Rankine, resulting in less loading on the shoring system. The coefficient of passive pressure gives greater passive resistance. The assumptions for Coulomb earth pressure are as follows:

1. The shoring wall is not smooth and has friction with the soil.
2. The failure surface of the soil is a plane, not curved.
3. There must be movement of the shoring wall.

The angle of wall friction δ is a function of the soil type and wall material; Table 6.4 shows typical values of wall friction for sheet piles.

The value for painted steel may be a degree or so less than for bare steel.

Steel sheet piles against the following soils:	(degrees)
Clean gravel, gravel sand mixture, well graded rock with spalls	22
Clean sand, silty sand-gravel mixture, single size hard rock fill	17
Silty sand, gravel or sand mixed with silt or clay	14
Fine sandy silt, nonplastic silt	11

Source: NAVFAC, *Foundation and Earth Structures,* Design Manual 702, p. 7.2.61–7.2.67, Naval Facilities Engineering Command, Alexandria, Va., 1971.

TABLE 6.4 Angle of Wall Friction for Steel Sheet Piles against Various Soils

The values of active and passive Coulomb pressure can be calculated using Eqs. (6.5) and (6.6)

$$K_a = \frac{\cos^2 \phi}{\cos\delta\left[1 + \sqrt{\dfrac{\sin(\phi+\delta)\sin(\phi-\beta)}{\cos\delta\cos\beta}}\right]^2} \tag{6.16}$$

$$K_p = \frac{\cos^2 \phi}{\cos\delta\left[1 - \sqrt{\dfrac{\sin(\phi+\delta)\sin(\phi-\beta)}{\cos\delta\cos\beta}}\right]^2} \tag{6.17}$$

where ϕ = angle of internal friction of soil
δ = angle of wall friction, positive for active pressure and negative for passive pressure
β = angle of backfill with respect to horizontal

If there is no angle of wall friction, a smooth frictionless wall surface, the Coulomb values are the same as the Rankine values. Tables 6.5 and 6.6 have calculated Coulomb values for K_a and K_p for level backfill. Figure 6.3(a) and (b) gives the Coulomb active and passive pressure diagrams. The calculations for passive pressure with large angles of wall friction give passive pressures that will create a curved failure surface which will fail before the full Coulomb passive pressure is achieved (Fig. 6.4). Table 6.6 has been corrected for this condition.

The angles of the Coulomb failure planes for level backfill are given by the following equations

$$\alpha_a = 90^\circ - \phi - \arctan\left[\frac{-\tan\phi + \sqrt{\tan\phi(\tan\phi + \cot\phi)(1+\tan\delta\cdot\cot\phi)}}{1+\tan\delta(\tan\phi+\cot\phi)}\right] \tag{6.18}$$

Rankine and Coulomb K_a, $\beta = 0$ (no back slope)					
Angle of Internal Friction ϕ (deg)	δ, steel wall friction to soil (deg)				
	0	22	17	14	11
	Rankine K_a	Coulomb K_a			
0	1.00	1.00	1.00	1.00	1.00
10	0.70	0.60	0.61	0.62	0.63
12.5	0.64	0.55	0.56	0.57	0.58
15	0.59	0.51	0.51	0.52	0.53
17.5	0.54	0.46	0.47	0.48	0.49
20	0.49	0.43	0.43	0.44	0.44
22.5	0.45	0.39	0.39	0.40	0.41
25	0.41	0.36	0.36	0.36	0.37
27.5	0.37	0.33	0.33	0.33	0.34
30	0.33	0.30	0.30	0.30	0.31
32.5	0.30	0.27	0.27	0.27	0.28
35	0.27	0.24	0.25	0.25	0.25
37.5	0.24	0.22	0.22	0.22	0.23
40	0.22	0.20	0.20	0.20	0.20

TABLE 6.5 Coulomb Active Earth Pressure

$$\alpha_p = 90^\circ + \phi - \arctan\left[\frac{\tan\phi + \sqrt{\tan\phi(\tan\phi + \cot\phi)(1 + \tan\delta \cdot \cot\phi)}}{1 + \tan\delta(\tan\phi + \cot\phi)}\right] \quad (6.19)$$

The Coulomb active soil pressure at depth is given by

$$p_a = k_a \gamma H \quad (6.20)$$

Total pressure $$P_a = k_a \gamma \frac{H^2}{2} \quad (6.21)$$

Total horizontal pressure $$P_{\text{active horizontal}} = P_a \cos\delta \quad (6.22)$$

The Coulomb passive soil pressure at depth is given by

$$p_p = k_p \gamma H \quad (6.23)$$

Total pressure $$P_p = k_p \gamma \frac{H^2}{2} \quad (6.24)$$

Rankine and Coulomb K_p, $\beta = 0$ (no back slope)					
Angle of Internal Friction ϕ (deg)	δ, steel wall friction to soil (deg)				
	0	22	17	14	11
	Rankine K_p	Coulomb K_p			
0	1.00	1.00	1.00	1.00	1.02
10	1.42	2.23	1.99	1.87	1.76
12.5	1.55	2.54	2.24	2.10	1.96
15	1.70	2.89	2.53	2.34	2.18
17.5	1.86	3.29	2.85	2.62	2.43
20	2.04	3.76	3.21	2.94	2.71
22.5	2.24	4.31	3.63	3.31	3.02
25	2.46	4.95	4.12	3.73	3.39
27.5	2.71	5.73	4.70	4.21	3.80
30	2.99	7.00	5.38	4.79	4.29
32.5	3.31	7.50*	6.00*	5.46	4.85
35	3.68	8.00*	7.00*	6.00*	5.52
37.5	4.10	8.80*	8.00*	7.00*	6.31
40	4.59	9.50*	9.00*	8.00*	7.27

*Corrected to eliminate false failure surfaces at toe.

Table 6.6 Coulomb Passive Earth Pressure

and

$$\text{Total horizontal pressure} \quad P_{\text{passive horizontal}} = P_p \cos\delta \tag{6.25}$$

6.1.4 Log-Spiral Theory

This theory assumes that failure surfaces are curved instead of planar which is fairly evident if one looks at the intact soil after a landslide. Major assumptions for this theory are as follows:

1. The shoring wall is not smooth and has friction with the soil.
2. The failure surface of the soil is curved.
3. There must be movement of the shoring wall.

The results yield active and passive K values that would be considered less conservative yet arguably more accurate than the Rankine

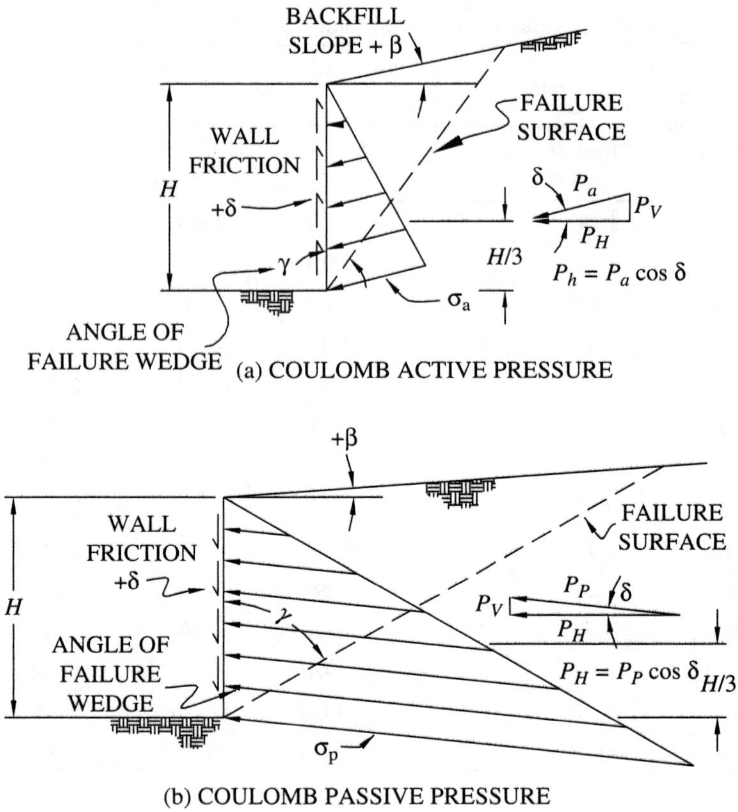

Figure 6.3 (a) Coulomb active soil pressure diagram, (b) Coulomb passive soil pressure diagram.

and Coulomb values. This method is most often used in the design of permanent retaining walls. Calculations, tables, and graphs for this method were first developed by A. Caquot and J. Kerisel, *Tables for the Calculation of Passive Pressure, Active Pressure, and Bearing Capacity of*

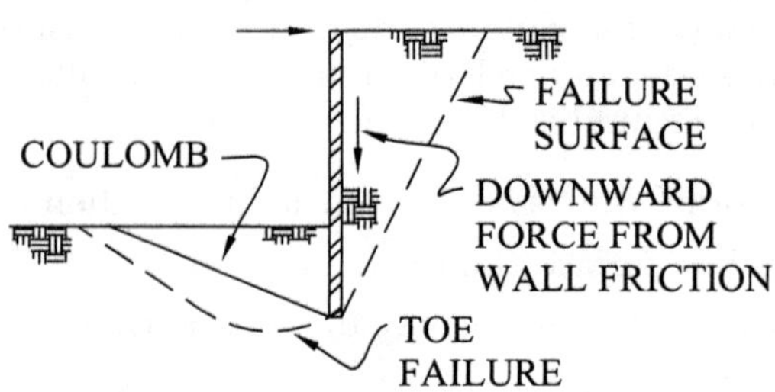

Figure 6.4 Toe failure due to large angle of wall friction.

Foundations and can be found in NAVFAC, *Foundation and Earth Structures*, Design Manual 7.02, pp. 7.2.61–7.2.67.

6.2 Use of Earth Pressure Theories

The lateral earth pressures given by the Rankine, Coulomb, and log spiral theory are theoretical and should not be used in most cases of shoring design because the excavation process does not replicate the theoretical condition (see next section). It is reasonable to develop the theoretical active and passive pressure coefficients K_a and K_p provided the assumptions are met. The most conservative assumption would be the Rankine coefficients because the theory does not assume wall friction which is present to some degree in all shoring applications that contact the soil. The triangular shape of the active soil pressure distribution in braced walls is the assumption that is most often incorrect; however, in the case of cantilevered shoring walls it is reasonably accurate. In braced toed-in shoring walls, the passive pressure distributions given by the theories are used to develop toe resistance, generally with additional factors of safety. It is also important to note that in shoring design the K_a and K_p values only apply to (assumed) noncohesive soils where the angle of internal friction is at work.

Theoretical earth pressure distributions are most often used in the design of cantilever retaining walls and permanent structures because of control of the backfill against them and the long-term effect of time. Calculation of Coulomb and log spiral loading is slightly more complicated than Rankine so the reason for using them would be to obtain a more accurate and cost-effective design. Example 6.1 calculates the lateral pressure using both the Rankine and Coulomb methods in order to see what the difference is.

Example 6.1 Calculate the lateral soil force P_a on the braced wall shown in Fig. 6.5, using Rankine and Coulomb earth pressure theories. Assume that there is enough movement in the wall for active pressure to develop.

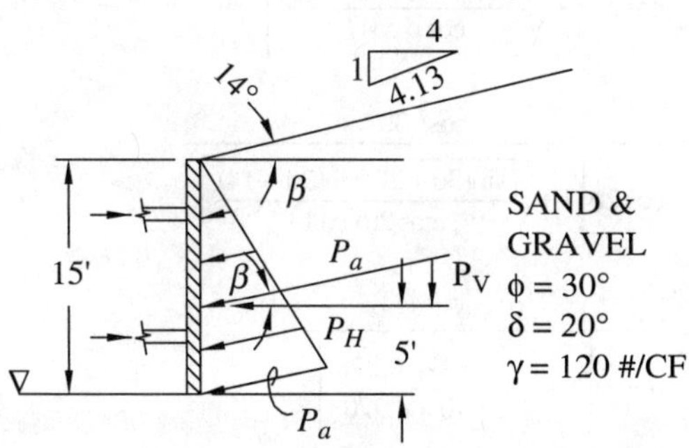

FIGURE 6.5 Braced wall with lateral soil pressure distribution.

Rankine lateral pressure

Find the coefficient of active pressure.

$$K_a = \cos\beta \frac{\cos\beta - \sqrt{\cos^2\beta - \cos^2\phi}}{\cos\beta + \sqrt{\cos^2\beta - \cos^2\phi}}$$

$$K_a = \cos 14° \frac{\cos 14° - \sqrt{\cos^2 14° - \cos^2 30°}}{\cos 14° + \sqrt{\cos^2 14° - \cos^2 30°}} \tag{6.11}$$

$$K_a = 0.970 \frac{0.970 - \sqrt{0.970^2 - 0.866^2}}{0.970 + \sqrt{0.970^2 - 0.866^2}} = 0.970 \times \frac{0.533}{1.407}$$

$$K_a = 0.368$$

Find the soil pressure and horizontal force.

$$p_a = K_a \gamma H = 0.368 \times 120 \times 15 = 661 \text{ psf}$$

$$P_a = 661 \text{ psf} \times \frac{15 \text{ ft}}{2} = 4961 \text{ lb}$$

$$P_{\text{horizontal}} = P_a \cos\beta = 4961 \times \cos 14° = 4961 \times 0.971$$

$$P_{\text{horizontal}} = 4847 \text{ lb}$$

Coulomb lateral pressure

Find coefficient of active pressure.

$$K_a = \frac{\cos^2\phi}{\cos\delta \left[1 + \sqrt{\frac{\sin(\phi+\delta)\sin(\phi-\beta)}{\cos\delta\cos\beta}}\right]^2} \tag{6.16}$$

$$K_a = \frac{\cos^2 30}{\cos 20 \left[1 + \sqrt{\frac{\sin(30+20)\sin(30-14)}{\cos 20 \cos 14}}\right]^2}$$

$$K_a = \frac{0.866^2}{0.940 \left[1 + \sqrt{\frac{0.7660 \times 0.276}{0.940 \times 0.970}}\right]^2}$$

$$K_a = 0.365$$

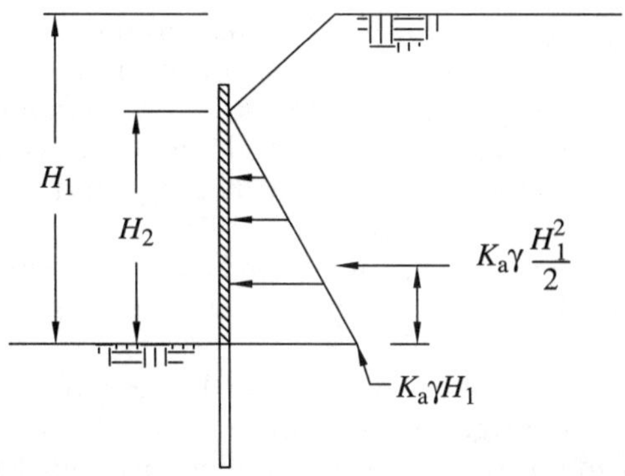

FIGURE 6.6 Loading diagram for sloped bank cantilever shoring wall.

Find the soil pressure and horizontal force.

$$p_a = K_a \gamma H = 0.365 \times 120 \times 15 = 657 \text{ psf}$$

$$P_a = 657 \text{ psf} \times \frac{15 \text{ ft}}{2} = 4928 \text{ lb}$$

$$P_{\text{horizontal}} = P_a \cos\delta = 4928 \times \cos 20° = 4928 \times 0.940$$

$$P_{\text{horizontal}} = 4632 \text{ lb}$$

∴ The difference in soil loading between Rankine and Coulomb theory is

$$\Delta = 4847 - 4632 = 215 \text{ lb}$$

Sloped Embankment behind Shoring Wall

It is a common practice to slope the embankment at the top of shoring walls so that the shoring height is lowered. Reasonable cantilever heights using sheet piles or pile and lagging are roughly 12 ft. By sloping the embankment this sheeting height can be achieved with excavations to approximately 18 ft. For cantilever walls the simplest way to approach the loading diagram is to calculate the full height pressure and total force and then apply them to the cantilever wall height, as shown in Fig. 6.6.

6.3 Apparent Soil Pressure Diagrams for Braced Cuts

Braced cuts in temporary excavations have to meet three criteria. They must

1. Protect workers from cave-in
2. Provide a stable excavation in which the production work can to be constructed
3. Control surrounding settlement

Even though there has been a decrease from the past, today failure on one of or all these requirements still occur at an alarming rate. Failure in the past, say until the late 1960s, was partially due to the lack of understanding about how the soil interacted with the shoring during the construction process. Strut loadings in deep excavations were measured and studied from the 1930s to the 1960s, and soil loading design envelopes were developed from these studies. Today's failures are most likely due to lack of understanding and adherence to design standards.

Prior to development of apparent earth pressure diagrams, shallow trench excavations generally considered to be less than 20 ft deep were shored using mostly 2- to 4-in timber sheeting, 4 × 6 to 4 × 10 timber waling, and 4 × 6 to 8 × 8 struts depending on the trench width. The early 1917 construction trench safety orders prescribed minimum vertical and horizontal strut spacing to be 4 ft vertical by 8 ft horizontal in hard compact soil and 5 ft horizontal in softer soils. Sizing was based on observation of the soil, and calculating loading was not standard practice. When OSHA developed the *Draft Construction Safety Standards for Excavations,* they formalized this in greater detail, resulting in OSHA Appendix C, Timber Shoring.

With the advent of manufactured shoring equipment, such as aluminum trench jacks and shoring shields, starting around 1950, there was a need to adapt soil loading theories that applied to the equipment. There was no standard practice, and OSHA left loading and use tabulation up to the manufacturers. Each manufacturer has been trying to make the "mine is stronger, lighter, and will go deeper than any other similar product" claim ever since. In this game there are two different ways to parse the data: change the soil loading assumption and change the allowable working material strength assumption. Assumed high soil loading and low allowable material strength does not work in the manufacturer's favor. The result of this race has been a constant push in the direction of less conservative tabulated data. Fortunately today we have a history of use for manufactured shoring equipment. Although it has not been studied in depth, it appears that the tabulated data assumptions are safe and reasonable.

The excavation industrywide adoption of apparent soil pressure diagrams for shoring systems would lead to a level playing field for shoring manufacturers and greater versatility for the competent person and shoring design engineer when adapting shoring equipment to special situations. Even though the apparent pressure diagrams presented below were derived from deep excavations, they have been accepted to a large extent as the standard of the industry. They were first developed by Peck, Hanson, and Thornburn among others and further developed by the Naval Facilities Engineering Command, NAVFAC, and are used extensively throughout the excavation industry today.

6.3.1 Important Information Resulting from Studies of Deep Excavations

Loading, deflection, and settlement measurements were taken on struts in major, 26-ft (8-m) to 60-ft (19-m) deep excavations in the 1940s through the 1960s. Study results indicate that in braced cuts

- Loads are distributed to wale levels evenly, as if the sheeting between the wales were pin-connected at the wales.
- Strut loads toward the top of the excavation increase as the excavation proceeds to depth.
- Within the overall depth of the excavation loads are highest in the middle.
- Earth at the bottom of a cut acts to some extent as a strut.

Of major importance is the finding that the total load on the system was what would be calculated using a triangular Rankine earth pressure; however, the resultant does not act at one-third times the height of the excavation. Instead it acts roughly in the middle of the excavation. Researchers also found that individual strut measurements at each level could vary as much as ±60 percent of the average and that the total load at a level could vary ±30 percent in different cuts in the same area. There are several reasons for this:

1. There are variations within the soil such as slightly different soil types, densities of noncohesives, cohesiveness of noncohesives, and moisture content.
2. Both the influence from construction loading and the effects from surrounding structures are a factor.
3. Timing of construction sequence is important. In strutted excavations the first strut is usually placed at a shallow depth and fairly rapidly. As the excavation depth increases, the time between each level of strut placement increases as does the movement of the corresponding soil.
4. In places the soil pressure is somewhere between at rest and active due to limited wall movements.
5. Distribution of soil load to sheeting, wales, and struts plays a role. Deflections around each of these elements are different, setting up different soil interface interaction and mobilization of wall friction.
6. Lateral movement and settlement in cohesive soil can cause tension cracks that lose the cohesive strength.
7. In excavations strutted near the top, the soil does not fail in a triangular wedge, resulting in a triangular pressure distribution. The soil fails in a curved section, and the pressure

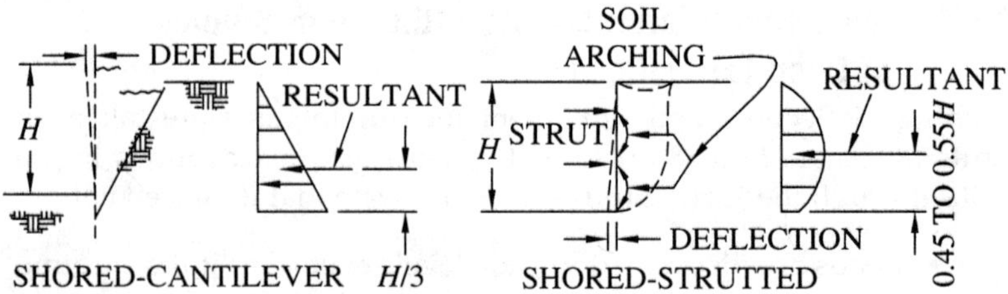

FIGURE 6.7 Rankine triangular and strutted failure modes.

distribution is roughly parabolic with the resultant acting more toward the center than the bottom third of the excavation (Fig. 6.7).

Struts are the critical element in a shoring system; while sheeting or a wale could conceivably fail, the entire shoring system would not fail. However, the failure of one strut would more than likely set up a progressive failure in all the struts, and the entire shoring system would fail.

Apparent earth pressure diagrams were developed as a result of measured strut loads and predict maximum loads that the struts will see. The diagrams do not necessarily represent the distribution of earth pressure on sheeting and wales. Also the excavations studied were dewatered outside the excavation to the bottom, and the wale levels and struts were fairly evenly spaced; anything different from this should be studied carefully before relying on these diagrams.

The above information is thoroughly discussed in Terzaghi, Peck, and Mesri (1996, Article 46).

6.3.2 Apparent Earth Pressure Diagram for Cuts in Noncohesive Soils

Figure 6.8 was thoroughly discussed in Terzaghi and Peck (1967) and given by Peck, Hanson, and Thornburn (1974), and developed by NAVFAC (1971) and has become the standard of the industry for noncohesive soils, sands, and gravels. The diagram gives total strut loads 30 percent higher than the Rankine Triangular load area would be.

$$p_a = .65k_a\gamma H \tag{6.26}$$

where k_a = coefficient of active pressure
γ = unit weight of soil

$$P_1 = (p_a)\left(a_1 + \frac{a_2}{2}\right)$$

$$P_a = 0.65k_a\gamma H \qquad (6.26)$$

where

k_a = coefficient of active pressure

γ = unit weight of soil

$$P_1 = (P_a)(a_1 + \frac{a_2}{2})$$

$$P_2 = (P_a)(a_2 + \frac{a_3}{2})$$

$$P_3 = (P_a)(a_3 + \frac{a_4}{2})$$

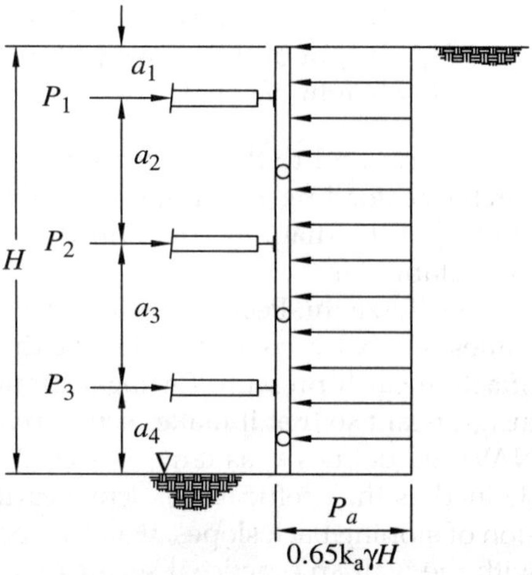

FIGURE 6.8 Apparent earth pressure diagram for strut loading in noncohesive soils.

$$P_2 = (p_a)\frac{(a_2 + a_3)}{2}$$

$$P_3 = (p_a)\frac{(a_3 + a_4)}{2}$$

The Terzaghi, Peck, and Mesri article does not discuss water except to state that the excavations studied were dewatered to the bottom and that seepage pressure needs to be accounted for. The diagram also does not state that it only applies to sheeting that terminates at the base of the excavation. Sheeting that stops at the base of an excavation is generally assumed to allow seepage water to flow into the bottom of the excavation and relieve the hydrostatic pressure. With this assumption, dewatering is usually planed around pumping from sumps inside the excavation, thereby eliminating the cost of setting and maintaining dewatering wells outside the excavation. Water can easily back up against sheeting that terminates at the bottom of the excavation or sheeting that has spaces between it in soils with low permeability. If there is going to be a problem with a shoring system, misjudgment of dewatering capability is most likely where it started from. In practice, if the water table is above the base of the excavation, use the natural unit weight of the soil above the water table and the buoyant unit weight of the soil σ' below the water table so that

$$p_a = 0.65K_a(\gamma H_1 + \gamma' H_2) \qquad (6.27)$$

where p_a = earth pressure at bottom of excavation as shown in Fig. 6.8

K_a = coefficient of active pressure

γ = unit weight of soil above water table

γ' = buoyant unit weight of soil
H_1 = height of soil above water table
H_2 = height of soil above water table

Always add to the soil loading diagrams the water loading and surcharge loading diagrams with their resultants at the center of gravity of the diagram, one-third up for triangular and one-half up for rectangular.

The Terzaghi, Peck, and Mesri article discusses the fact that given values of φ between 30° and 45°, the difference between the coefficient of active earth pressure for logarithmic spiral and Rankine values is insignificant so that it makes sense to use the easier Rankine formula. NAVFAC defines k_a as $\tan^2(45 - \varphi/2)$, and in other publications it is defined as the coefficient of active earth pressure. There is no discussion of sloping back slopes, wall friction, and the angle of the resultant with the wall. In practice it seems to make sense to use the coefficient from the Rankine, Coulomb, or logarithmic spiral theory that makes the best sense to the designer. The article also does not discuss the design of the sheeting and wales except to state that they will be lower.

6.3.3 Apparent Earth Pressure Diagram for Cuts in Cohesive Soils

In shoring design in cohesive soils, a $\phi = 0$ assumption is made and K_a becomes 1. Equation (6.3)

$$p_a = \gamma z K_a - 2c\sqrt{K_a}$$

becomes

$$p_a = \gamma z - 2c \qquad \text{(cohesive, } \phi = 0\text{)} \tag{6.28}$$

Without support a bank of soil is on the verge of collapse when the active pressure becomes the value of the unconfined compression strength or $p_a = q_u$. Since $q_a = 2c$, we can substitute into Eq. (6.28) to get the critical height H_c.

$$2c = \gamma H_c - 2c$$

$$H_c = \frac{4c}{\gamma} \tag{6.29}$$

where H_c = critical height prior to bank failure
c = cohesion of in situ soil
γ = unit weight of in situ soil

In theory a bank of cohesive soil should be able to stand without support to a height of H_c. The bank of soil is said to be stable when $\gamma H/c = 4$ and $\gamma H/c$ is referred to as the *stability number.*

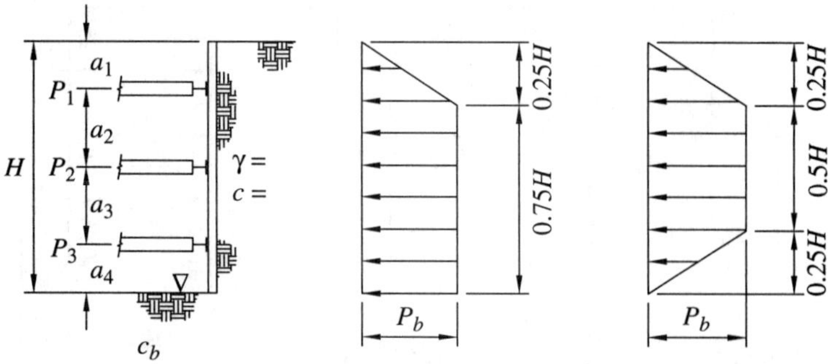

(a) BRACED CUT (b) for $\frac{\gamma H}{c} > 4$ (c) for $\frac{\gamma H}{c} \leq 4$

(a)

For for $\frac{\gamma H}{c} > 4$: The greater of $p_b = \gamma H - 4mc$ or $p_b = n\gamma H$

For for $\frac{\gamma H}{c} \leq 4$: $p_b = n\gamma H$

where γ = unit weight of soil

H = depth of excavation

$$m = \begin{cases} 4, \text{ for } \frac{\gamma H}{c} > 4 \\ 1, \text{ otherwise} \end{cases}$$

c = cohesion of soil beside cut

c_b = cohesion of soil below excavation level

$$n = \begin{cases} 0.2, \text{ if loading time is short and sheeting movement restricted} \\ 0.4, \text{ long loading time and flexible sheeting} \end{cases}$$

For struts: Assume pinned connections between struts and calculate tributary load. For example:

$$P_2 = p_a \frac{a_2 + a_3}{2}$$

(b)

Figure 6.9 Pressure diagrams for braced cuts in clay (after Peck, Hanson, and Thornburn, 1974).

For design of struts in braced cuts in clay, Peck, Hanson, and Thornburn have recommended Fig. 6.9. NAVFAC Design Manual 7.02 (Fig. 6.10) has a slightly different version utilizing the stability number concept. These two similar methods have become standard in the industry because they take the following factors into consideration:

- The amount of time the cut will be open
- The stability of the bottom of the excavation
- The cohesion of the soil

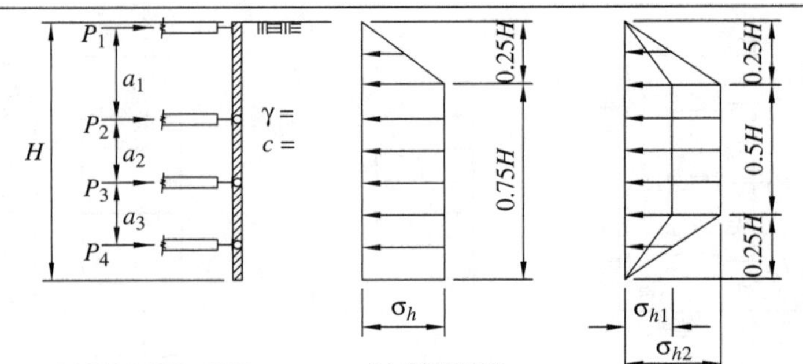

(a) BRACED CUT (b) SOFT TO MEDIUM CLAY (c) STIFF CLAY

For soft to medium clay, $N_0 > 6$

$$N_0 = \frac{\gamma H}{c} \quad \text{and } \sigma = k_a \gamma H$$

where N_0 = stability number
γ = unit weight of soil
c = cohesion of soil
H = depth of excavation

$$k_a = 1 - m\frac{4c}{\gamma H}$$

$m = 1$ or $0.4F_{sb}$ where cut is underlain by deep soft normally consolidated clay
F_{sb} = factor of safety against bottom instability
$1 \le F_{sb} \le 1.5$ see bottom stability formulas.

For stiff clay, $N_0 > 4$

$$N_0 = \frac{\gamma H}{c} \quad \text{and } \sigma = 0.2\gamma H \text{ or } = 0.4\gamma H$$

where $\sigma_{h1} = 0.2\gamma H$ use when movements are minimal and short construction period
$\sigma_{h2} = 0.4\gamma H$ use when movements occur and long construction period

For stiff clay, $4 < N_0 < 6$
Use largest of diagrams (b) and (c)

For struts
Assume pinned connections at struts and calculate tributary load.
For example: 3

$$P_3 = \sigma_x \frac{a + a}{2}$$

FIGURE 6.10 Pressure diagrams for braced cuts in clay [NAVFAC Design Manual 7.2 (after Fig. 26), 1971.]

The Navy version was developed similar to the Peck, Hanson, and Thornburn version and was revised last in 1986. There are some differences in the diagrams; however, the major difference is that the Navy version requires a bottom stability calculation. Tables 6.7 and 6.8 show the results of calculations at different depths and cohesions

p_b/σ_h for Braced Cuts in Clay—Hard Bottom Results of Peck, Hanson, and Thornburn/NAVFAC Formulas						
Depth H (ft)	p_b (psf) $c_b = 2000$, $n = 0.4$ Long Term Cohesion c (psf)					
	250	500	750	1000	1500	2000
5	260	260	260	260	260	260
10	520	520	520	520	520	520
15	950	780	780	780	780	780
20	1600	1040	1040	1040	1040	1040
25	2250	1300	1300	1300	1300	1300
30	2900	1900	1560	1560	1560	1560
35	3550	2550	1820	1820	1820	1820
40	4200	3200	2200	2080	2080	2080

Notes:

1. Below dark line use chart for $\frac{\gamma H}{c} > 4$.
2. Above dark line $p_b = 0.4\gamma H$; use chart for $\frac{\gamma H}{c} \leq 4$.

TABLE 6.7 p_b/σ_h for Cohesive Soils

using both formulas. In these calculations with the assumptions made the results are exactly the same for both formulations. The tables are provided to show what is the range of cohesive pressures at different depths and how they change with a soft and hard bottom. Another conclusion that can be made from these tables is that in most cases the horizontal soil loading in cohesive soils is $0.2\gamma H$ to $0.4\gamma H$, depending on how long the excavation was open.

Although the discussions and titles on the diagrams use the word *clay*, industry practice has been to apply the theory to cohesive soils as defined by the Uniform Soils Classification System—fine-grained soils with more than 50 percent passing the No. 200 sieve. Clays give up their cohesion over time and rely on the internal angle of friction for strength after that. The closer a soil is to noncohesive, more silt and less clay, the sooner the cohesion will disappear. On shoring projects that will be in the ground for a few months, it is prudent to run a noncohesive, $c = 0$, analysis and use the worst case.

Always add to the soil loading diagrams the water loading and surcharge loading diagrams with their resultants at the center of gravity of the diagram, one-third up for triangular and one-half up for rectangular. Use γ' below the water table.

Sloping back surfaces are not addressed in these calculations, and because there is no angle of internal friction involved it is not possible

p_b/σ_h for Braced Cuts in Clay—Soft Bottom Results of Peck, Hanson, and Thornburn/NAVFAC Formulas						
	p_b (psf) c_b = 250, n = 0.4 Long Term					
Depth *H*	**Cohesion *c* (psf)**					
(ft)	250	500	750	1000	1500	2000
5	260	260	260	260	260	260
10	900	520	520	520	520	520
15	1550	1150	780	780	780	780
20	2200	1800	1400	1040	1040	1040
25	2850	2450	2050	1650	1300	1300
30	3500	3100	2700	2300	1560	1560
35	4150	3750	3350	2950	2150	1820
40	4800	4400	4000	3600	2800	2080

Notes:
1. Below dark line use chart for $\frac{\gamma H}{c} > 4$.
2. Above dark line $p_b = 0.4\gamma H$; use chart for $\frac{\gamma H}{c} \leq 4$.

TABLE 6.8 p_b/σ_h for Cohesive Soils, Soft Bottom

to work them in through the determination of K_a (in NAVFAC they define a k_A that has nothing to do with φ). One method for dealing with back slopes is to treat them as surcharge loads. Since back slopes typically are not steep and usually flatten off eventually, another conservative method is to use the full height of the back slope as the depth of excavation (Fig. 6.6).

These diagrams were developed to determine worst-case strut loads, and sheeting and wale loads may be less. They also depend on the idea that the struts are spaced relatively evenly. It would probably be a good idea for the use of these diagrams to require that the variation in distance between the struts be not more than 30 percent.

Another important point that is made in the Terzaghi, Peck, and Mesri article is that in stiff clays with stiff sheeting and rigid supports, the soil never gets a chance to relax from the at rest condition to the active condition, causing loads to be much higher. Use K_0 for design in this case.

6.3.4 Pressure Diagrams for OSHA Appendix A Soil Types

As shown in Table 5.6b, Recommended Soil Classification System for the Standard Practice, there is a column for W_e, the effective lateral weight of the soil. In that column W_e is defined as 20 psf for type A, 40 psf for type B, and 80 psf for type C. The final field soil classification system, Appendix A, does not mention W_e. The primary reasons for this are as follows:

- Soil loading is dependent on more than soil properties. It is essentially a different problem that is to be considered after the soil is classified. The type of shoring system, protective shield, support such as trench jack, and preload all affect the loading. Cohesive soils load differently than noncohesive soils, and yet both types can fall into the type B or type C categories. The depth of the excavation and the type of soil below the bottom of the excavation can have an effect.
- The "if I use this and a problem develops, it is someone else's responsibility" mentality is promoted.

The only way that a competent person can determine soil loading is if a method is described in the tabulated data for the shoring system that he or she is using. In OSHA Appendix C, Timber Shoring, $P = 25H$ +2 ft depth for surcharge is set out for type A soil, and 45 and 80 for type B and C. Tabulated data for shoring shields usually supply an allowable depth for each soil type and a maximum psf rating that is not to be exceeded. The manufacturer's anticipated soil loading cannot always be backed out of the psf rating, and the shape of the loading diagram is not evident. Tabulated data for trench jacks are based on soil type and depth, again with no indication of how the tabulation was arrived at. The problem for the competent person lies in determining how to back off allowable depth when large surcharge loads are encountered. Tabulated data should be more specific on this issue.

By using Table 5.3, SPT *N/f* to angle of internal friction, then calculating the Rankine K_a, and finally applying the Peck, Hanson, and Thornburn apparent earth pressure formulas, horizontal loading from these noncohesive soils can be calculated. Table 6.9 shows the results of this exercise for noncohesive soils. Of particular interest is the fact that in type C noncohesive soils, the lateral loading never gets above 39*H*. OSHA Appendix A type C soils are typically associated with horizontal loading of 80*H* and 60*H*. The option to use apparent noncohesive earth pressure is only available with OSHA option 4 design by a registered engineer. Shoring systems designed using this option will be far more cost-effective due to the lighter soil loads. The best example of this is in the selection of shoring shields. In a C-60 soil at 25 ft deep, the shoring shield would need to have a minimum strength of $25 \times 60 + 72 = 1572$ psf. Most 6-in wall shoring shields 20 ft and longer are rated 1300 psf and less. An 8-in wall shield or a shorter stronger 6-in wall shield would have to be used. Using design by an engineer, the shield would need to be rated for $25 \times 39 + 100 = 1075$ psf.

Critical Height for OSHA Cohesive Soil Types—By using Table 5.1, SPT *N* to unconfined compressive strength for cohesive soils, Table 6.10 can be developed to determine critical height for OSHA

Approximate Horizontal Soil Pressure P_b for OSHA Appendix A, Noncohesive Soils					
Relative Density					
Sands and Gravels	**SPT, *N***	**Angle of Internal Friction, ϕ (deg)**	**Rankine* K_a**	**OSHA Soil Type**	**P_b^{3} (psf)**
Very loose	0–4	18–28	0.52–0.36†	Type C	39*H*
Loose	5–10	28–30	0.36–0.33	Type C-60	27*H*
Medium dense	11–30	30–36	0.33–0.26	Type B	25*H*
Dense	31–50	36–41	0.26–0.21	↓	19*H*
Very dense	50+	41	0.21	↓	16*H*

*From Peck, Hanson, and Thornburn, apparent earth pressure diagram for strut loading, noncohesive soils, Fig. 6.8, $P_b = 0.65k_a\gamma H$, $\gamma = 115$ to 120 pcf.

†$k_a = \tan^2\left(45 - \frac{\phi}{2}\right)$

Table 6.9 OSHA Noncohesive Soil Types and Horizontal Loading

OSHA Appendix A Soil Type and Critical Height Based on SPT *N* Blow Counts					
Consistency					
Silts and Clays	**SPT, *N***	**Unconfined Compression Strength (tsf)**	**Cohesion* c (psf)**	**OSHA Soil Type**	**Critical Height† H_c (ft)**
Very soft	0–2	0.05–0.25	50–250	Type C	2.2
Soft	3–4	0.25–0.50	250–500	Type C-60†	10
Medium stiff	5–8	0.50–1.00	500–1000	Type B	16
Stiff	9–11	1.00–1.5	1000–1500	Type B	32
Stiff	12–16	1.5–2.00	1500–2000	Type A	48
Very stiff	17–32	2.00–4.00	2000–4000	↓	64
Hard	32+	> 4.00	4000+	↓	128

*q_u = unconfined compression strength; $c = \frac{q_u}{2}$

†Soil should stand vertical without failure (due to cohesion); $H_c = \frac{4c}{\gamma}$

Table 6.10 OSHA Soil Types and Critical Height for Cohesive Soils

Hoizontal Force Factor, P_b for OSHA Cohesive Soils—Rectangular Loading Diagram*						
OSHA Soil Type	**Unit Weight (pcf)**	**Cohesion (psf)**	**From $\gamma H - 4mc$**	**From† $n\gamma H$**		
				$n = 0.2$	$n = 0.3$	$n = 0.4$
A	125	1500		19*H*	28*H*	38*H*
B	125	500	49*H*			
C-60	115	250	60*H*			
C	100	50	80*H*			

*From Peck, Hanson, and Thornburn, formulas for braced cuts in clay, Fig. 6.09. Developed for use with a rectangular loading diagram.

†For short loading time, (pipeline construction, footings, etc.), $n = 0.2$. For long loading time (burried structures, retaining walls, etc.), $n = 0.4$.

Table 6.11 OSHA Soil Type and Apparent Soil Loading for Cohesive Soils Using Rectangular Loading Diagrams

soil types. This is useful only for estimating the depth to which an excavation can be taken prior to installing the shoring. Use of trench jacks and shoring shields is dependent on the soil standing up long enough to insert the shoring. The critical height calculation assumes no tension cracks, fissures, surcharge loads, different soil layers, etc.

Hoizontal Force Factor, P_b for OSHA Cohesive Soils for Figure 6.09 Loading Diagrams*						
OSHA Soil Type	**Unit Weight (pcf)**	**Cohesion (psf)**	**From $\gamma H - 4mc$**	**From† $n\gamma H$**		
				$n = 0.2$	$n = 0.3$	$n = 0.4$
A	125	1500		25*H*	38*H*	50*H*
B	125	500	49*H*			
C-60	115	250	73*H*			
C	100	50	82*H*			

*From Peck, Hanson, and Thornburn, pressure diagram for braced cuts in clay, Fig. 6.09.

†For short loading time (pipeline construction, footings, etc.), $n = 0.2$. For long loading time (buried structures, retaining walls, etc.), $n = 0.4$.

Table 6.12 OSHA Soil Type and Apparent Soil Loading for Cohesive Soils Using Trapezoidal Loading Diagrams

Critical height should never be used to justify putting workers inside unshored excavations.

Table 6.11 gives the apparent soil loading for a rectangular loading diagram. This was developed so that a direct comparison to OSHA required rectangular loading can be made. Table 6.12 uses the Peck, Hanson, and Thornburn trapezoidal diagrams. Note that the generally accepted OSHA type A-25, B-45, C-60, and C-80 effective loadings are a close fit to the apparent soil pressure developed by Peck, Hanson, and Thornburn for cohesive soils.

6.4 Soil Arching Theory

The use of arch construction has been around since before the Romans made it famous in their architecture. The unique aspect of an arch is that it eliminates tensile forces created when a structure spans a space, such as in an arch bridge. All the forces are resolved into compressive forces that are carried through the arch to an unyielding base. As the arch is applied to soil mechanics, Karl Terzaghi (in Terzaghi, Peck, and Mesri, 1996) described the soil arching process as "the transfer of pressure from a yielding mass of soil onto adjoining stationary parts." In the case of shoring applications, the unyielding parts would be wales and struts in cofferdams, hydraulic cylinders in trench jacks, the bottom edges of an excavation, and the inside corners of vertical excavated walls. Soil arching is the underlying principle behind the fact that lagging can be removed from a small portion of shored excavation wall without the soil behind it flowing out. Arching is responsible for the fact that trench walls shored with trench jacks and no sheeting between will stand without wall failure. Arching is part of the reason that the apparent pressure on strut and wale supported shoring walls has a rectangular or trapezoidal distribution. Soil arching does not reduce the total soil load, it just distributes it to the shoring elements differently than one would expect. The primary function of soil arching in shoring systems is to redirect the loading away from the excavation wall face and concentrate it onto supporting struts.

In theory, for arching to work, there has to be some sort of infinitesimal movement of the soil. At the location of the supports the soil cannot move, and in order to fail the remaining soil must flow around the supports. As shown in Fig. 6.11, when this happens, the soil jams into the neck of the opening and compresses, thereby setting up the arch. Just as soil will not flow out of the neck of an upturned bottle, it will not flow between excavation wall supports or the corners and bottoms of excavations (Fig. 6.12).

In shoring applications arching soil movement can happen in two ways. Trench jacks, hydraulic struts, and anything driven into the soil, such as a pile or even a trench box being pounded in with

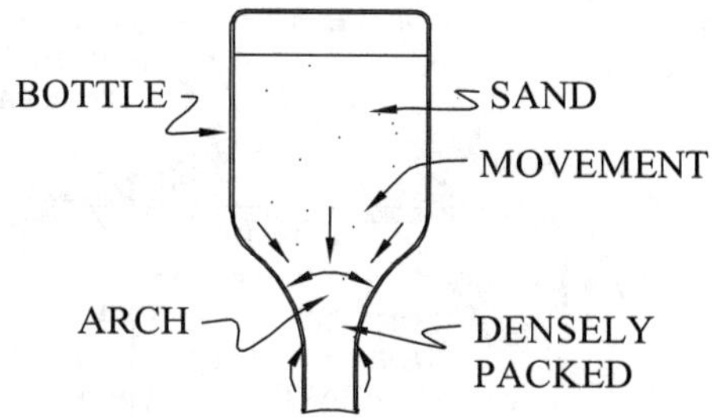

Figure 6.11 Soil arching.

an excavator, will add a compressive force and infinitesimal movement into the soil, setting up an arched stress path between the supports. The major concept behind preloading shoring struts is to set up the arching capability without allowing the soil to move outward. In the case of drilled piles and sheeted shoring systems, the soil moves outward and sets up arching to the more rigid parts of the system, thereby relieving some of the loading on the more flexible parts (Fig. 6.13). The key element in this process is to have rigid supports for the soil to arch to or, in other words, something that the soil has to flow between for failure to occur.

Another important element of soil arching is the consistency of the soil. Both cohesive and noncohesive soils at any strength will exhibit arching to some degree. The closer the supports are placed, the better the system works. Due to viscosity, even water flowing between two rocks experiences arching, as is evident by the arch on the water surface in front of the rocks.

Even though a considerable amount of observation and experimentation with arching has been done, there are no reliable

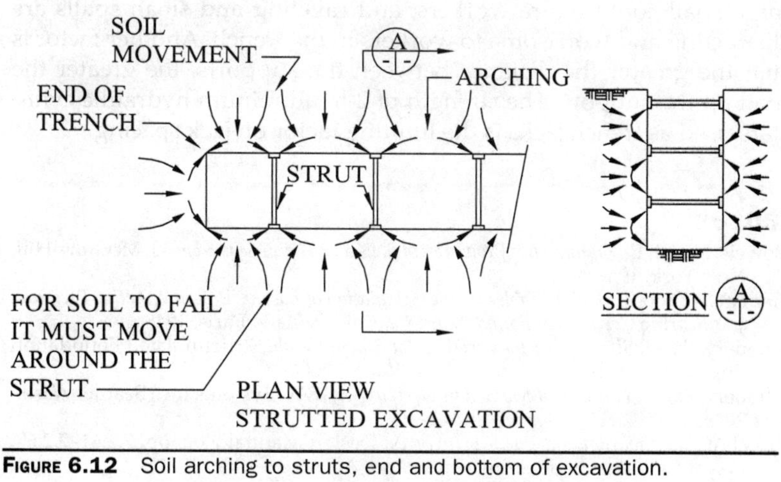

Figure 6.12 Soil arching to struts, end and bottom of excavation.

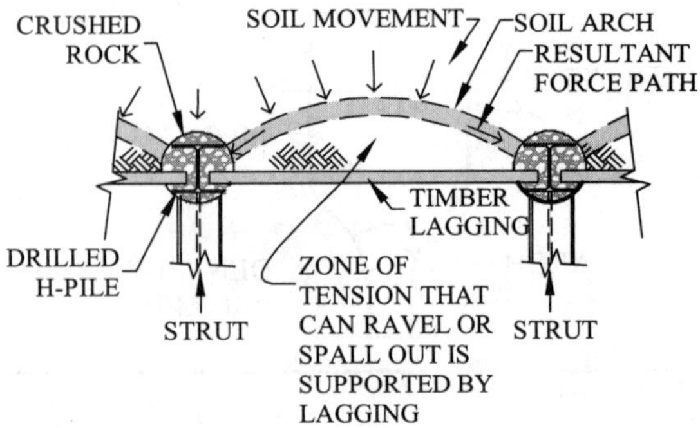

FIGURE 6.13 Soil arching in H-pile and lagging shoring. Note that if the struts are preloaded, soil movement toward the shoring wall is not required for arching to occur.

methods of calculating maximum spans or the radius of the curve of the theoretical arch in the soil. In the case of H-pile with timber lagged walls, there is no way to calculate how much of the soil load is distributed to the timber and how much goes directly to the pile. However, it can be assumed that the pile and ultimately the struts will see the entire apparent pressure predicted from the soil.

In the case of trench jacks, throughout the history of their use they have never been tabulated for use at spacing greater than 8 ft on center. There is no arching formula that says this is the correct or maximum distance. The bank of soil in front of the theoretical arch can also have an effect on the spacing decision, because that soil is not supported by the arch and can still ravel out or spall off. The farther apart the supports, the deeper the unsupported tension area (Fig. 6.13). A large spall could injure workers, and raveling and small spalls are distracting and worrisome to workers in the trench. Another factor is that the greater the distance between the supports, the greater the load on the support. The strength of 2-in aluminum hydraulic cylinders used as trench jacks is the limiting factor on jack spacing.

References

Bowels, Joseph E., *Engineering Properties of Soils and Their Measurement*, McGraw-Hill, New York, 1986.

Caquot, A., and Kerisel, J., *Tables for the Calculation of Passive Pressure, Active Pressure, and Bearing Capacity of Foundations*, Gauthier-Villars, Paris, 1948.

Chouery, Farid, *Slip Surface by Variation for Smooth Wall*, Structural and Foundation Engineer, FAC Systems Inc., Seattle, Wash., 2006.

Chouery, Farid, *Variational Method in Deriving* K_o, FAC Systems Inc., Seattle, Wash., 2006.

NAVFAC, *Foundation and Earth Structures*, Design Manual 7.02, pp. 7.2.61–7.2.67, Naval Facilities Engineering Command, Alexandria, Va., 1971.

Occupational Safety and Health Administration, "Regulatory Review of 29 CFR 1926, Subpart P: Excavations," *Federal Register,* March 2007.

Peck, R. B., Hanson, W. E., and Thornburn T.H., *Foundation Engineering,* John Wiley & Sons, Inc., New York, 1974.

Sowers, George B., and Sowers, George F., *Introductory Soil Mechanics and Foundations,* Macmillan, New York, 1970.

Terzaghi, K., and Peck, R., *Soil Mechanics in Engineering Practice,* John Wiley & Sons, Inc., New York, 1967.

Terzaghi, K., Peck, R., and Mesri, G., *Soil Mechanics in Engineering Practice,* 3d ed., John Wiley & Sons, Inc., New York, 1996.

Yokel, Felix Y., and Stanevich, Ronald L., *Development of Draft Construction Safety Standards for Excavations,* vol. 1, National Bureau of Standards, Department of Commerce, Washington, April 1983.

CHAPTER 7

Surcharge Loading, Base Stability, and Surface Settlement

7.1 Surcharge Loads

Shoring systems resist three major types of loads: soil, water, and surcharge loads. There are other types of lateral loads that might occur such as ice, soil swelling from expansive clays, thermal expansion, pressure from compaction, seismic forces, pressures generated from lateral movement of nearby structures, and horizontal jacking pressures. Due to the short-term nature of shoring systems, the possible occurrence of these forces is not routinely included in their design; however, if the shoring is going to be in the ground for a long time or it is obvious that the construction process, backfill compaction, microtunnel thrust backboards, etc., will involve these forces, then they should be included in the design.

Manufactured shoring equipment is designed and tabulated around anticipated soil and surcharge loads. Water loading, or hydrostatic loading, is never included in their tabulation because typical manufactured shoring equipment (trench jacks, shoring boxes, slide rail, and hydraulic equipment) does not hold back water under normal circumstances. Water flows out below, around, or through this type of shoring equipment. The assumption is that always the water table will be maintained below the bottom of the shoring. Even though there may be seepage of water through or under the equipment, the seepage forces are generally neglected by the manufacturer and the competent person. Typically the equipment tabulation gives allowable usage depths in OSHA classified soils, types A, B, C-60, and C, with a note that requires the competent person to evaluate surcharge loads of spoil piles or construction equipment greater than 20,000 lb. The tabulated data also give an allowable load rating in pounds per square foot, or maximum strut or cylinder load in pounds. Use of the soil classification method is generally intended for the competent person, and the maximum load rating is intended for the shoring design engineer to use in designing a shoring system that utilizes that specific piece of equipment. One method or the other can be used, and there is no requirement that both conditions—allowable depth in soils and maximum psf rating—be satisfied. The soil classification use method is unclear and confusing because the manufacturer does not state how much surcharge load was considered in the tabulation formula, and the competent person is given little or no instruction in the data on how to evaluate surcharge loads.

In OSHA tabulations for worker protection, there are references to surcharge loads, but they give little instruction on how to evaluate them. With open cut systems (Appendix B) the reference to surcharge loads is as follows:

> *Subpart P Appendix B, Sloping and Benching (c)(3)(iii)* When surcharge loads from stored material or equipment, operating equipment, or traffic are present, a competent person shall determine the degree to which the actual slope must be reduced below the maximum allowable slope, and shall assure that such reduction is achieved. Surcharge loads from adjacent structures shall be evaluated in accordance with 1926.651(i).

This particular note says nothing about weight, configuration, or proximity to the excavation. The competent person is given a requirement with absolutely no guidelines about how to carry it out. On the subject of proximity, an additional element of confusion is present here because in shored trenches it is generally thought that surface encumbrances that are farther away than the depth of the excavation have little effect on the excavation. However, with sloped excavation walls 1:1 and greater, the surcharges loads by necessity fall into that category. The reference to 1926.651(i) is with regard to existing structures and leads to a requirement that an engineer evaluate their effects. Surcharge loads do have an effect on slopes. See Chap. 8 for more on slope stability and open cut protection systems. In timber shoring, OSHA Appendix C, the references to surcharge are as follows:

> Appendix C to Subpart P of Part 1926—Timber Shoring for Trenches—(2) *Limitation of application.* (ii)
>
> (A) When loads imposed by structures or by stored material adjacent to the trench weigh in excess of the load imposed by a two-foot soil surcharge. The term "adjacent" as used here means the area within a horizontal distance from the edge of the trench equal to the depth of the trench.
>
> and
>
> (C) When surcharge loads are present from equipment weighing in excess of 20,000 pounds.

In the first reference there is information about load and proximity but nothing on the configuration, weight of the spoil pile, or how far it is set back from the trench. In the tabulation the lateral force effect of the spoil pile is defined as 72 psf. There is still confusion about the configuration of the spoil pile: Is it 2 ft high, or is it any height spoil pile set back 2 ft from the edge of the excavation? Due to setback requirements to keep soil from rolling down the bank of an open cut or shored trench, it is sometimes assumed that this spoil pile is set back 2 ft. The definition of the term *adjacent*, as it applies to this appendix only, eliminates that possibility unless the trench is only 2 ft deep. If the spoil pile is 2 ft high, there is still no indication about how far away from the edge the pile extends. In the second reference there is still no information about the setback and configuration of the equipment.

In aluminum hydraulic shoring (Appendix D), the only reference to surcharge is this:

> (C) When surcharge loads are present from equipment weighing in excess of 20,000 pounds.

Again there is nothing about proximity or configuration of the equipment.

In all these tabulations there is nothing said about traffic loads, railroads, cranes, K-Rail, trees, and the spectrum of construction equipment that operate adjacent to excavations. Aside from rendering the tabulations useless and requiring the competent person to take the problem to an engineer, there is no guidance about estimating lateral loads or setback distances that will reduce the lateral loads, and there is still the lingering question about how much surcharge load was included in the design of the shoring equipment.

In practice all the above-mentioned limitations are ignored to some extent by everyone involved. Turning a blind eye to the subtleties and confusion on the subject has not led to a spate of accidents; if it had, OSHA would have done more to clarify the subject; however, more practical information and knowledge would make the job of the competent person a lot easier. The fact is that calculating surcharge effects is inexact, complicated, and time-consuming. However, once the calculation work is done, it is easy to understand how surcharge loads affect the loading on the shoring system. This section will provide a method for calculating the lateral pressures from surcharge loads, how to control their effect on the shoring through location and setback, and guidance for the competent person on how to make decisions regarding surcharge and tabulated shoring equipment.

7.1.1 Calculating Surcharge Loads

Surcharge loads are due to the effect of things placed on the surface of the earth. Unlike a mass of soil, they have a definable length, width, and height. They transmit loads through the soil with little regard to water table, soil type, relative density, or consistency—at least the formulas used to calculate these loads work better with that assumption. In fact soil properties do have an effect, but there is little agreement on how to calculate it. Generally soft, loose, and gravely soils transmit greater surcharge pressure than dense fine-grained soil, cohesive or noncohesive. Even though there are always variations in the soil mass (layers of different soil types, range in particle size, moisture content, etc.), in surcharge calculations, once the prominent soil type is determined, it is assumed that the soil is infinitely the same in all directions. For the purposes of shoring design it is more important to have a reasonable value for the surcharge loading and an idea about the sensitivity of that number.

The generally accepted equation for calculating surcharge loads is accredited to Joseph Valentin Boussinesq (1842–1929), a French mathematician. The equation is referred to as the *Boussinesq equation* and is

$$\sigma_r = \frac{P}{2\pi}\left[\frac{3r^2z}{R^5} - \frac{1-2\mu}{R(R+z)}\right] \tag{7.1}$$

where s_1 = lateral force at some distance away resulting from a point load on surface of earth, psf

P = point load on surface of earth, lb

r = radial distance, ft

$$r = \sqrt{x^2 + y^2}$$

R = radial hypotenuse

$$R = \sqrt{r^2 + z^2}$$

x = surface coordinate, ft

y = surface coordinate, ft

z = depth coordinate, ft

m = Poisson's ratio, ratio of axial compression to volumetric expansion. In cases where the surcharge area is considered infinite in one direction, use m'

Figure 7.1 shows the geometric parameters for Eq. (7.1). Any consistent set of units will work.

Poisson's ratio m is the only factor in the Boussinesq equation associated with the soil properties; the rest are strictly geometric. Simply stated, m has to do with the idea that when a cube is compressed on two sides, the other sides expand, as if one were squeezing a marshmallow. In the case where there is an infinite amount of soil in one direction, such as parallel to a shoring wall, there can only

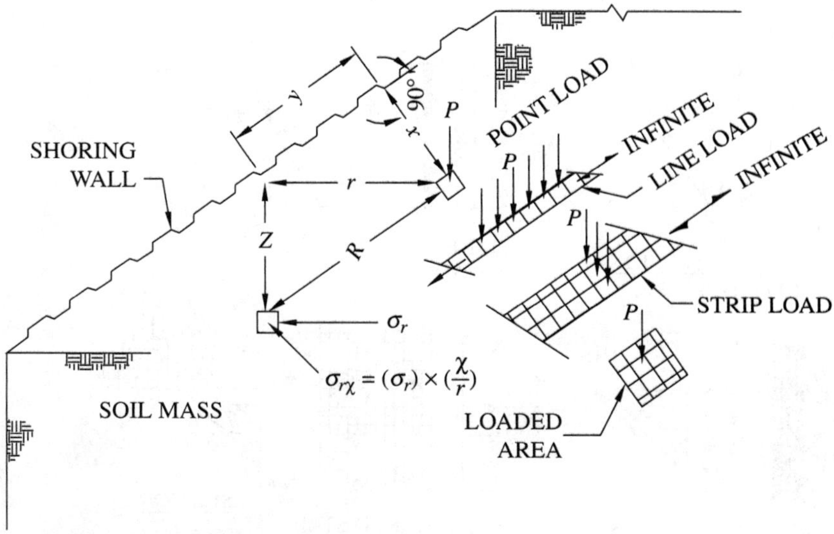

FIGURE 7.1 Boussinesq geometric parameters.

be expansion in one direction, sometimes referred to as *plane strain*. In this case plane strain is

$$\mu' = \frac{\mu}{1-\mu} \tag{7.2}$$

Table 7.1 gives commonly used values for *m* and *m′* for soils.

Figure 7.2(a), (e), and (f) would use *m* and the rest would use *m′*. In another special case, the surcharge load is considered to be infinite in both the direction perpendicular to the wall and parallel to the wall. In this case the surcharge load can be treated as if it were soil with the unit weight of the surcharge, and the normal Rankine and Coulomb methods for calculating lateral pressure can be used instead of the Boussinseq equation. If $m = 0.5$ which is normal for most clay soils, the third factor in the Boussinesq equation, $1 - 2\mu/[R(R+z)]$, becomes zero, leaving $\sigma_r = 3Pr^2z/2\pi R^5$, making the equation easier to solve on a hand calculator.

Soil Type	Volumetric Strain μ	Plane Strain μ'
Moist clay soils	0.4–0.5	0.67–1
Saturated clay soils	0.45–0.5	0.82–1
Cohesionless, medium, and dense	0.3–0.4	0.42–0.67
Cohesionless, loose to medium	0.2–0.35	0.25–0.54

TABLE 7.1 Common Values of a Poisson's Ratio for Soils

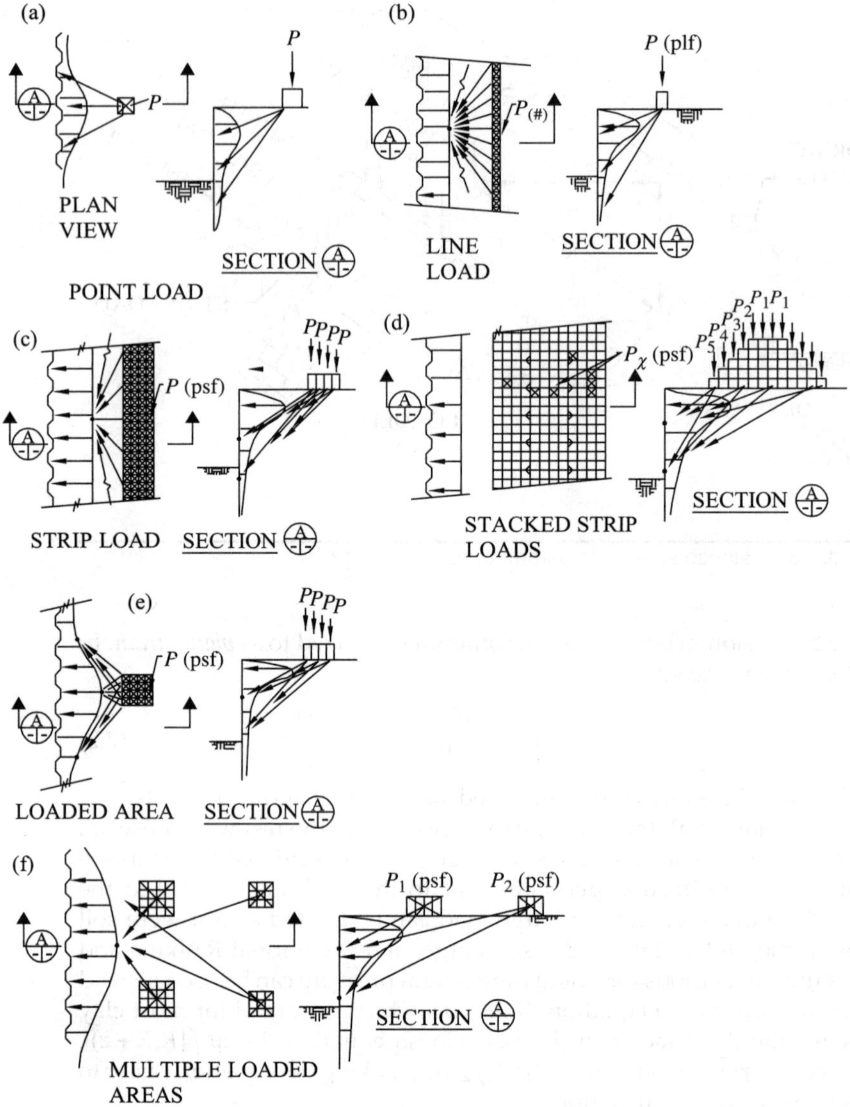

FIGURE 7.2 Common construction loading surcharge configurations.

As shown in Fig. 7.2(a), the Boussinesq equation gives the lateral stress at a point below the surface resulting from a point load on the surface. To get a profile of the loading on the wall from the surface to some distance below the excavation line, it is necessary to solve the equation for all locations of interest. If a load profile from surface to 40 ft deep is needed, then the equation needs to be solved 40 times and the results plotted. If there is another point load in the vicinity, it will also have an effect on the wall at the points just calculated, adding another 40 Boussinesq calculations and 40 additions to get the loading on that line. By positioning point loads and adding up the cumulative effects, the distribution of lateral loading at any point on the wall can be obtained. Figure 7.2(a) to (f) shows common cases encountered in construction problems.

These problems are fairly easy to set up and calculate on a spreadsheet program using feet for distance and pounds or kips per square foot as the load, or any equivalent set of dimensions. As a general rule, surcharge loads more than the depth of the excavation away from the edge are ignored, and normal construction equipment loads become less than 75 psf after approximately 20 ft of depth.. For practical purposes the effect of surcharge loading usually becomes insignificant at a distance in any direction approximately 2 times the depth of the excavation. There are several programs on the market that calculate surcharge loading; however, one of the advantages of setting it up on a spreadsheet is that the person setting it up has a better understanding of what went into getting the results and how to model specific problems.

7.1.2 Surcharge Load Cases

Surcharge loading problems come to the competent person or engineer in two forms:

1. The load and position are fixed such as the location of traffic, railroads, footings, and trees.
2. The size of the load is adjustable, and the location can be spotted so that it meets the criteria of the shoring strength. Cranes, concrete trucks, loaders, spoil piles, etc., can be set back a distance that will prevent the surcharge loading from exceeding the strength of the shoring. Crane pads and wheel locations can also be located at the corners or strut locations of shored walls so that bending and deflection are minimized at the center of the span.

In the first case, the shoring capacity has to be designed to include the surcharge loads delivered. Where the shoring has a fixed capacity such as a shoring shield or trench jacks, the surcharge load has to be subtracted from the capacity, and the strength that is left can be used to resist the soil. The result is that allowable shoring shield depth or the trench jack spacing is reduced. In the second case, the load can be moved away from the excavation until the load is within the capacity of the shoring.

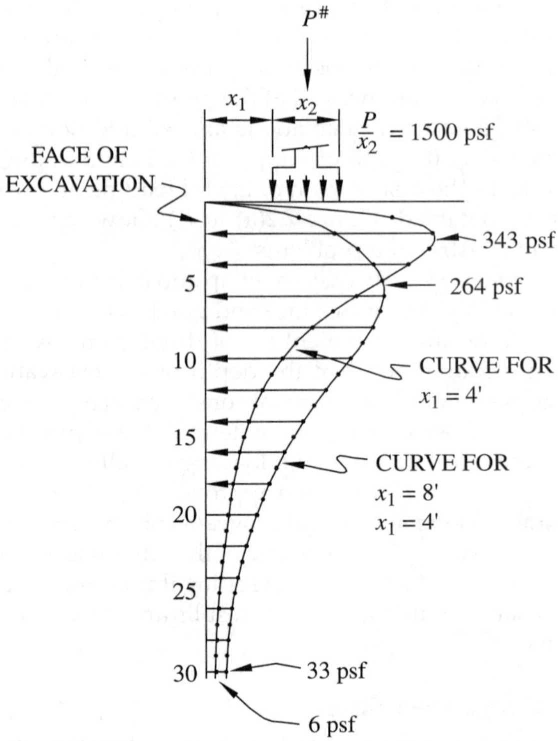

FIGURE 7.3 Surcharge loading curves for 4-ft-wide × 1500 psf footing. Setback = 4 ft and 8 ft.

In the load cases shown below a setback that will result in less than a 100 psf loading on the shoring is used, which is the normal amount of surcharge load allowed in the tabulated data. There is also a calculation that shows the surcharge load when it is placed a normal construction distance, 2 to 4 ft back from the shoring wall.

In tabulating and working with the results of surcharge calculations, it is common to flatten off the curves into average rectangular loads where the resultant is equivalent to the resultant area of the curve on the theory that the shoring will distribute the spiked part of the load over the rest of the system. This is a reasonable assumption except in the case of horizontal wood lagging because the lagging cannot distribute loads to the lagging pieces above and below, and also in the case of trench jacks because in soft clays and loose sands and gravels, spike loads can exceed the arching capacity of the soil. A very convenient mitigating factor for surcharge loading is the fact that soil loading increases with depth of excavation while surcharge loading does the opposite. Surcharge loads are P-shaped (Fig. 7.3), resulting in high loads near the surface and smaller loads after about

20 ft deep. The closer the surcharge is to the edge of the excavation, the closer to the surface the high point of the load is. As a general rule, in the top 5 to 10 ft set the rectangular loading so that the spike load does not exceed it by more than 30 percent, and set the 10- to 20-ft depth rectangular loading so that the spike load does not exceed it by more than 50 percent, and ignore deep loads less than 75 psf.

The following load cases were calculated using Eq. (7.1) and illustrate these principles. The specific example is given to show the layout and assumptions, and the tables that follow were developed by changing variables in the original assumptions.

Foundation Loads

Most foundation design parameters do not allow more than 1500 psf dead load to the soil. Spread perimeter footings are usually 2 to 3 ft wide, and column footings for tilt-up concrete and low-rise steel structures average 3 to 6 ft square. Structures founded on piles do not deliver appreciable lateral loads to shoring walls. If the soils below the foundation are soft clays or loose sands and gravels, structures will be founded on piles. It is a reasonable assumption to use $m = 0.4$, resulting in a plane strain of $m' = 0.67$ (Table 7.1). Table 7.2 shows the results of various footing setbacks and widths for continuous footings and Table 7.3 shows the results for square footings.

Excavation Spoil Pile

Pipeline excavation operations in cross-country (Fig. 7.4), commercial building projects, new subdivision infrastructure, and small pothole-type excavations are the most likely to generate spoil piles at the side of the excavation. In excavations in city streets and underground structure excavations such as pump stations, water treatment plant work, and basements, the spoils generated are usually hauled off or shuffled by loader to piles that are set a large distance away from the excavation. OSHA requires that the spoil pile be set back a minimum of 2 ft from the edge, but does not set limits on the height or length of the pile. The width of the pile is generally governed by the height, normally the result of a 1:1 slope. In the case of an infinitely long spoil pile, the problem becomes plane strain and requires the use of m', where in a fixed short-length problem, say less than 30 ft, m should be used. Calculations indicate that for this problem the difference in going from $m = 0.5$ to $m' = 0.67$ is an approximately 25 percent increase. Table 7.4 was calculated using $m' = 0.67$.

Concrete Truck Perpendicular to Excavation Wall

Placing concrete in structure excavations is normally done through a pump truck and does not necessarily require that the concrete truck and a pump truck be close to the edge of the excavation. The use of

Poisson's Ratio[§] μ'	Surcharge Load[*] (psf)	Setback from Shoring Wall (ft)	Footing Width (ft)	Lateral Surcharge Load[†] (psf) s_1	s_2	s_3	s_4	s_5
0.67	1500	2	2	250	56	20	5	5
		4		180	85	35	10	5
		6		130	95	45	15	10
0.67	1500	2	3	340	95	36	10	10
		4		250	130	55	20	10
		6		180	140	70	25	15
0.67	1500	2	4	420	140	55	15	10
		4		300	175	80	25	15
		6		220	180	100	30	15
		Setback for 100 psf max.						
0.67	1500	8	2	95	90	50	20	10
0.67	1500	10	3	100	100	80	30	20
0.67	1500	14	4	80	100	100	50	35

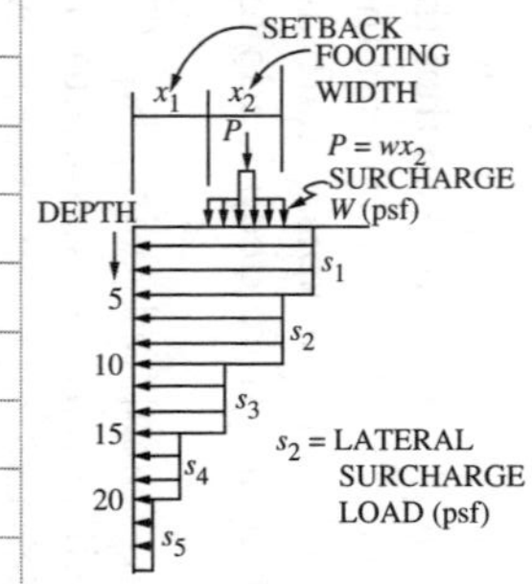

[*] For soft clay and loose sand and gravel increase surcharge loading by 20%.

[†] For higher or lower surcharge loads use direct proportion to get new lateral surcharge. For example, for 1200 psf load on 3-ft-wide footing setback 4 ft-S_1= 1200/1500 × 250 = 200 psf.

[‡] Table is based on Boussinesq equation $\sigma_r = \frac{P}{2\pi}\left[\frac{3r^2z}{R^5} - \frac{1-2\mu}{R(R+z)}\right] \quad \mu' = \frac{\mu}{1-\mu}$.

[§] μ' is used for plane strain cases such as continuous footings. Footing lengths used in this table are 60 ft.

TABLE 7.2 Surcharge Loading for Continuous Footing[‡]

Poisson's Ratio[§] μ	Surcharge Load[*] (psf)	Setback from Shoring Wall (ft)	Footing Width × Length (ft)	Lateral Surcharge Load[†] (psf) s_1	s_2	s_3	s_4	s_5
0.50	1500	2	4 × 4	190	25	5	0	0
		4		90	30	10	0	0
		6		50	30	10	0	0
0.50	1500	2	6 × 6	300	60	15	5	0
		4		160	70	20	10	0
		6		90	70	25	10	0
0.50	1500	2	8 × 8	375	110	30	10	5
		4		210	115	40	15	5
		6		125	100	50	20	10
		Setback for 100 psf max.						
0.50	1500	8	4 × 4	90	30	10	0	0
0.50	1500	10	6 × 6	90	70	25	10	0
0.50	1500	14	8 × 8	75	85	50	25	10

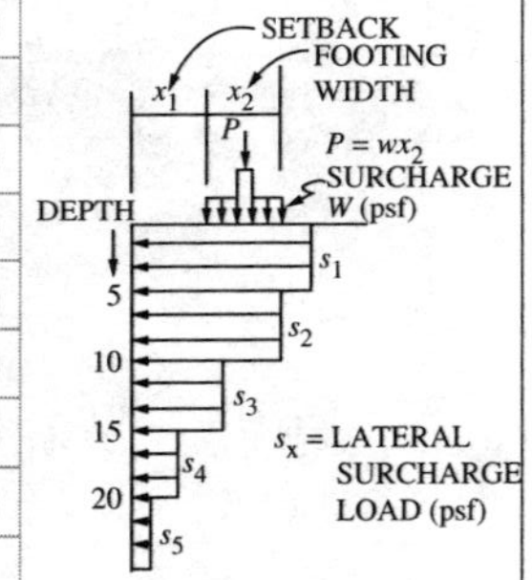

[*] For soft clay and loose sand and gravel, increase surcharge loading by 20%.

[†] For higher or lower surcharge loads use direct proportion to get new lateral surcharge. For example, for 1200 psf load on 3-ft-wide footing setback 4 ft-S_1 = 1200/1500 × 250 = 200 psf.

[‡] Table is based on Boussinesq equation $\sigma_r = \frac{P}{2\pi}\left[\frac{3r^2z}{R^5} - \frac{1-2\mu}{R(R+z)}\right]$ $\quad \mu' = \frac{\mu}{1-\mu}$.

[§] μ' is used for plane strain cases such as continuous footings. Footing lengths used in this table are 60 ft.

Table 7.3 Surcharge Loading for Square Footing[‡]

FIGURE 7.4 Cross-country excavation spoil pile for approximately 16 ft deep excavation. Pile is approximately 12 ft high x 20 ft wide.

concrete as pipeline backfill and in electrical duct construction has increased significantly over the past 25 years. Concrete placement for these operations is usually done by backing the truck up close to the edge of the trench and dumping the concrete down the chute. Accidents involving concrete trucks falling into excavations where the shoring or sloping has failed are not uncommon. The results can be deadly because workers are usually close to the truck and the failure is rapid. Table 7.5 shows the vertical distribution and horizontal distribution of surcharge load from a 9 cubic yard (cy) and 12-yd concrete truck with the rear axle set back 4 ft from the edge of the trench. The table also gives the setback needed to keep the lateral surcharge below 100 psf.

HS20-44 Traffic Loading

Figure 7.5 shows the configuration for a 12-cy concrete truck with the hydraulic axle down running parallel to a shored excavation. The load is factored up by 30 percent to allow for impact loading. This loading is approximately equivalent to an HS20-44 highway load. Table 7.6 shows gives surcharge loading for traffic parallel to the excavation.

Poisson's Ratio† μ'	Surcharge Load* (psf)	Setback from Shoring Wall X_1 (ft)	Spoil Pile X_2 by Height (ft)	Lateral Surcharge Load (psf)				
				s_1	s_2	s_3	s_4	s_5
0.67	100	2	4 × 8	120	80	40	20	15
		4		90	80	50	30	20
		6		65	70	50	30	20
0.67	100	2	6 × 12	160	130	70	40	25
		4		120	120	80	50	30
		6		90	110	80	50	35
0.67	100	2	8 × 16	210	190	120	70	45
		4		160	170	120	80	50
		6		125	150	120	80	60
0.67	100	4	12 × 24	230	280	225	160	120
		6		190	250	210	160	120
		8		150	210	200	160	120
		Setback for 100 psf max.						
0.67	100	8	4 × 8	90	90	50	30	20
0.67	100	6	6 × 12	100	100	80	50	35
	100	10	8 × 16	80	115	105	80	60
0.67	100	20	12 × 24	80	100	100	50	35

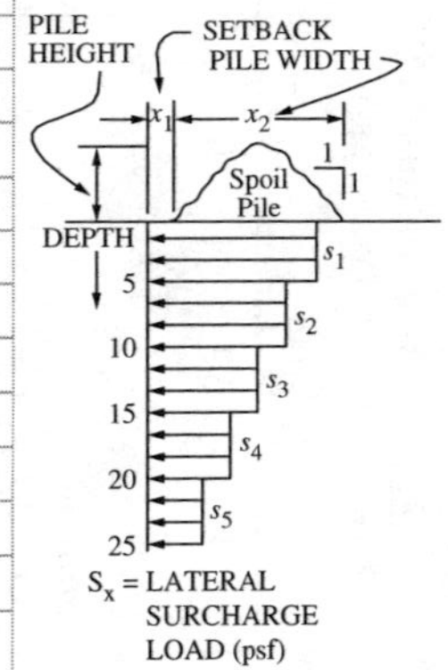

* For soft clay and loose sand and gravel, increase surcharge loading by 20%.

† Table is based on Boussinesq equation $\sigma_r = \frac{P}{2\pi}\left[\frac{3r^2 z}{R^5} - \frac{1-2\mu}{R(R+z)}\right]$ $\mu' = \frac{\mu}{1-\mu}$.

‡ μ' is used for plane strain cases such as continuous spoil piles. Spoil pile lengths used in this table are 60 ft.

TABLE 7.4 Surcharge Loading for Excavation Spoil Piles†

Poisson's Ratio μ	Rear Dual Tire Load (psf)	Dual Tire Area (ft)	Truck Yardage (cy)	Setback from Shoring Wall (ft)	Lateral Surcharge Load (psf)				
					s_1	s_2	s_3	s_4	s_5
0.50	4500	4	9	4	150	75	25	0	0
0.50	6000	4	12	4	200	100	50	0	0
μ	(psf)	(ft)	(cy)	Setback for 100 psf max.	s_1	s_2	s_3	s_4	s_5
0.50	4500	4	9	6	100	75	25	0	0
0.50	6000	4	12	7	100	80	35	0	0

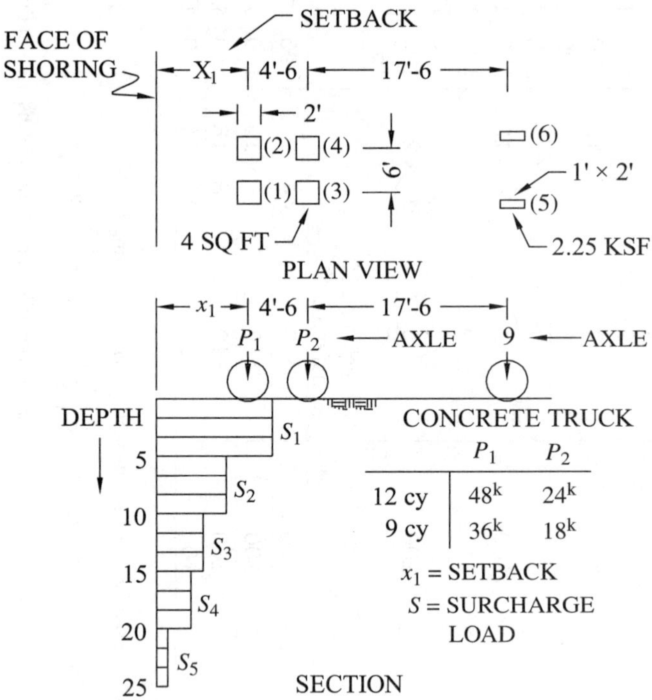

TABLE 7.5 Concrete Truck Surcharge Loading

Poisson's Ratio m	Dual Tire Load (psf)	Dual Tire Area (ft²)	Truck Axle + Impact k	Truck Setback from Shoring Wall (ft)	Lateral Surcharge Load (psf)				
					s_1	s_2	s_3	s_4	s_5
0.50	5000	2	9	4	230	108	40	20	8
m	(psf)	(ft)	k	Setback for 100 psf max.	s_1	s_2	s_3	s_4	s_5
0.50	5000	2	41.6	8	100	100	50	25	0

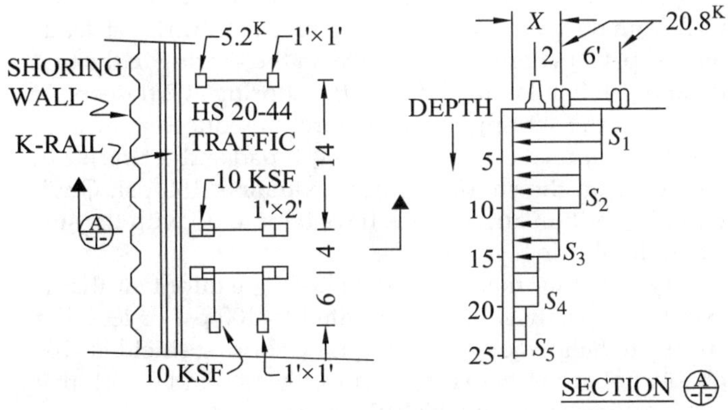

* Includes weight of K-rail.

Table 7.6 HS20-44 Traffic Loading* with Impact Factor = 0.3

7.1.3 Surcharge Load Decisions for the Competent Person

In the case of surcharge loading, the task at hand for the competent person is to make sure that the loads do not destabilize slopes or exceed the allowable loading of the shoring equipment. This is accomplished by requiring sufficient setback to limit the surcharge loading. For timber shoring, in OSHA Appendix C it is stated at the top of the tabulation that the shoring design load is calculated by using soil type load (25, 45, or 60) × depth + 75 psf for surcharge. Setting equipment back so that the average surcharge loading effect is around 75 psf ensures that the timber shoring design strength is not exceeded.

Shoring equipment such as shields, manhole boxes, trench jacks, wale rails, and slide rail systems are usually designed around the allowable strength of their materials plus a one-third increase for the assumption that short-term usage and shoring loads will lead to less overall stress on the equipment. Tabulated data providing allowable depth for this equipment are developed by subtracting 75 psf for surcharge loads from the calculated strength of the equipment before calculating the allowable depth. In OSHA tabulated data for open cut systems, timber shoring, and trench jacks, there is the statement that the data are not applicable "when surcharge loads are present from equipment weighing in excess of 20,000 lb." Manufacturers place similar language in their tabulated data. The use of the word *present* in that statement instead of a certain setback distance such as 2 ft from the edge would lead one to believe that the equipment can be set as close as possible to the trench edge without overloading the shoring equipment, and there is nothing in the statement about the distribution of the 20,000-lb load. A 20,000-lb load placed on a 2-ft × 2-ft pad set 2 ft from the edge, similar to a boom truck outrigger, leads to an average 230 psf surcharge lateral force on the shoring in the first 5 ft of depth, and a 450-lb lateral force on the shoring at 2 ft deep. At 4 ft from the face of the shoring the same force is reduced to around 100 psf. If the load is spread out over a 4-ft × 4-ft pad and set back 2 ft, the surcharge force on the shoring is approximately 150 psf. Obviously the load distribution and setback have a lot to do with the surcharge effect on the shoring.

The surcharge load on manufactured shoring equipment that is used at maximum capacity should be limited to 100 psf or less. For instance, if the stated allowable psf rating for a shoring shield is 1100 psf, it is probably not accurate to more than ±100 psf , and therefore if a 100 psf load is added, there will be little effect.

There are three other surcharge load mitigating factors that help reduce the consequences of surcharge loading on shoring systems:

1. After approximately the first 10 ft, the surcharge load decreases with depth while the soil loading is usually small in relationship to the strength of the shoring equipment being used to that depth.
2. Wheel loads from construction equipment and traffic are intense at the location of the wheel but become insignificant after approximately 10 ft in any lateral direction; therefore the overall effect is averaged out over the length of the shoring. It is reasonable to average wheel loading in the horizontal and vertical directions. A special case where this does not apply is with trench jacks since they are not laterally connected to one another.
3. Shoring equipment is not normally used at maximum allowable depth or psf rating. The residual strength not used to support the soil can be used to support surcharge effects. A shoring shield with an 1100 psf capacity used at a depth requiring 900 psf capacity has an additional 200 psf to support

Normal Construction Surcharge	Setback Required for Maximum Surcharge Load (ft)	
	$s_1 = 100$ psf, $s_2 = 50$ psf	$s_1 = 200$ psf, $s_2 = 100$ psf
K-rail	1	1
Spoil pile 2 ft high × 4 ft base	2	2
Spoil pile 4 ft high × 8 ft base	3	2
Less than 20,000 lb equipment	3	2
Backhoe 30,000 lb	4	2
Footing 4 ft square	4	3
3-cy loader	4	2
12-in round tree	4	3
10-cy dumptruck	4	3
5-cy loader	4	3
325 excavator parallel to trench	4	3
9-cy concrete truck	6	4
24-in round tree	6	6
Spoil pile 6 ft high × 12 ft base	6	2
Footing 2 ft wide × 1500 psf	8	4
Footing 3 ft wide × 1500 psf	10	6
HS-20-44 traffic with 30% impact factor, distance to closest wheel	8	4

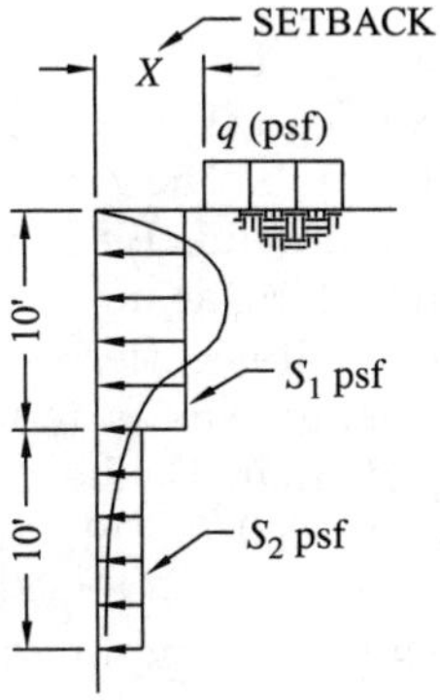

Notes
Crane and boom truck outrigger pads are not normal construction loads and should be analyzed separately.

TABLE 7.7 Setback Table to Achieve 100 psf and 200 psf Surcharge Loading on Shoring

surcharge. The competent person can factor this in when deciding about setback.

Table 7.7 gives setback distances required to maintain a 100 psf and a 200 psf surcharge effect on shoring equipment.

7.1.4 Railroad Cooper E-80 Loading

Loading from trains originates from axle loads at the engines and cars, by far the most intense loading being at the engines. Unlike highway loading where the truck tires distribute the load directly to the road surface or bridge deck, the train axles distribute the load to track and subsequently to the ties creating a distributed strip load to the ballast and soil below. Rail ties are normally 8 ft 6 in to 9 ft long, and the major engine axles are 5 ft on center. The name *Cooper* in the term "Cooper E-80 Loading" came from Peter Cooper's locomotive engine of 1830, the first steam-powered railroad locomotive made in the United States. Today a "Cooper's Train" consists of two locomotives and an undefined number of cars. The E number refers to the axle load; E-10 is 10,000 lb on the axle and E-80 is 8 times that, or 80,000 lb. The American Railroad Engineering Association (AREA) recommends E-80 loading for mainline railroad structures (Fig. 7.5). In looking at surcharge loading on shoring structures parallel to railroads, it is easier to not get caught up in the amount and spacing of the engine axles and to simply consider it as a uniformly distributed load that is the width of the ties. For Cooper E-80 this is $q = 80{,}000/(9 \times 5) = 1779$ psf for 9-ft ties and 1882 psf for 8-ft 6-in ties. For lateral loading (Fig. 7.6), the equation specified by AREA is the Boussinesq modified by experiment (after Teng)

$$P_s = \frac{2q}{\pi}(\beta - \sin\beta\cos 2\alpha) \tag{7.3}$$

where P_s = lateral pressure at point of interest

$$\phi = A\tan\frac{x_1}{H_2}$$

$$\beta = A\tan\frac{x_2}{H_2} - \phi$$

$$\alpha = \phi + \frac{\beta}{2}$$

α and β in radians

$$\text{Radians} = \frac{\text{degrees} \times \pi}{180}$$

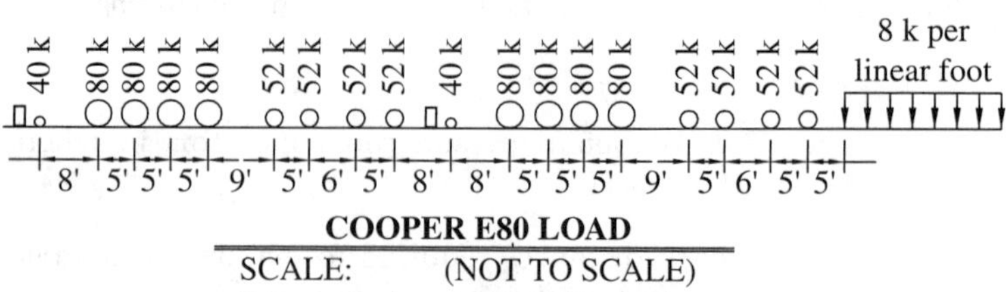

Figure 7.5 Cooper E80 track loading.

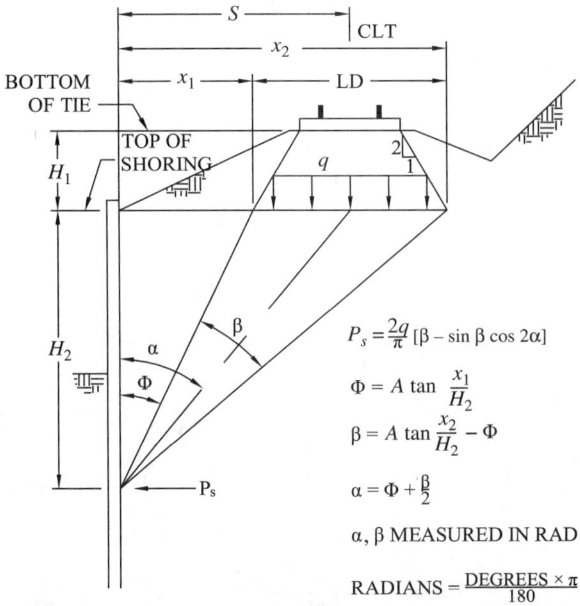

FIGURE 7.6 Surcharge loading from railroad.

$$q = \frac{P}{LD \times AS}$$

where P = Cooper E axle load
LD = tie length + H_1
AS = axle spacing

Most railroad companies and agencies have adopted this formulation and publish it in their requirements for temporary shoring, usually along with tables and graphs for the case where H_1, the distance from the bottom of the tie to the top of the shoring, is equal to zero. In actuality there is always approximately 2 ft of ballast under the ties and some distance from the bottom of the ballast to the top of the shoring. If that combined distance is less than 3 ft, it is reasonable to assume that the calculated tables are within a reasonable margin of error. The equation is easy to set up using a spreadsheet program, and once it is set up, it can be used for any configuration. Figure 7.7 shows the results of a calculation for 7.5 ft from the centerline of the track. Due to the high lateral pressures generated the pressure curve should not be averaged to less than 10 percent of the peak, and the averaged areas should exceed the total area under the curve.

There are other elements of railroad loading that are not usually discussed in the railroad shoring requirements. These items may not

Cooper E-80 Loading **CLT = 7.5' H_1 = 0'**

CLT = S	7.5 ft
Axle load = P =	80,000 lb
Axle spacing = AS	5 ft
Tie length = TL	9 ft
$x_1 = S - (TL + H_1)/2$	3
$x_2 = x_1 + TL + H_1$	12
$x_2 - x_1 = Ld$	9
$H_1 =$	0
$q = \dfrac{P}{LD \times AS} =$	1778 lb

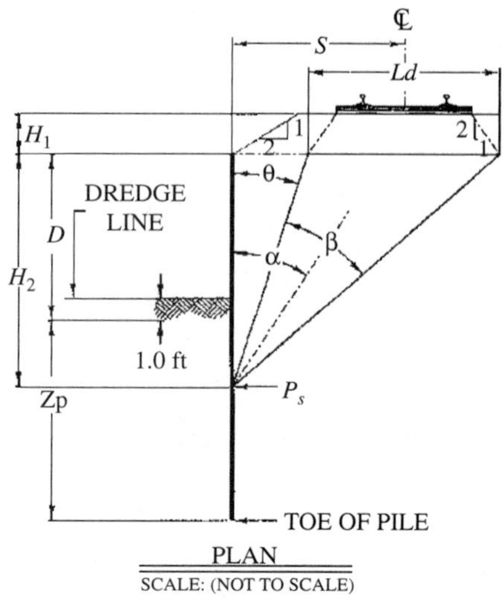

$$P_s = \frac{2q}{\pi}(\beta - \sin\beta\cos 2\alpha)$$

$$\phi = A\tan\frac{x_1}{H_2}$$

$$\beta = A\tan\frac{x_2}{H_2} - \phi$$

$$\alpha = \phi + \frac{\beta}{2}$$

α and β in radians $\qquad$ $\text{Radians} = \dfrac{\text{degrees} \times \pi}{180}$

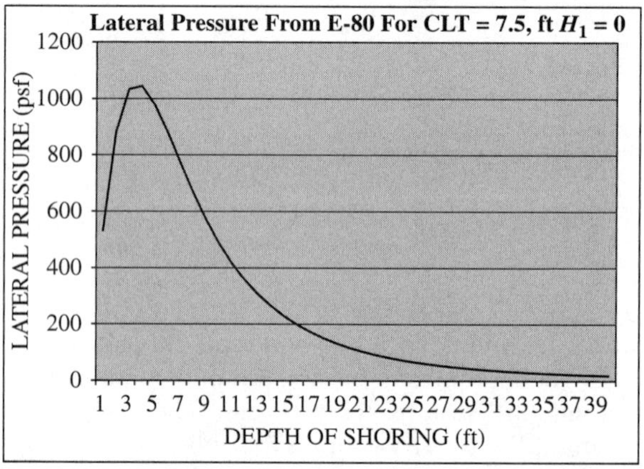

FIGURE 7.7 Surcharge loading for Cooper E-80 loading.

be significant; however, even not knowing what thoughts went into the adoption of the formula by AREA, the shoring designer must still be aware that they exist.

- If there is more than one set of tracks, there is the possibility that two engines will line up on the tracks directly in front of the shoring, and for sure two train cars will pass each other in front of the shoring. The full Cooper E-80 loadings for each set of tracks should be added together and used in the shoring design.

- On curved tracks there is an additional lateral load on the track and ties that is resisted by the soil and delivered to the shoring. Given train speed and long curves, this is most likely negligible until the track is very close to the shoring.
- There is an additional impact force that is taken up by the soil and transmitted to the shoring. As the impact force is dissipated perpendicular to the train on both sides, the shoring side would see one-half of the impact loading that a bridge would have to carry. On a bridge less than 80 ft long, the railroad impact force as a percent of the live load is $I = 40 - 3l^2/1600$. If the shoring wall is considered similar to a bridge, a 60-ft line of shoring would see an impact of roughly a 15 percent increase in lateral force.
- The Boussinesq equation, Eq. (7.3), does not have an input for Poisson's ratio m. When Eq. (7.1) is set up and solved for a strip load using $m' = 1$, the results are very close to the results of Eq. (6.43). Therefore it seems that one could conclude that Eq. (7.3) does consider the plane strain case for a long strip load using $\mu' = 1$. Research by Rehnman and Broms (1972) indicates that when loose sands and gravels are present, as is often the case along riverbeds, the horizontal surcharge load is higher than with dense sands and gravels and that gravely backfill generates higher pressures than fine-grained soils. It also makes sense that soft clays would transmit higher pressures to the shoring wall than stiff clays because soft clays act more as water does, transmitting equal pressures in all directions.

Perhaps the largest mitigating factor to these potential imperfections in the formula is that it seems to have survived the test of time. If there were evident problems, AREA would have altered the formula. Also the Cooper E-80 engine is said to be a rare animal on the United States railroad trackway; E-60 to E-70 is said to be more common on mainline tracks.

Lateral Surcharge Pressure on Shoring Walls Perpendicular to the Track

Virtually all utilities that cross under railroad tracks are installed inside a steel pipe casing. The casing is bored and jacked or microtunneled under the tracks, and then the utility is pushed inside the casing and the annular space is filled with cement grout. The bore pits that are used to install the casings have walls parallel and perpendicular to the track (Fig. 7.8). The surcharge loading on the walls

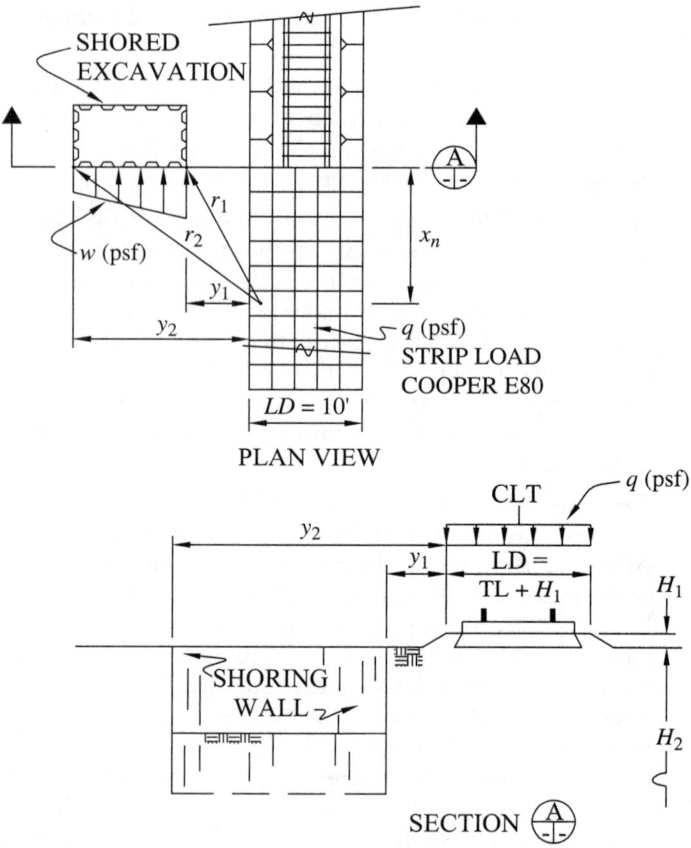

FIGURE 7.8 Cooper E80 load on wall perpendicular to tracks.

perpendicular to the tracks is different from the loading on the shoring walls parallel to the tracks. The AREA approach to solving this problem is given by

$$P_s = K_a \times q \tag{7.4}$$

where P_s = surcharge pressure at point of interest, psf
K_a = Rankine active soil coefficient

$$K_a = \tan^2\left(45 - \frac{\phi}{2}\right)$$

ϕ = angle of internal friction, deg

$$q = \frac{P}{LD \times AS}$$

P = Cooper E axle load
LD = tie length + H_1
AS = axle spacing

CLT	Variable	
Axle load = AL	80,000	lb
Axle space = AS	5	ft
Tie length = TL	9	ft
H_1	1	ft
LD = TL + H_1	10	ft
q = AL/(AS + LD)	1600	psf

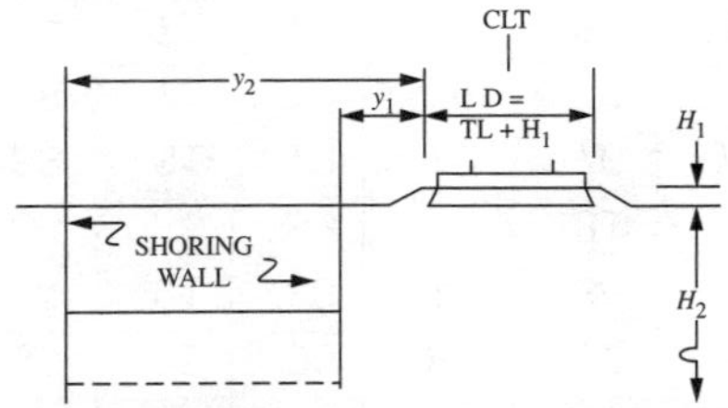

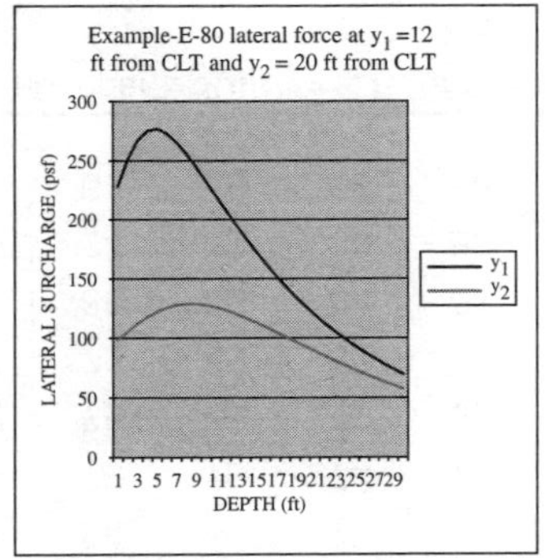

	Distance from CLT or Edge of Track to Point of Interest on Shoring Line																
	CLT	10	11	12	13	14	15	16	17	18	19	20	21	22	23	24	25
	y_1	5	6	7	8	9	10	11	12	13	14	15	16	17	18	19	20
	H_2	Surcharge Pressure at Depth H_2 *(psf)															
Depth (ft)	1	328	270	228	197	172	157	151	135	121	109	98	89	82	75	69	63
	2	361	298	251	216	187	169	165	146	130	117	106	96	87	80	73	67
	3	376	314	267	230	200	178	176	156	139	124	112	102	92	84	77	71
	4	375	320	275	239	209	185	184	163	146	131	118	107	97	88	81	74
	5	365	317	277	242	214	189	189	168	151	136	122	111	101	92	84	77
	6	349	309	273	242	215	191	192	171	154	139	126	114	104	95	87	80
	7	330	296	265	238	213	190	191	172	156	141	128	116	106	97	89	82
	8	311	282	256	231	209	187	189	171	156	141	129	118	107	98	90	83
	9	291	267	244	223	203	182	185	169	154	141	129	118	108	99	91	84
	10	272	252	232	214	196	176	180	165	152	139	128	117	108	99	92	85

TABLE 7.8 Surcharge Loading on Shoring Line Perpendicular to Track—Cooper E-80, H_1 = 1 ft

	Distance from CLT or Edge of Track to Point of Interest on Shoring Line															
CLT	10	11	12	13	14	15	16	17	18	19	20	21	22	23	24	25
y_1	5	6	7	8	9	10	11	12	13	14	15	16	17	18	19	20
H_2	Surcharge Pressure at Depth H_2 *(psf)															
11	254	237	220	204	189	162	174	161	148	137	126	116	107	99	92	85
12	237	222	208	194	180	156	168	155	144	133	123	114	106	98	91	84
13	221	208	196	184	172	149	161	150	139	130	120	112	104	97	90	84
14	206	195	184	174	163	142	153	144	134	125	117	109	102	95	88	82
15	192	183	173	164	155	135	146	137	129	121	113	106	99	93	87	81
16	179	171	163	155	147	129	139	131	124	116	109	103	96	90	85	79
17	167	160	153	146	139	122	132	125	118	112	105	99	93	88	83	78
18	156	150	144	138	131	115	125	119	113	107	101	96	90	85	80	76
19	146	141	135	130	124	109	119	113	108	102	97	92	87	82	78	73
20	136	132	127	122	117	103	112	107	102	98	93	88	84	79	75	71
21	128	124	119	115	111	98	106	102	97	93	89	85	80	76	73	69
22	120	116	112	108	105	92	101	97	93	89	85	81	77	74	70	67
23	112	109	106	102	99	87	95	92	88	84	81	77	74	71	67	64
24	105	102	99	96	93	82	90	87	84	80	77	74	71	68	65	62
25	99	96	94	91	88	78	85	82	79	76	74	71	68	65	62	59
26	93	90	88	86	83	74	81	78	75	73	70	67	65	62	60	57
27	87	85	83	81	79	70	76	74	72	69	67	64	62	60	57	55
28	82	80	78	76	74	66	72	70	68	66	64	61	59	57	55	53
29	77	76	74	72	70	62	68	66	64	62	60	58	56	55	53	51
30	73	71	70	68	66	59	65	63	61	59	58	56	54	52	50	48

TABLE 7.8 Surcharge Loading on Shoring Line Perpendicular to Track—Cooper E-80, $H_1 = 1$ ft (*Continued*)

This approach is quick and simple. With $K_a = 0.3$, the result is a rectangular load distribution with $P_s = 533$ psf. The problem is that the formulation does not consider depth or distance away from the track.

Another approach is to use Eq. (7.1) and sum up a series of point loads over an area of LD × infinity (after 40 ft back the influence is negligible), as shown in Fig. 7.8. Table 7.8 can be used to determine the surcharge load for various distances from the track and depths when $H_1 = 1'$. Poisson's ratio $m' = 1$ was used in the analysis.

7.2 Base Stability of Excavations and Surface Settlement

Removal of the earth from a location will always result in movement of the surrounding ground. The walls of the excavation will move toward the center, the surface will settle, and the bottom will move upward due to elastic rebound and possible heave from the inability of the excavation base soil strength to hold up the weight of the surrounding earth walls. The extent of these movements is based on the strength properties of the soil. If water is present, seepage forces can transport soil particles toward the shoring walls and up through the bottom. Ground loss in one area results in settlement at the surface and ground increase in the walls or bottom of the excavation. Prediction of these movements—determining how much movement is allowable, how to control them, and the associated cost—is part of the analysis needed to decide on a shoring system.

Surface settlement will be the combined result of wall movement and base uplift. Wall movement is controlled by the stiffness of the shoring system. Base uplift can be controlled by the stiffness and depth of sheeting below the bottom of the excavation and by changing the weight and strength of the base material, such as adding a concrete plug. Both wall movement and base uplift can also be highly dependent on quality and timing of the construction procedure.

There are three major types of base stability problems in excavation work:

1. Structural base deterioration during the construction process
2. Base instability in granular soils due to seepage and high water table
3. Bottom heave due to soft cohesive soils

The approach to determining base stability is to divide the forces that resist failure by the forces that contribute to failure to arrive at a factor of safety. Generally a factor less than 1 predicts imminent failure, and 1.5 is an acceptable risk. This is only as accurate as the underlying assumptions. Variations in weight and strength properties of the soil are easy to analyze and judge sensitivity; however, water table fluctuations and surcharge limitations can sometimes be hard to predict and control. Pumps fail, rainstorms and river levels are unpredictable, cranes and material stockpiles are moved in after the shoring job without regard to the shoring design assumptions, as the factor of safety decreases the anticipated load on the shoring system increases. All these factors can add up to disastrous failure of the shoring system and should be carefully considered in the analysis.

7.2.1 Structural Base Deterioration during the Construction Process

The ground surface at the bottom of an excavation has to be stable so that production work can take place. The exact grade has to be maintained, and the base has to be solid. Water, wind, and heat over time will always have a detrimental effect on an earth surface. Workers and equipment moving on the surface will have an effect on the surface. Wet ground also presents a safety risk to workers. The tried and true solution to base deterioration problems is to maintain the water table a minimum of 2 ft below the bottom of the production work and add a minimum of 6 in of crushed rock over the ground surface. A third element to this solution is to provide drainage sloping and a sump pumping pit so that surface drainage and rainfall can be removed. In the case of microtunnel work pits where weeks and months of construction activity will take place, a 12-in-thick concrete work slab is advisable. In deep excavations often the base work slab can act as the bottom wale of the shoring system and add some protection from bottom heave. Whether these base stabilization elements are shown on the project plans or not, the estimator should get ample amounts of money in the estimate to cover them. In construction, especially underground, a safe, clean, and stable working environment will always pay for itself.

7.2.2 Bottom Stability in Noncohesive Soils

Base instability problems in granular soils can be broken down into three different cases:

1. Water table at excavation depth below the base
2. Water table at the base
3. Water table above the base

Figure 7.9(a) and (b) shows cases 1 and 2. Equation (7.5), proposed by Terzahgi and Peck (1967) show a method for calculating base stability in a granular soil for these two cases. This equation applies only when the sheeting extends only to the bottom of the excavation. The factor of safety can be increased by extending sheeting below the base of the excavation.

$$F_s = 2N\gamma_2\left(\frac{\bar{\gamma}_2}{\gamma_1}\right)k_a \tan\phi \qquad (7.5)$$

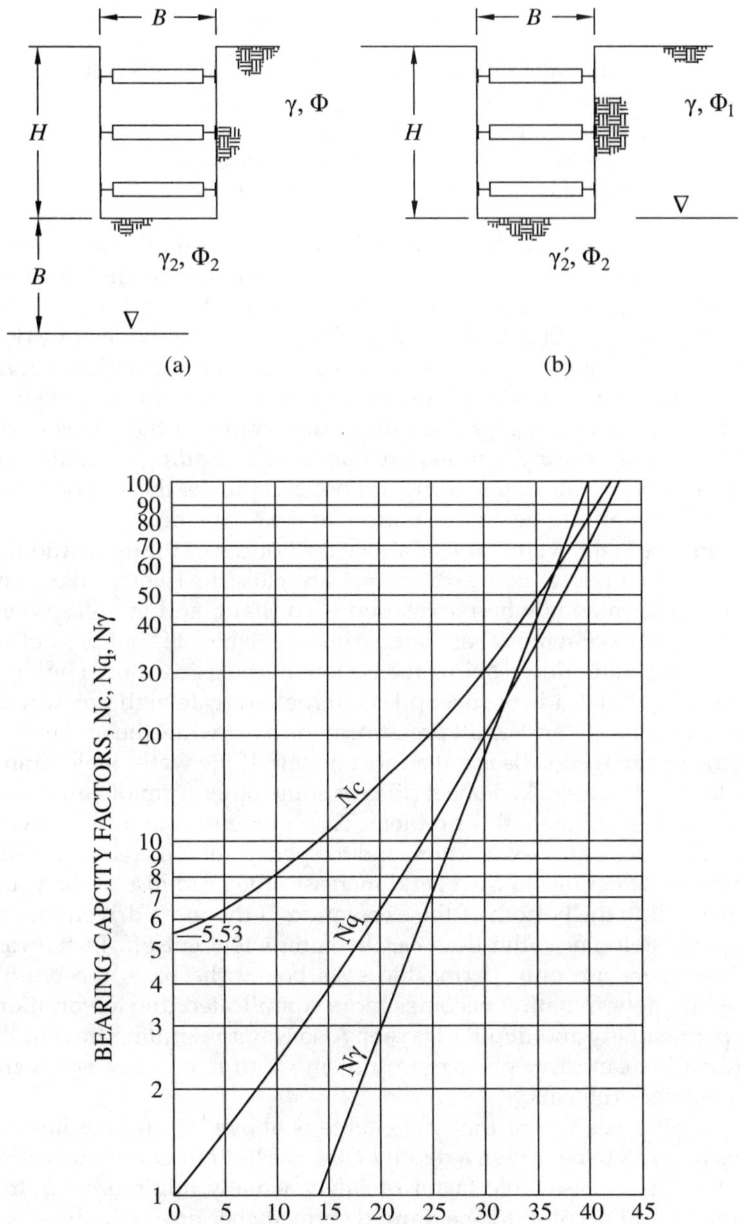

Figure 7.9 Bottom stability for noncohesive soils: (a) Water table B below bottom of excavation, (b) water table at base of excavation, (c) bearing capacity factor N_g (Meyerhof, 1955).

where $N\gamma_2$ = bearing capacity factor from Fig. 7.9 for soil below bottom of excavation

$\bar{\gamma}_2$ = average unit weight of soil within depth **B** below bottom of excavation

$\underline{\gamma}'_2$ = buoyant unit weight of soil

γ_1 = average unit weight of soil above excavation

ϕ = angle of internal friction of base material

Equation (7.5) can be solved to determine the minimum angle of internal friction ϕ given the unit weight for the soil needed to obtain a factor of safety of 1.5. Calculations using $\gamma = 120$ and $\phi = 22.5°$ give $F_s = 1.5$ for Fig. 7.9(a) and $\phi = 26°$ for Fig. 7.9(b). On the relative density chart these are very loose sands and gravels with blow counts less than 4. The result of a loose noncohesive base will be settlement at the surface and a slight rise up of the bottom at the edges of the shoring. The shoring will also see additional loading. If settlement and additional shoring loading are not acceptable, the solution is to extend the shoring below the bottom of the excavation.

In case 3, the water table is above the bottom of the excavation. In many situations where excavations are close to rivers, lakes, and bays, in farmlands where irrigation is constant, and in areas where settlement from removal of water is unacceptable, it is not possible to pump the water down below the bottom of the excavation. High water heads on the side of sheet pile cofferdams create high pressure on the shoring walls and uplift pressure at the base, while the unit weight of the submerged soils is cut almost in half. If the water table cannot be lowered, interlocked sheet piles or some other form of cutoff wall is the only solution to this problem. They prevent water from flowing into the sides of the excavation, and toe penetration increases the distance the water has to travel and increases the head loss of the water flowing into the bottom of the excavation. If the toe is driven into an impermeable layer, the flow can be completely cut off. In the case where there are only permeable soils below the base, the bottom stability determination becomes more complicated due to variations in permeability and depth of layers. Also water migrating toward the excavation can carry fine sands and silts with it, creating voids and settlement problems.

In all cases where the water table is above the dredge line the sheeting has to be driven a depth below the bottom of the excavation that yields a reasonable factor of safety, usually minimum 1.5. It is critical that the cutoff toe be figured correctly because once the sheeting is driven and the excavation is finished to the bottom, a great deal of money and time have been spent in getting there. If the water cutoff and base stability problems are not controlled, there is practically no other solution than to start over with a deeper sheet pile toe. Figures 7.10 and 7.11 use the USCS terms *dense sand* and *loose sand, depth of toe, height of water, the half width of the excavation,* and the *distance*

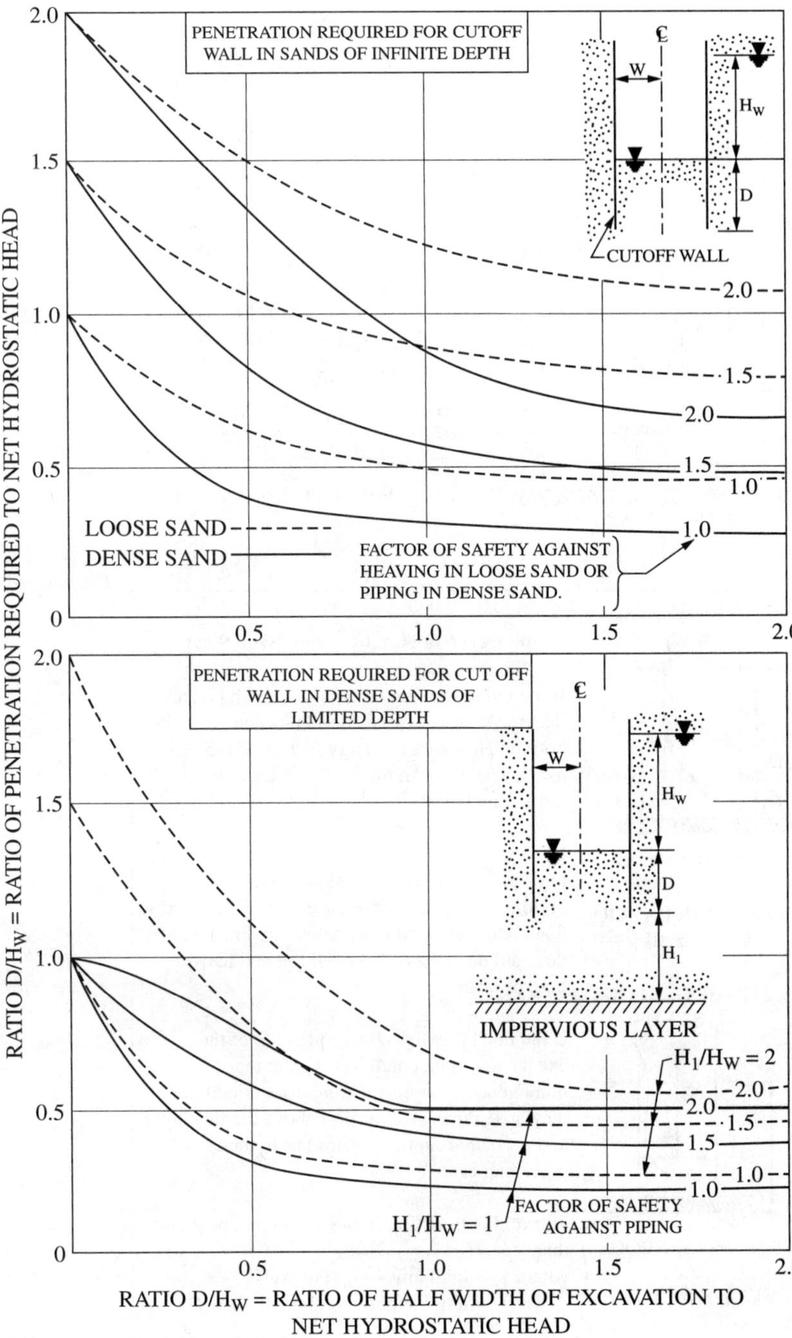

FIGURE 7.10 Penetration of cutoff wall to prevent piping in sand that is consistent in all directions (NAVFAC DM-7).

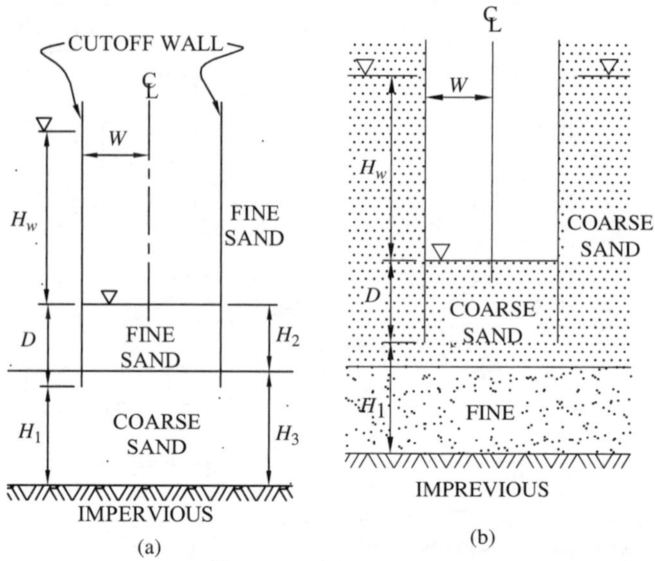

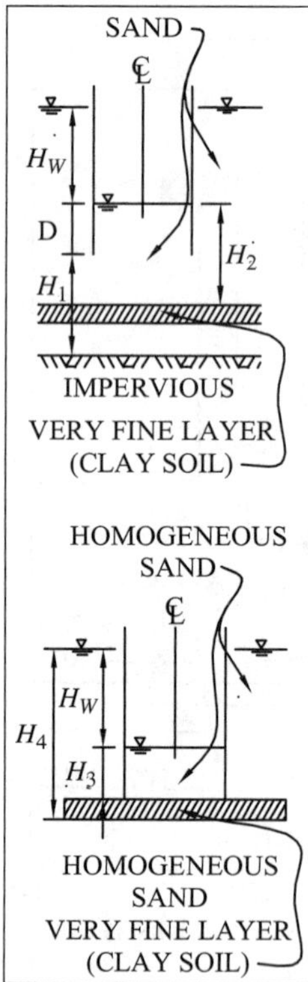

Fine Layer in Homogeneous Sand Stratum

If the top of the fine layer is at a depth greater than the width of excavation below cutoff wall bottom, $H_2 - D \geq 2W$, safety factors of Fig. 6.22 apply, assuming an impervious base at the top of the fine layer.

If the top of the fine layer is at a depth less than the width of excavation below cutoff wall bottom, $H_2 - D \geq 2W$, pressure relief is reduced so that the unbalanced head below the fine layer does not need exceeded height of soil above base of layer.

If the fine layer lies above subgrade of the excavation, final condition is safer than homogeneous case. But dangerous condition may arise during excavation above the fine layer, and pressure relief is required as in the preceding case.

To avoid bottom heave $\gamma_T \times H_3$ should be greater than $\gamma_w \times H_4$

where γ_T = total unit weight of soil

γ_w = unit weight of water, 62.4 pef

Figure 7.11 (a) Coarse sand underlying fine sand; (b) fine sand underlying coarse sand; (c) and (d) fine sand underlying coarse sand.

below the toe to an impervious layer to determine the factor of safety for piping. On this chart the soil is assumed to be sand and the same density in all directions.

7.2.3 Bottom Stability in Cohesive Soils

In the case of soft cohesive soils in the layer below the bottom of the excavation, the potential problem is the bottom of the excavation rising up while the sides settle. This is caused by the weight of the surrounding soil column exceeding the shear strength of the base soil. Bottom heave can be very subtle and can take a bit of time to happen. In the case of pipeline construction where excavating, pipe laying, and backfill are taking place in rapid series during the day, there may be no evidence of bottom heave while the excavated and laid section that is not backfilled at the end of the shift may be off-grade the next morning. Large structure excavations always take more time than a day to complete, and usually bottom heave is not discovered until grades are checked before the final pour. In very soft clays, cohesion less than 500 psf, the bottom failure can happen more rapidly, most often as it is being excavated. The cost of reworking grades and reconstructing at the base of the excavation can be dwarfed by the cost of problems associated with settlement at the surface. The solution to bottom heave problems in soft clays is to drive sheet piles below the bottom of the excavation, or remove surrounding soil at the surface. In cases where the problem is discovered after the shoring and excavation are complete, several methods of heave control have been tried. Placing a surcharge load at the base of an excavation such as a concrete work slab can tip the balance; however, considering that concrete weighs about 25 lb more than soil, it would take a 4-ft-thick slab to add a 100 psf surcharge and as the extra 4 ft is being excavated, the depth is increased and the factor of safety is further decreased.

Stacking K-Rail or steel plate in areas of the excavation where the structure does not go has also been used successfully. Drilled and anchored tension rods have also been used.

Bottom heave is a soil bearing capacity problem and is affected by the depth and shape of the structure and the weight and strength of the surrounding soils. Terzaghi developed a method of analysis based on bearing capacity factors, and NAVFAC further developed the approach shown here (Fig. 7.12).

For $T \geq (B/\sqrt{2})$

If sheeting terminates at base of cut,

$$\text{Safety factor } F_s = \frac{cN_c}{\gamma_T H + q} \tag{7.6}$$

where c = unit cohesion of soil in failure zone beneath and surrounding base of cut

N_c = bearing capacity factor, Fig. 7.12(b)

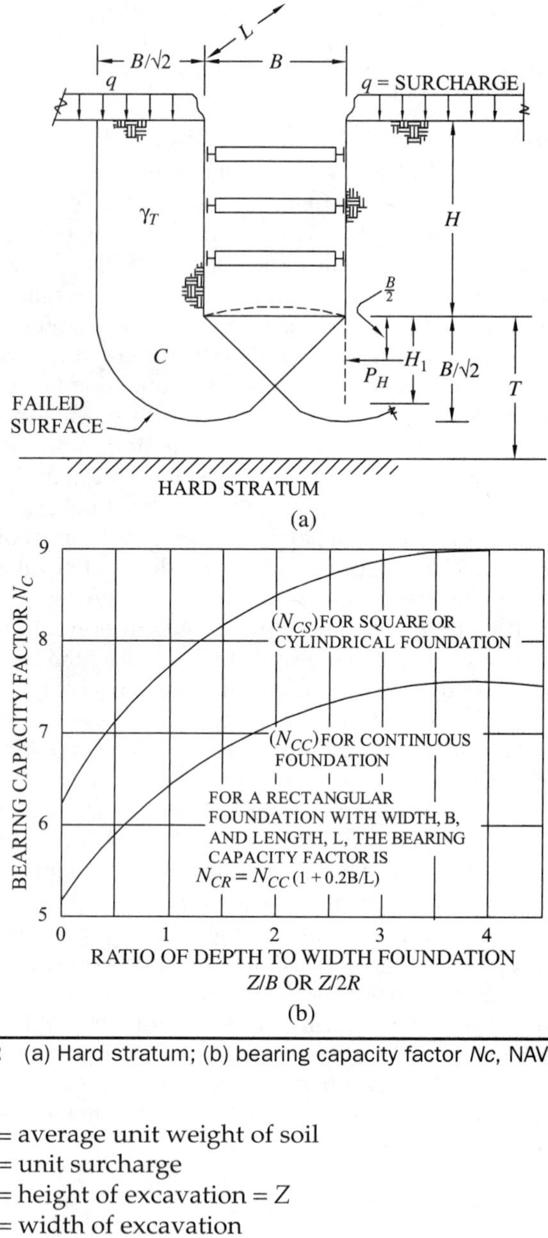

FIGURE 7.12 (a) Hard stratum; (b) bearing capacity factor *Nc*, NAVFAC.

γ_T = average unit weight of soil
q = unit surcharge
H = height of excavation = Z
B = width of excavation
L = length of excavation

If safety factor is less than 1.5, extend sheeting below excavation until $F_s \geq 1.5$, using Eq. (7.7).

$$F_s = \frac{c}{\gamma H + q}\left(N_c + \frac{2d}{B}\right) \tag{7.7}$$

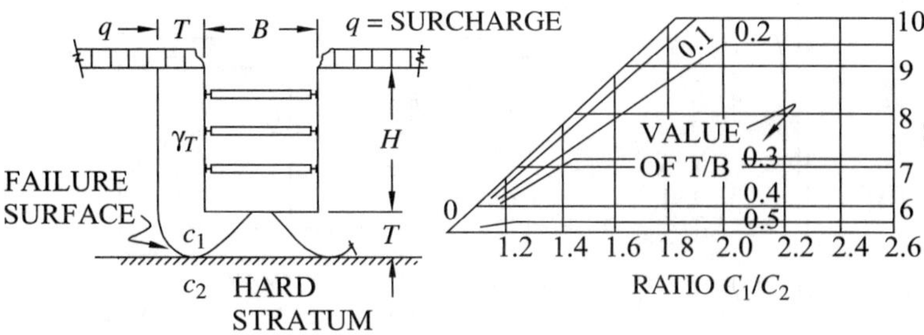

	Effect of *D*					
	D/B					
	0	0.5	1	2	3	4
N_{CD}/N_C	1	1.15	1.24	1.36	1.43	1.46

FIGURE 7.13 Cut in clay with hard stratum at $T \geq B/\sqrt{2}$.

For forces on buried length,

If $H_1 \geq 2B/3\sqrt{2}$ $\quad P_H = 0.7(\gamma_T HB - 1.4cH - \pi cB)$

If $H_1 < 2B/3\sqrt{2}$ $\quad P_H = 1.5H_1(\gamma_T H - [(1.4cH)/B] - \pi c)$

For $T \leq (B/\sqrt{2})$

The approach for a bottom in soft clays with a hard stratum less than $\leq (B/\sqrt{2})$ below is shown in Fig. 7.13. If sheeting terminates at base of cut,

Continuous excavation $\quad F_s = N_{CD}\dfrac{c_1}{\gamma_t H + q}$ (7.8)

Rectangular excavation $\quad F_s = N_{CR}\dfrac{c_1}{\gamma_t H + q}$ (7.9)

where $D = H$ = depth of excavation
B = width of excavation
L = length of excavation
c_1 = cohesion of upper layer of soft clay
c_2 = cohesion of lower layer of stronger clay
N_C = bearing capacity factor for continuous footing with $D = 0$
N_{CD} = factor for continuous footing with $D > 0$

$$N_{CD} = N_C \times \frac{N_{CD}}{N_C}$$

Determine $\dfrac{N_{CD}}{N_C}$ from the effect of the D table.

N_{CR} = factor for rectangular footing with $D = 0$

$$N_{CR} = N_{CD}\left[1 + 0.2\left(\frac{B}{L}\right)\right]$$

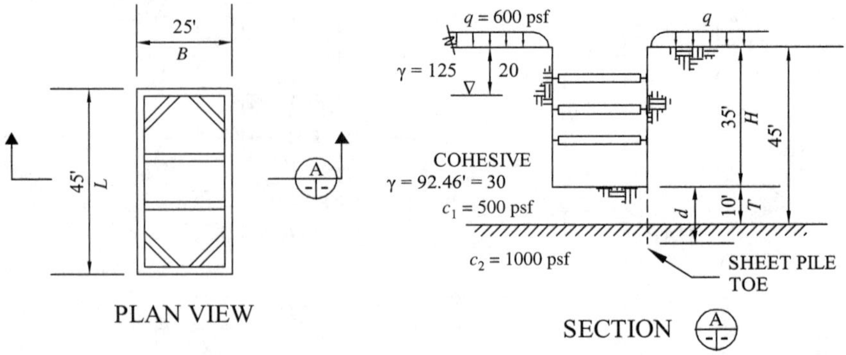

FIGURE 7.14 Determine factor of safety for bottom heave in Example 7.1.

If factor of safety < 1.5, extend sheeting as required to achieve

$$F_s = \frac{c_1}{\gamma H + q}\left(N_{CR} + \frac{2d}{B}\right) \tag{7.10}$$

where d = depth of piling below dredge level.

Example 7.1 Determine the factor of safety for bottom heave in the excavation shown in Fig. 7.14.

Step 1. Geometric properties and ratios

$B = 25$ ft	$\frac{B}{\sqrt{2}} = 17.6$ ft	$\therefore T < \frac{B}{\sqrt{2}}$
$H = D = 35$ ft	$B/L = 0.625$	
$L = 45$ ft	$D/B = 1.4$	
$T = 10$ ft	$T/B = 0.4$	

Water table depth = 20 ft

Strength and force properties and ratios

$\gamma_1 = 125$ pcf	
$c_1 = 500$ psf	$c_2/c_1 = 2$
$\gamma_2 = 92.4$ pcf	
$\gamma_2' = 30$ pcf	
$c_2 = 1000$ psf	(consider hard stratum)
$q = 600$ psf	(crane pad)

Step 2. From Fig. 7.13 using the ratios $c_2/c_1 = 2$ and $T/B = 0.4$, find $N_c = 6$.
From Fig. 7.13, the effect of D table, interpolate D/B = 1.4 to get

$$\frac{N_{CD}}{N_C} = 1.28$$

$$N_{CD} = N_C \times \frac{N_{CD}}{N_C} = 6 \times 1.28 = 7.68$$

$$N_{CR} = N_{CD}\left[1 + 0.2\left(\frac{B}{L}\right)\right] = 7.68[1 + 0.2(0.625)] = 8.64$$

From Eq. (7.9), factor of safety for rectangular excavation,

$$F_s = N_{CR}\frac{c_1}{\gamma_t H + q} = 8.64 \times \frac{500}{(20 \times 125) + (15 \times 30) + 600} = 1.22 \textit{ No Good}$$

Drive sheets below base to get additional safety factor, and use Eq. (7.10) to determine the required depth.

$$F_s = \frac{c_1}{\gamma H + q}\left(N_{CR} + \frac{2d}{B}\right)$$

where d = depth of piling below dredge level. Set $F_s = 1.5$ and solve for d.

$$1.5 = \frac{c_1}{\gamma H + q}\left(N_{CR} + \frac{2d}{B}\right) = \frac{500}{3550}\left(8.64 + \frac{2d}{25}\right)$$

$$d = 17.6 \text{ ft}$$

Note: This length would require the sheet to be driven 7.6 ft into the hard stratum; 3- to 5-ft embedment into a hard stratum would be adequate. Use a 35 + 15 = 50 ft pile length. Use proper design and construction procedure to maintain factor of safety.

7.3 Prediction and Control of Deflection and Settlement from Shoring and Excavation Operations

The impact of construction operations on surrounding private and public property is a major source of lawsuits in excavation work. Surface settlement causes strain and visual damage to structures. To property owners, cracks and out of plumb and level structures are worrisome, measurable, and calculable, the stuff of lawsuits. Change and movement will occur. A reasonable expectation of what will occur and how it will be handled prior to start of construction is the best method of diffusing problems and making equitable reparations to the affected parties. Surface settlement will come from many sources such as

1. Vibrations from driving production piles
2. Surface loading from new structures
3. Surface loading from equipment, construction materials, and soil stockpiles
4. Dewatering
5. Vibrations from general construction operations

6. Vibrations from driving shoring piles
7. Movement of open cut and shored excavations

All the items on this list are under the direct control of the design engineer in terms of setting limitations on them and should be dealt with as part of the design process. Limitations on frequency and size of production pile driving equipment, dewatering levels, surface stockpile loading levels, and admissibility of driving shoring piles should be defined in the contract. Settlements from vibrations, even with shoring piles, usually occur prior to excavation and have a different ground loss dynamic than shoring movements. The cost increase associated with restricting types of shoring and installation methods should be fully evaluated in the design process. The predicted movements from non-shoring-related operations should be evaluated so that they can later be separated from shoring-related movements. The last two items on this list, after design criteria and limitations on means and methods are set by the design engineer, are decided and controlled by the contractor; consequently the contractor has to have a method of predicting movements and surface settlement.

This section will look at different types of excavations and shoring systems and the factors that contribute to ground loss at the surface. Ground loss in open cut systems will be discussed in the section on open cut worker protection systems.

7.3.1 Ground Loss in Excavations to 20 ft Deep Shored with Trench Jacks and Shoring Shields

Medium Stiff to Stiff Cohesive Soils

In medium stiff to stiff cohesive soils, trench jacks and shoring shields are by far the most commonly used shoring systems at depths up to 20 ft. The unique aspect of this type of excavation is that usually the excavation has to be made to the bottom prior to installing the shoring, and as a result cave-in is the major ground loss potential. In cohesive soils the critical height as calculated by Eq. (6.29), $H_c = 4c/\gamma$, indicates that a cohesion of 625 psf, medium stiff on the consistency chart (Table 5.1), is required for the soil to stand to a depth of 20 ft. Surrounding conditions such as surcharge loads and existing buried facilities will have an effect on the reliability of this calculation; however, if there is no cave-in prior to properly installing the shoring, there should be no measurable settlement. If shoring boxes are used to support these excavations, there has to be sand or excavated soil placed between the box wall and the cut wall, or else there is still potential for cave-in. With the use of trench jacks and shoring shields, more often than not the ground below the bottom of the excavation and the potential for bottom heave are not looked at. Cave-in at the zone of existing utilities usually means complete failure and disruption of the system. If the excavation is in a roadway, it also means a complete loss of road base and pavement structure within the cave-in area. The caved in area usually extends out to a 3/4:1 slope from the bottom. Even small amounts of bottom heave will stress existing utilities and degrade the road base and pavement structure in the area.

The alternative in terms of expense level to this type of shoring system is first slide rail, then H-pile with sheeting, and finally sheet pile. For the contractor the risk cost of replacing street and utilities has to be compared to the cost of the next expensive shoring system. Often if the shoring system is still cost effective after replacing street and utilities, it still makes sense to take the risk of cave-in. If the bottom cannot be reached and shored without extensive damage, the time and expense of the failed shoring system have to be added to the cost of replacing street and utilities. For the contracting agency the risk decision should take into account the impact of disruption of services, the potential for degradation of street and utilities even if the installation is ultimately successful for the contractor, and the contractor's ability to meet the expense and time constrictions related to a failed attempt at shoring the excavation. A factor of safety of 1.5 on cohesion, say 1000 psf minimum, and a 1.5 factor of safety for bottom heave should be a minimum starting point for this risk evaluation.

Noncohesive Soils

In noncohesive soils cave-in is still the major ground loss potential. If these soils are properly shored, there will be negligible settlement. Making the right call becomes even more critical because noncohesive soils stand up vertical to far lower heights than cohesive soils. In pipe laying operations it is important that the bottom portion of the trench wall stand vertical to a depth equal to the diameter of the pipe due to pipe bedding and backfill requirements. If there is a soft or loose bottom, additional excavation up to 2 ft can be required to provide a stable base. The cost of replacing street and existing utilities can make sloping and random "glory holing" far too expensive an alternative to take a risk on. On large pipeline projects with thousands of feet of sewer line, the wrong assumption about whether the soil will stand up long enough to get trench jacks or shoring shields into the trench can turn the project into a very expensive loss.

From previous experience with large pipeline projects in noncohesive soils with maintenance of the water table a minimum of 2 ft below the excavation bottom, and shoring installation following directly after excavation, the author has observed that temporary stand-up is in a range of depths from 2 to 20 ft. Blow counts, grain angularity, and the amount of cohesive material in the soil are the major factors contributing to "stand-up ability." Coarse-grained soils, at any level of the excavation, with less than 5 percent fines, GW, GP, SW, SP, will not stand up more than about 5 ft regardless of angularity or blow count. In this type of layer with cohesive soils above, the layer ravels out and then the cohesive soil above then collapses.

Coarse-grained soils with greater than 12 percent fines, GM, GC, SM, SC, and a minimum blow count of 6 will stand to approximately 8 ft prior to collapse. At blow counts of 12 and above, the reliability of the soil standing seems to be greatly increased. With increased blow count, angularity, and clayey fines these soils can stand as much as 20 ft. Variations in these soils can be subtle and yet exhibit drastic differences in their ability to stand up. Meandering riverbeds, alluvial fans, existing parallel pipelines, and sewer line replacement in previously open cut trenches are some of the sources of these variations. The cost/risk assessment is generally the same as for cohesive soils except that in these soils collapsed walls fall back to 1:1 and even 1.5:1, making utility, street, and sidewalk repairs even more expensive. On large projects, even though it may be very expensive, the best method for the contractor to determine "stand-up ability" is to perform test excavations prior to the bid.

Equipment size and installation procedure can also have a profound effect on stand-upability. The stand-up time for cohesive soils can be hours and days while for sands and gravels it can be minutes to hours. Large excavators that can move more dirt rapidly combined with shorter shoring shields, 16 ft versus 24 ft, can mean the difference between success and failure. Other change-ups such as digging the last few feet when the last dump truck is in place and the shoring shield is ready to be pulled ahead, putting a front end plate on the shoring shield and digging almost vertical at the face of the trench, or digging the shield into the excavation can shorten the required stand-up time. Steel plates extended below the sides of the shield are often used successfully. More often than not in pipeline excavations to about 12 ft deep adjusting equipment and installation procedures can solve the problem.

Alternate shoring systems can also have downsides that make it worthwhile to try to make shoring shields and trench jacks work. Heavy excavators required to handle slide rail equipment can tear up road base and pavement. Vibrations from driving sheet piles or pounding in slide rail systems can destroy surrounding property. Vibrations in sands and gravels will cause settlement, repetitive moving loads, and heavy equipment will cause movement and surface cracking in noncohesive soils.

7.3.2 Ground Loss in Excavations Shored with Slide Rail, H-Pile and Lagging, and Sheet Piles

Ground loss in this type of shoring comes from incremental movements of shoring elements and in the extreme from progressive failure of the entire system, starting with one single failed strut. In the latter case multiple losses of life and entire city blocks including roads, utility structures, and surrounding buildings have fallen into the excavation. Fortunately the rate of occurrence of these incidents has decreased; however, they still happen today. Due to advanced equipment, technology, and surface land use demands, larger and deeper structures are being constructed below the surface, and consequently the impact of shoring failure today is greater in the single event and overall.

Unlike with steel, concrete, and timber, there are no exact calculations using material properties of the soil to predict movements in shoring systems. Soil and strength properties vary throughout the soil structure, and the contributing reasons for movement are numerous. Terzaghi and Peck and others have approached this problem empirically by closely measuring surface settlements, wall movements, and construction procedures in past excavations and using the information to predict reasonable expectations toward future excavations. They looked at dewatered cuts shored with H-pile with lagging and sheet pile in depths from 20 to 70 ft. They found that ground movement is easier to control in sands and gravels and in medium to stiff cohesive material than it is in loose sands and gravels and in very soft to soft clay. They also found that construction procedure, the amount of time an excavation is open, the depth of the excavation, the stiffness of the shoring system, and the factor of safety from bottom heave have an effect on surrounding settlement.

In theory the ground volume loss at the surface of an excavation should be the same as the ground volume loss to excavation, wall deflection, and bottom heave. Terzaghi and Peck found that "for well constructed and supported cuts"

$$\frac{V_{ss}}{V_{wm}} \approx 0.5 \qquad \text{for drained sands and stiff clays}$$

and

$$\frac{V_{ss}}{V_{wm}} \approx 1 \qquad \text{for soft to medium saturated clays}$$

where V_{ss} = volume of surface settlement and V_{wm} = volume of wall movement. The first conclusion is most likely due to the fact that there is some elastic rebound inside and around an excavation when the soil weight is removed, and as arching is set up between wales and struts, soil movement may not extend beyond the arched zones. They concluded that minimizing wall movements is the best approach to eliminating settlement problems.

Regarding the settlement range, depth at the edge, and distance away from the face, they found that for medium to dense sand without surcharge the settlement at the edge is about 0.5 percent of the depth, the distance back is about one-half the depth of the excavation, and for poor construction, loose sands, or heavy surcharge this can double. For very soft to soft clays, edge settlement can be 1 to 2 percent of the depth and can extend back as much as 2 to 4 times the depth. In cases where there is a soft bottom without a hard stratum below it, the settlement can extend 2 to 4 times the depth and for hard bottom it is closer to the edge. They found that these are reasonable expectations and that even under the best of conditions movements could be limited but not eliminated with these types of structures.

Given a reasonable set of limitations, a shoring system can be designed that will meet these criteria. There are several methods that can be used to limit movement in these shoring systems and an associated cost escalation that goes with them.

Fortunately it is possible to calculate loads and deflections on shoring systems and therefore to approximate the volume associated with the movement. Today's structural analysis programs easily calculate beam deflections that can be used to make simple approximate volume calculations which are adequate for this purpose. By calculating and adding the volumes contributed by different types of movements, a prediction of surface settlement can be made.

Surface Volume Loss Formula

As a result of years of observing shored excavations and surface settlement at the perimeter, the author has developed an empirical approach to predicting ground loss. It is important to note that any method of ground loss prediction is at best a rough estimate and should always be followed up in the field to sharpen the input assumptions on future projects. Approximation and simple calculations provide the same level of accuracy as detailed analysis. Equation (7.11) provides an estimate of ground loss in cubic feet of soil per lineal foot of shored excavation.

$$V_{ss} = (V_1 + V_2 + V_3 + V_4 + V_5 + V_6) \times F_1 \times F_2 \times F_3 \qquad (7.11)$$

where V_{ss} = volume of surface settlement
V_1 = volume from pile mass void
V_2 = volume from drill hole loss
V_3 = volume from sheeting movement to wales
V_4 = volume from deflection of sheeting and perimeter creep
V_5 = volume from deflection of wales
V_6 = volume from bottom heave
F_1 = factor for soil type, Table 7.11
F_2 = factor for construction procedure quality
F_3 = factor for movement restriction elements

Figure 7.15 illustrates some of these volume loss conditions, and a discussion focusing on ways to calculate and minimize their effect follows.

Pile Mass Void V_1 Even though at the time of driving the pile there is soil densification to make room for the pile, by the time the excavation is opened up and subsequently backfilled, if the pile is removed, there is a definite void left that soil will eventually move into. In the

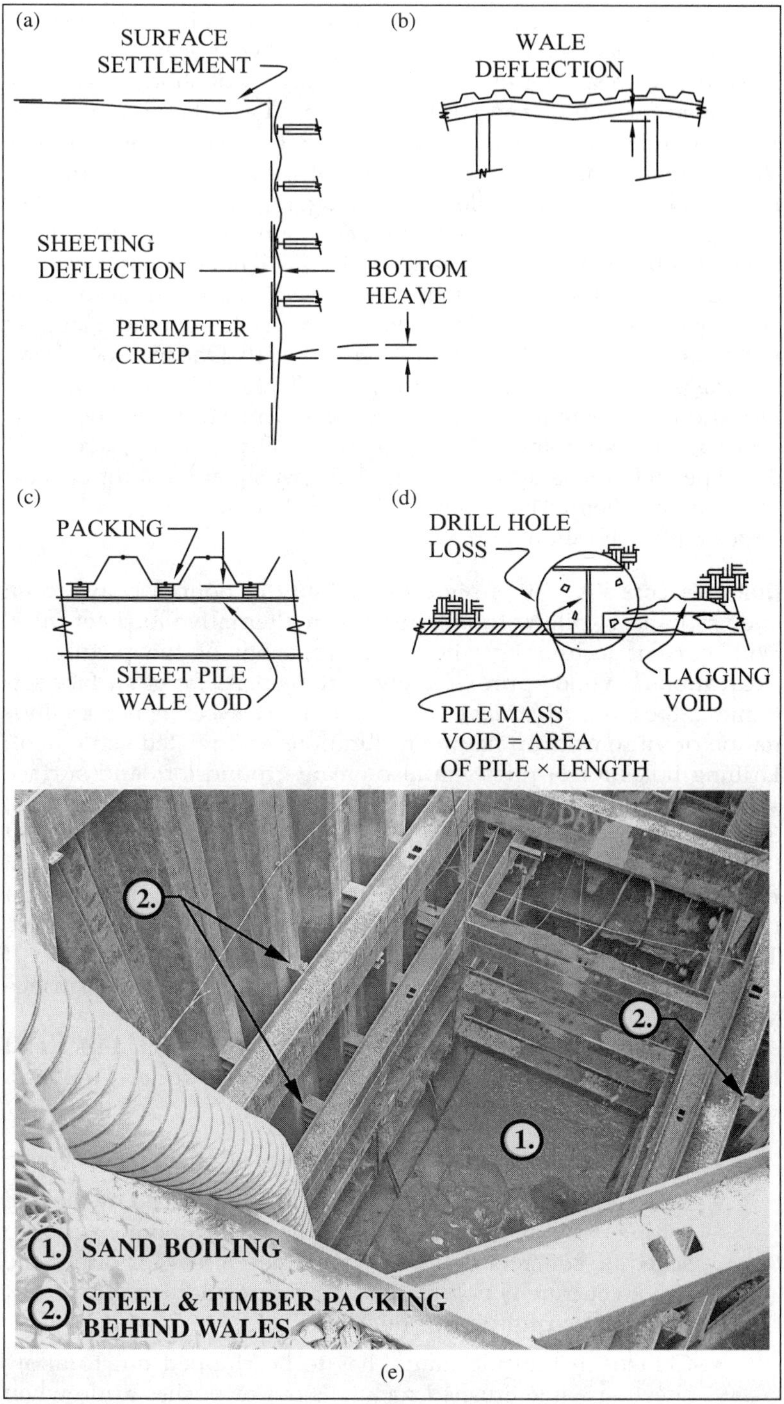

Figure 7.15 Volume loss contributors.

case of sheet piles, the pile mass void is the sheet pile area times the length of the pile. A 50-ft-long, 3/8-in-thick sheet pile will leave a 1.5 ft^3 void per lineal foot or 1 yd^3 per 17 lineal ft of shoring. The pile mass void for a 50-ft-long HP 14 × 117 spaced at 8 ft on center is also 1.5 ft^3 per lineal foot of wall. If the sheeting for the H-pile is 1-in-thick steel plate, the mass void is 3 ft^3/lin ft of wall, making the approximate total for 35-ft-deep pile and steel lagging 4.5 ft^3/lin ft of wall or 1 yd^3 per 6 lin ft of wall. Usually timber lagging is left in place. If it is left below the water table or in wet soils, it will not decompose; however, if it is open to the air as in noncohesive soils it will decompose and leave a void. If left in place and decomposed, 4-in timber lagging will void approximately 30 ft^3 per 100 ft^2 of wall. One method of preventing pile mass void is to cut the piles off a few feet below the surface and leave them in place. The cost of leaving steel sheeting in the ground, such as 3/8-in-thick sheet piles at $0.60 per pound, is roughly $1300 per 100 ft^2 less approximately $250 per square foot for the cost of removing them. This amounts to a cost of $75,000 for a 35-ft × 35-ft × 35-ft-deep shored excavation.

Drill Hole Loss V_2 If there is no possibility of bottom heave, then slide rail and H-pile with lagging are an alternative to sheet piles. With current technology slide rail drops out of the picture for excavations beyond approximately 35 ft deep. As far as surface settlement goes, the advantage to going to these systems is the elimination of vibro or drop hammer vibrations and related settlement. Drilling holes for H-piles can also create ground loss and surface settlement. In cohesive soils with $q_u > 750$ tons per square foot (tsf), 30-in round drilled holes will not collapse. In areas of soft clays and loose sands and gravels, it is possible to lose large volumes of soil in the drilling process. This can be detected by observing the amount of soil coming out of the drill hole versus the calculated amount. This can be prevented by adding drilling mud or a casing pipe. If the voids created are not filled, eventually there will be surface settlement.

There are two popular types of H-pile hole fill, crushed rock and lean cement. Crushed rock, such as ¾ minus drain rock, is the least expensive for several reasons:

- The cost per cubic yard is about 6 times less than for concrete.
- Crushed rock can be stored on the site and placed by the pile setting crew as they are setting the pile without having to schedule concrete delivery. Sometimes access for out of the shoot concrete is not available, creating additional time lost and cost for pumping or bucketing.
- Cement in the pile flange has to be chipped out to allow sheeting while crushed rock falls out or pushes aside when sheeting is advanced through it.

- H-piles set in crushed rock can be retrieved at the end of the job.

Problems associated with crushed rock fill include these:

- The rock can ravel out of the hole during the excavation and lagging placement operation. One solution is to flush a mixture of water and cement down the hole to add cementation properties to the rock.
- Sand and silt can wash out of the surrounding soils over time into the crushed rock, creating ground loss and settlement. A solution to this problem is to flush a sand/cement mixture down the hole at the time the pile is being removed.
- The reliability of pile flange bearing through crushed rock versus concrete is debatable. Certainly there has to be some movement to mobilize the rock to the passive condition and some indentation of the rock particles into the soil at the face of the pile hole. However, the extent of that movement may still be insignificant (unmeasurable). For the same reasons the arching capability of the soil around the piles is called into question. There should be no argument that lean concrete pile fill represents greater quality; however, it is difficult to calculate the benefit.

In addition to the items mentioned above, the benefits of H-piles set in lean concrete are as follows:

- The cement fill is more likely to flow into all voids of the drill hole.
- There will not be any raveling during lagging installation.
- The pile will not move within the fill mixture and will mobilize the passive resistance of the surrounding soil better.
- There will be more positive soil arching between the piles.
- A slightly stiffer beam element is produced.

There are two major problems with lean cement H-pile fill that are extremely hard to solve:

- It is impossible to advance steel plate sheeting ahead of the excavation down the flange of the pile because the flange is cemented in.
- Removal of H-piles in cement is hard to accomplish. For shorter piles to about 35 ft deep it is possible in less than two sack concrete to break them loose by first vibrating or driving them downward to break the bond and then pulling them up.

Slide rail is a unique alternative to sheet pile and H-pile shoring because it is advanced with the excavation and does not require pile driving or drilling, and it is removed as the backfill progresses, leaving no sheeting or H-pile mass void. When installed properly, it provides the same excavation wall support as predriven sheet piles. The installation time and cost of slide rail are substantially less than those of the other two alternatives. The downside to slide rail is that it works best to about 25 ft and is limited to approximately 35 ft in depth. The other limitation, the same as with H-pile and sheeting, is that there is no way to drive the sheeting below the base of the excavation to prevent bottom heave or cutoff water.

Sheeting Movement to Wales V_3 Sheet piles are almost always driven against a template; usually the first wale frame is set in place, and the piles are started at the perimeter face. The pile is driven as plumb as possible but not perfectly. Sheet piles also get stretched or compressed along the length of the frame. As a result of these variations, the piles do not fit perfectly against the wales [Fig. 7.15(c)], and the accuracy is decreased with every succeeding wale level. Quality pile driving is represented by about 1 percent vertical accuracy about the x-x axis, 6 in for a 50-ft pile, and 0.25 percent about the y-y axis. The pile line is somewhat rigid in the plane of its face. Like a piece of thin sheet metal that is not perfectly flat, if the edges and some middle points are fixed, it becomes impossible to push the curves out. This is why under full calculated load sometimes the sheet pile does not move to the wale face. The way to prevent movement is to add packing between the pile and wale (metal or wood shims). In areas where settlement is not an issue, sheet pile shoring to about 30 ft deep without continuous packing between the sheet and wale is common. In this type of construction if the sheets are packed with a steel spacer welded to the sheet and the wale at about 10 ft on center, it represents good quality and economical construction practice. Even if the piles do not totally move to the wale, there is some movement. When the piles are packed at 10 ft on center to the wale with no packing between, the author has observed in sands and gravels and medium stiff clays about 50 percent of the piles between the fixed point moving about ½ in. This calculates to approximately 2 ft^3 of ground per 100 ft^2 of wall. In soft clays, unpacked piles will rotate until some point on the pile touches the wale. One way to control this without packing every sheet pile is to specify that any pile more than some minimum distance from the wale shall be blocked.

If there is no soil movement, there will be no load on the shoring system from the soil. Limited movement means lighter loads on the system, greater safety factor, and less surface settlement. On all projects where settlement is an issue and on excavations over 30 ft deep, all the piles should be packed to the wales, even if it is only for the purpose of limiting the loading on the system. In H-pile with steel plate lagging it is still possible to get a gap between the pile flange and the plate because two piles can be out of plane, creating a three-point bearing; or there can be soil and rocks between the plate and the flange. These gaps should also be packed. In the case of slide rail, the sheeting panels ride in a channel on the post with little room to get off track. There is no need for packing with this system.

There are two common types of packing, steel and wood wedge. Wood wedge is far easier and cheaper to install. With this system, timber blocks and bridge falsework wedges are used to shim between the wale and the close face of a z-pile. The wedges should be set with a sledge hammer. With this system it is easier to get the shim under the web of the wale instead of out at the flange edge. Critics of this system say that the wood can crush and still allow movement, and it will not take tension should the pile want to move away from the wale. Rarely does the pile move away from the wale, and there is enough surface area on the wedge to prevent all but infinitesimal crushing. Wood packing can be installed at a rate of approximately 6 worker-hours per 100 lin ft of wale using a two-person crew. Even with wood packing the wale should be welded to the sheets at some interval to prevent compression flange buckling. Steel packing is more costly due to the amount of cutting and welding involved. This can require approximately 20 worker-hours per 100 lin ft of wale to install and 6 worker-hours per 100 lin ft to remove. Good construction practice would mean that the wales are welded at required intervals and 50 percent packed prior to excavating more than 2 ft below the wale.

Volume from Deflection of Sheeting and Perimeter Creep V_4 Loading and deflection on sheet piling are easy to calculate. The worst-case design condition for the sheet pile is just prior to installing the bottom wale or work slab. At this point the excavation is usually 2 ft below the wale. The pile is considered fixed at the wale and has a point of inflection, no bending moment, and approximately one-third of the required toe depth (Fig. 7.16). For instance, if the next wale is 10 ft below the last wale installed and a 12-ft pile toe is required, the span for moment and deflection calculations would be 10 ft + 2 ft + (12/3) ft = 16 ft. The bending calculation determines the minimum pile section modulus and therefore the moment of inertia which is required for deflection calculations. This sheet pile condition is equivalent to a beam fixed at one end and pinned at another, and the deflection formula is

$$\Delta_{max} = \frac{wl^4}{185EI} \tag{7.12}$$

where Δ_{max} = maximum deflection, in
w = load per unit length, k/in

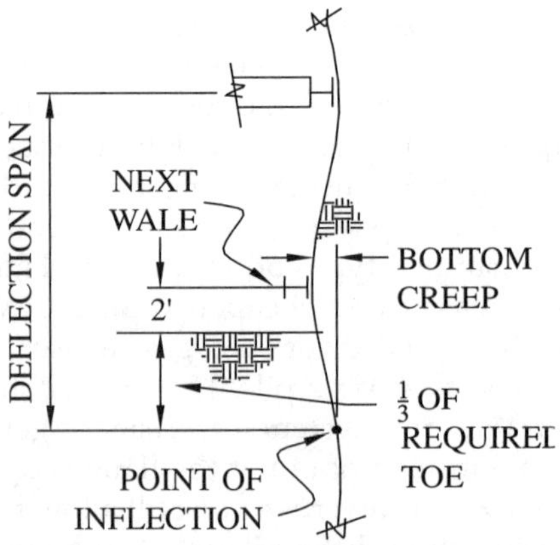

FIGURE 7.16 Deflection span for bottom creep calculation.

E = modulus of elasticity for steel, 29,000 ksi
I = moment of inertia for sheet pile, in^4/ft width of pile

This deflection, although it may be small, on the order of ¼ to ½ in for sheet piles designed close to their allowable bending strength, adds up so that at the forth wale or bottom work slab the deflection can be as much as 2 in. In a 30-ft-deep excavation the volume lost to these deflections could be approximated by a triangular wedge 2 in × 30 ft equaling 2.5 ft^3, or 8 ft^3 per 100 ft^2 of sheet pile wall. The deflection is permanent, even preloading the struts will not move the soil back any significant amount; however, soil movement and resulting deflection can take hours and even days. Therefore the sooner the next wale is installed, the less deflection there will be.

There is also deflection between the wales; however, since the pile is designed for much longer spans and most of the deflection is already there from the bottom creep movement, this volume contribution is usually insignificant. Most computer beam design programs will draw the deflected shape for the final condition of the beam, and the deflection volumes can be calculated from that or transferred to an AutoCAD drawing program that automatically calculates areas. The beam design program does not assume that the supports are allowed to settle, so the triangle calculated above should be added to the deflection volumes.

Another point to remember is that the apparent pressure diagrams are meant to encompass the worst-case loading for the struts. Sheeting and wale loads are usually approximately 20 percent less.

Slide rail shoring is always cantilevered at the bottom and top. The trick to good slide rail installation is to keep the bottom strut close—the excavator can move it up and down—to within about 4 ft of the bottom at all times during the excavation process. This eliminates bottom creep at the post because the strut cannot shorten, and due to the large moment of inertia of the post the 4-ft cantilever deflection is insignificant. Also there is no time lapse between excavation and strut installation. After the bottom is reached, the strut can be pulled up to allow an approximately 8- to 12-ft cantilever depending on soil type. Cantilever deflection is

$$\Delta_{max} = \frac{wl^4}{8EI} \tag{7.13}$$

A 2.5-in deflection with a 10-ft cantilever amounts to ground loss of 1 ft^3/lin ft of shoring system. Each strut frame has two struts spaced 4 ft apart, and more frames can be added to easily keep the strut spacing under 8 ft, resulting in insignificant deflection between struts.

Volume from Deflection of Wales V_5 As a general rule, wale beams are 14 in deep and less. Deeper beams are not common because deeper beams result in wider excavations. Pipes being installed have to clear the wales as the pipes are lowered in, and for concrete structure construction there should be about 2 ft clear between the wale and the wall to allow for formwork and worker access. Since deflection is an issue, strut spans are normally kept to less than 22 ft, to fit 20-ft pipes, and for structures about 18 ft. These spans do not necessitate the use of deep beams that have greater bending strength. Since the beams are set horizontal with gravity wanting to rotate them about their weak axis, short bulky beams versus deep thin beams are easier to handle. For the main purpose of limiting ground loss and for visual reasons (workers and onlookers get uncomfortable when they see movement in shoring), wale deflection should not exceed 1 in for any span. If strut spacing is fairly equal, the deflection for spans where struts do not fall on the end of the wale is

$$\Delta_{max} = \frac{wl^4}{384EI} \tag{7.14}$$

And for the span between a corner or a beam end and a strut that is not on the end of a wale, the deflection is

$$\Delta_{max} = \frac{wl^4}{185EI} \tag{7.15}$$

Figure 7.17 shows a W14 × 109 designed for maximum bending with a 20 klf load. This loading would be expected for an excavation approximately 32 ft deep in medium stiff clay with wales spaced about 8 ft on center. The area for the deflected shape is close to that of a parabola and is calculated by

$$Area = \frac{2bh}{3} \times \frac{1\ ft^2}{144\ in^2} \tag{7.16}$$

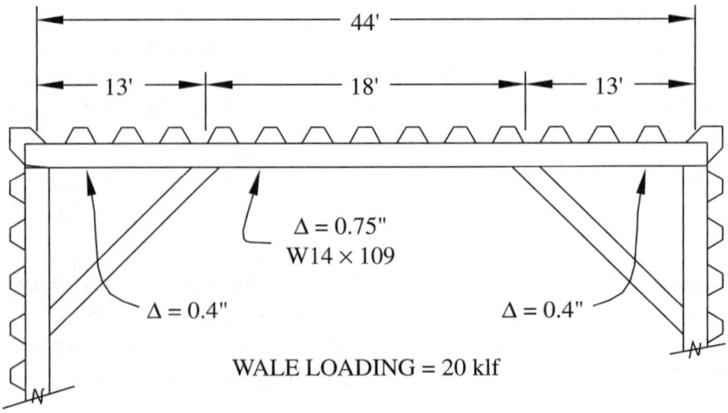

FIGURE 7.17 W14 × 109 wale deflection.

where b = wale length, in, and h = deflection, in. The volume of soil lost is the area times the wale spacing. For Fig. 7.17 the soil lost in the center is $2/3 \times [(18\,\text{in} \times 12\,\text{ft} \times 0.75\,\text{ft})/(144\,\text{in}^2/\text{ft}^2)] \times 8\,\text{ft} = 6\,\text{ft}^3$ $(0.33\,\text{ft}^3/$ lin ft of wale). It is also important to remember that steel deflection is extremely predictable so that if the wale deflection is greater than calculated, the soil loading is higher than assumed.

In the case of slide rail, the wall panel is the sheeting and wale. As shown in Fig. 7.18(a), during the excavation process if the excavation line is advanced slightly ahead of the panel, the panel deflects on the soil shear plane and prestresses the panel.

The panel deflection is due to prestress and not soil movement toward the excavation. Even though the panel deflection does not look good, prestress is helpful in preventing soil movement. In this case it is reasonable to figure that about 20 percent of the panel deflection comes from soil movement. In very soft clays, 4 blows per foot or less, and loose sands and gravels [Fig. 7.18(b)], the panel cuts the soil before the shearing takes place, and therefore there is no deflection. The panel deflects later as the soil loading and subsequent deflection progress. All the deflection in the panel will be from soil movement. The deflection can be calculated from Eq. (7.14). Table 7.9 shows allowable loads and deflections for a 3/8-in wall 6-in-thick slide rail panel when loaded to allowable stress.

Volume from Bottom Heave V_6 Even with a factor of safety of 1.5 there is bottom heave. Initially there is some sort of rebound from relieving the soil load above dredge line, and then just as there is some theoretical movement for soil arching, the same mechanism works in this instance as the soil tries to flow into the bottom of the hole. If in a 40-ft × 40-ft hole rebound and arching amounted to a 1-in rise in the bottom, it would still be 5 yd^3 or 1/3 yd^3 per 100 ft^2. With soft clays, blow count < 4, the movement is fairly rapid, 1 to 24 h, and the

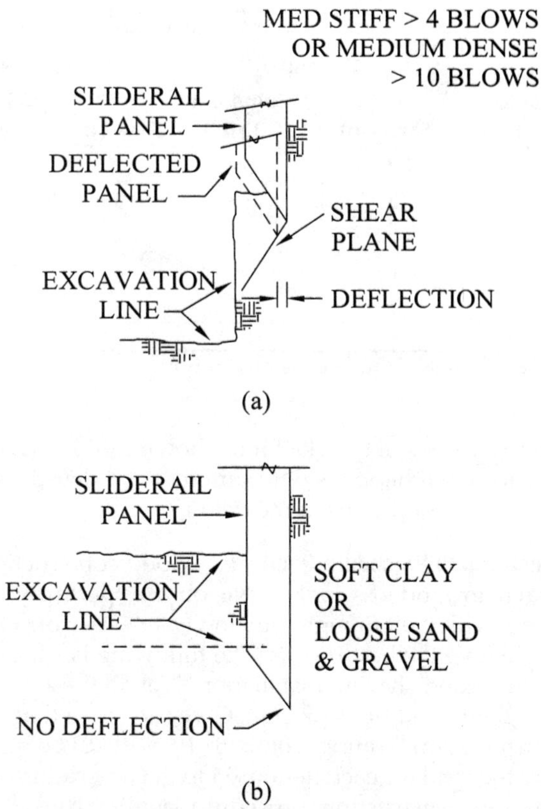

FIGURE 7.18 (a) Slide rail panel prestress; (b) slide rail panel in soft and loose soils.

excavation work at the bottom proceeds until the bottom stops moving. As much as a 1-ft-thick layer may have flowed into the hole before it locks up. As the factor of safety for bottom heave rises above 3, the potential for any loss beyond rebound is greatly reduced. Table 7.10 was developed from the author's experience from excavations 15 to 40 ft deep.

	Length (ft)				
Panel	**12**	**16**	**18**	**20**	**24**
Allowable load (klf)	3.9	2.2	1.7	1.41	1.0
Deflection (in)	0.81	1.45	1.79	2.26	3.27

1 – $F_y = 50$ ksi, $S_{y\text{-}y} = 25.7\ \text{in}^3$, $I_{y\text{-}y} = 77.2\ \text{in}^4$, $\Delta_{max} = \dfrac{Wl^4}{384EL}$

TABLE 7.9 Allowable Loading and Deflection for 3/8-in Wall × 6-in-Thick Slide

Factor of Safety from Bottom Heave	Soil Type			
	Cohesive		Noncohesive	
	Rise (in)	Volume per 100 (ft²)	Rise (in)	Volume per 100 (ft²)
1	12	100	6	50
1.5	3	33	2	16
2	1.5	16	1	8
2.5	1	8	1	8
3	0.5	4	1	4

TABLE 7.10 Approximate Ground Loss due to Bottom Heave

Soil Type Factor F_1 The soil type factor as shown in Table 7.11 is a way of accounting for what happens with different soil density or consistency under normal conditions in the field.

Factor for Construction Procedure Quality F_2 Good construction procedure is critical to ground loss control. No job goes perfectly; however, steady progress with well thought out and easy to perform repetitive procedures will give the best results. The following is a list of factors contributing to a good shoring installation. Not all these factors may be needed nor their conditions present. Give each item that applies to the shoring installation a rating from 1 to 10, with 10 being the best, and divide by the total for each item used to get a percent score. Considering that the construction procedure quality can change the ground loss quantity by 20 percent either way, Eq. (7.17) gives

$$F_2 = 0.8 + (1 - \%\ \text{score}) \times 0.4 \tag{7.17}$$

where $\%\ \text{score} = \dfrac{\text{score}}{\text{total}}$.

Soil*	Blow Count	F_1
Noncohesive	<4	1
	4–10	0.5–1
	>10	0.5
Cohesive	<2	2
	2–8	2–1
	>8	1

* Average soil type above dredge line.

TABLE 7.11 Soil Type Factor F_1

Quality of Installation Factors for Sheet Pile, H-Pile, and Slide Rail F_2

1. Crew size and experience
2. Equipment size appropriate for the job
3. Access, haul-off, and spoil pile storage available
4. Dewatering working and under control
5. Interlock sheet piles
6. Interlock corners when water is above dredge line
7. Template for sheet piles
8. Drive sheets in one-third increments (Fig. 7.19)
9. Excavate maximum of 2 ft below wale before installing
10. All packing and struts in place before excavating below 2 ft
11. Continuous operation, 10-h 7-day shift during excavation and waling process
12. Push slide rail panel below excavation line before next level of excavation
13. Slide rail panel pushed to dredge line

Factor for Movement Restriction Elements F_3 The more rigid the shoring system, the less ground movement there will be. Not only is deflection limited, but also the inertial forces that get the ground

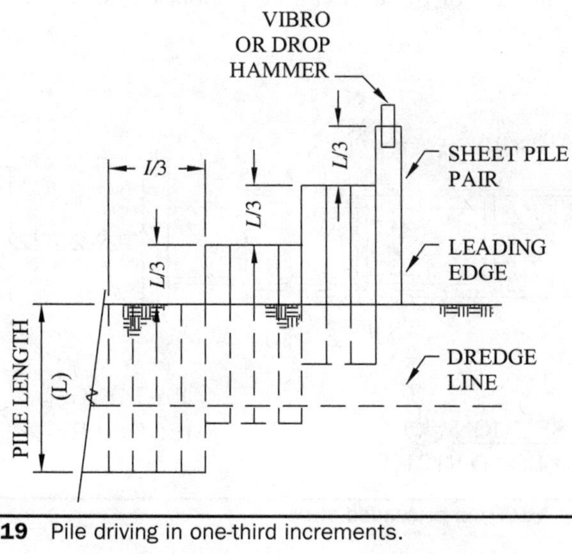

FIGURE 7.19 Pile driving in one-third increments.

moving in the first place are damped down with stiffer elements. Sheet piles and H-piles for pile and plate generally experience maximum bending just before the bottom wale is put in place. At that time the apparent load has most likely not developed. After the wale is in place, the pile bending load is often reduced to about 60 percent of allowable bending or less. If there is a long time before the wale is installed or the pile is designed for overstress prior to wale installation, there will be more deflection and ground movement. Unlike buildings that are designed for minimum steel weight and for the lifetime use of the structure, shoring is designed around what a contractor owns, can rent, or can purchase with future usage in mind. Through wale and strut spacing, bending and deflection can be controlled as much as is desired except when access considerations get in the way. Spatial requirements for overhead clearance for pipe, access for tunneling equipment, and strut spacing so that pipes can be dropped down between them usually require longer wale and sheeting spans that result in greater deflections.

Preloading wales prior to installing the struts causes soil arching to get set up and adds bending forces and deflection into the wale without having the soil move toward the excavation. In deep excavations over 2 wales deep, preloading should always be considered because it keeps the accumulated movements to a minimum (Fig. 7.20).

When excavations are close to manholes, utility vaults, and large pump station structures they turn into shoring elements. If these structures are not supported on the excavation side; they will experience bending and shear forces that they are not designed for and can slide if not toed in properly. In excavations with existing buried structures near the face of the excavation, preloading struts can help to

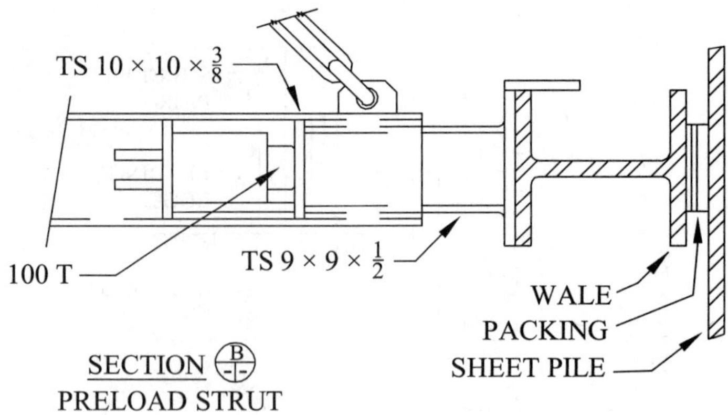

FIGURE 7.20 A 100-ton preloaded strut.

keep the soil stress distribution equal on both sides of the structure and prevent movement.

Preloading requirements are not very common because they add the cost of building a preload strut as well as an extra step in the strut installation procedure, both time loss and expenses that everyone would rather avoid. Preload struts can be built using a 100-ton jack, a steel column strut, and some welding (Fig. 7.20) and are easy to use. The preload step will add about 30 min to the installation time for each strut. Jacking and beam handling safety should be given special attention.

Packing between the wale and the sheet pile is discussed above. The volume of soil lost can be calculated; however, the effect that it has on restricting the initial ground movement can only be evaluated from the point of view that the more packing, the less likelihood of movement. Steel and welds that hold tension and compression are better than wood that compresses more easily and has no tensile capacity in this application.

Movement restriction or stiffness elements will have an overall effect, approximately 20 percent either way, on the potential for ground movement. Table 7.12 provides a rating system for determining F_3 where

$$F_3 = \frac{\text{rating total}}{\text{factor total}} \tag{7.18}$$

Element	% of Allowable Bending Stress Used	Rating
Sheeting	50	0.8
	75	1
	100	1.2
Wales	50	0.8
	75	1
	100	1.2
Preload struts	Yes	1.2
	No	1
Wale packing	Welded steel each pile	0.8
	Welded steel @ 10 ft O.C. w/wood at each pile	1
	Welded steel at 10 ft O.C.	1.1
	No packing	1.2

TABLE 7.12 Factor for Movement Restricting Elements F_3

Example 7.2 An experienced contractor is installing a 30-ft-deep, 108-in RCP sewer main next to an active light rail line. Estimate the amount and extent of settlement that will result from the proposed shoring system (Fig. 7.21).

Step 1. Determine settlement volume.

From Eq. (7.11),

$$V_{ss} = (V_1 + V_2 + V_3 + V_4 + V_5 + V_6) \times F_1 \times F_2 \times F_3$$

where V_{ss} = **volume of surface settlement**

V_1 = **volume from pile mass void.** The BZ20 sheet pile has an 0.551-in wall thickness. The pile volume is $v_1 = (0.556/2) \times 46 \times 1 = 2.13\ \text{ft}^3/\text{lin}$ ft of shoring wall.

V_2 = **volume from drill hole loss = 0**

V_3 = **volume from sheeting movement to wales.** Welded packing is at 10 ft on center with no packing between. Due to the medium stiff clay, figure 50 percent of piles will move ½ in or average ¼ in per square foot.

$$V_3 = (0.25/12) \times 30 \times 1 = 0.625\ \text{ft}^3$$

V_4 = **volume from deflection of sheeting and perimeter creep.** The first wale is set at 6 ft below grade; assume bottom creep is insignificant. The second wale is to be set 11 ft below the first wale. Assume that the required toe depth is 0.5 × the depth, and the point of inflection is 1/3 × the required toe depth.

The total deflection length is 11 + 3 + 3 = 17 ft (Fig. 7.22).
The soil is cohesive, $c = 750$ psf, and $g = 125$.

$$\frac{\gamma H}{c} = \frac{125 \times 20}{750} = 3.3 < 4$$

Use the NAVFAC formula

$$P_a = 0.3\gamma H = 0.3 \times 125 \times 20 = 750\ \text{psf}$$

Note that there is movement; however, the construction period is short so use 0.3.

For deflection

$$\Delta_{max} = \frac{wl^4}{185EI}$$

where $w = 0.75$ klf
$L = 17$ ft
$E = 29{,}000$ ksi
$I = 217.6\ \text{in}^4$

$$\Delta_{max} = \frac{0.75/12(17 \times 12)^4}{185 \times 29{,}000 \times 217.6} = 0.09\ \text{in}$$

For the bottom span deflection use

$$H = 30\ \text{ft} \qquad \text{Fig. 7.21}$$

$$16 \times 1/3 = 5.33\ \text{ft} = \text{point of inflection}$$

$$L = 13 + 5 = 18\ \text{ft} = \text{deflection length}$$

$$N_0 = \frac{\gamma H}{c} = \frac{125 \times 30}{500} = 7.5 > 4$$

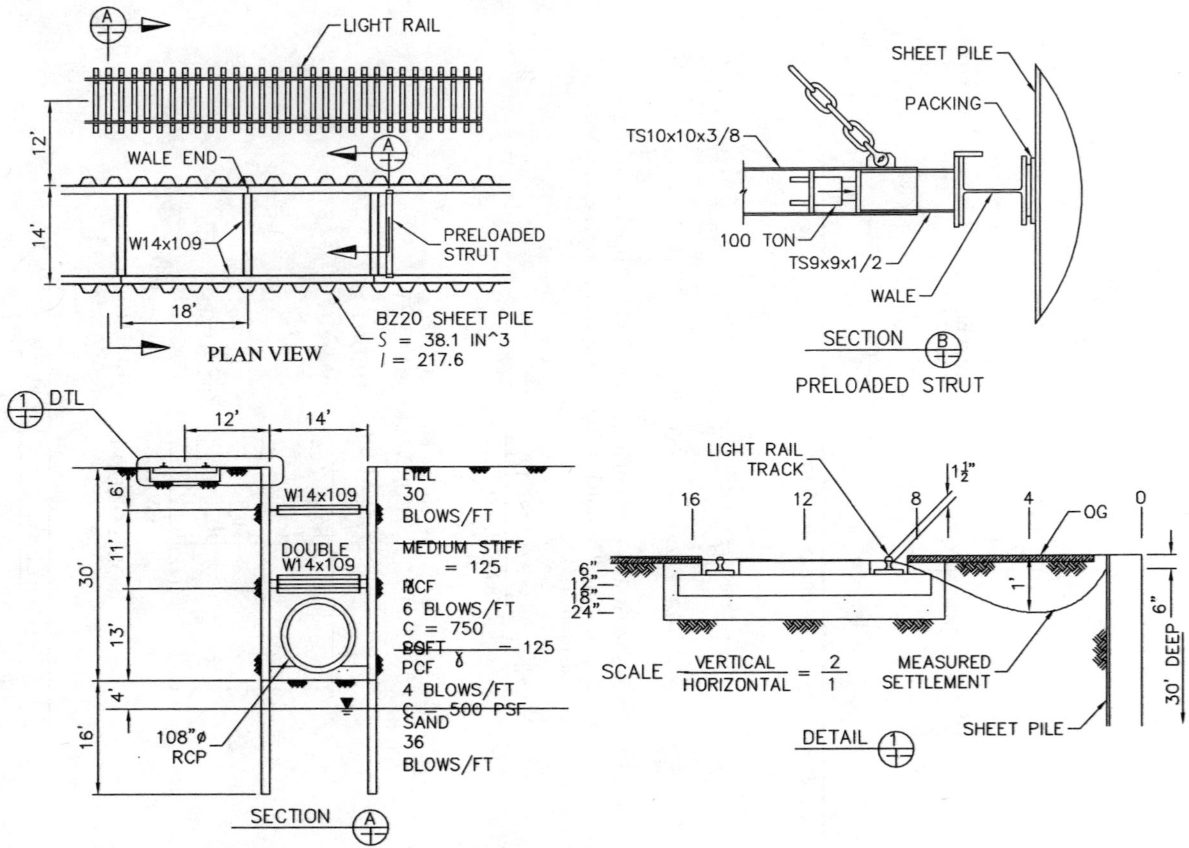

FIGURE 7.21 Example 7.2 ground settlement estimate for pipeline excavation.

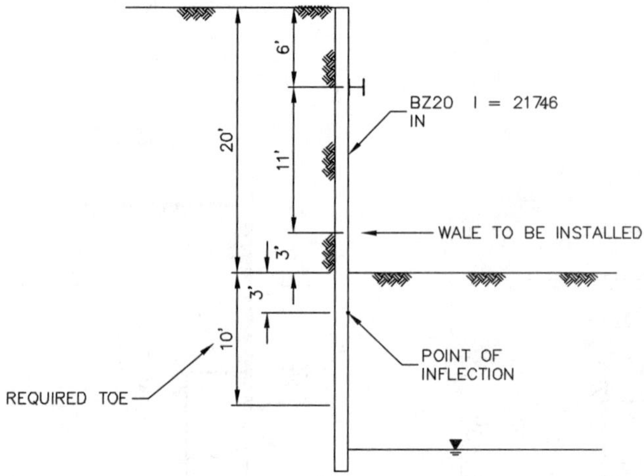

FIGURE 7.22 Second wale pile deflection.

Use $$k_A = 1 - m\frac{4c}{\gamma H} = 1 - 1\times\frac{4\times 500}{125\times 30} = 0.47$$

$$P_a = 0.47\times 125\times 30 = 1763 \text{ plf}$$

Figure 7.23 shows the results of tributary area loading calculations from the soil on the wales and sheeting. The unit loading tabulated is necessary to perform the deflection calculations.

$$\Delta = \frac{1.8/12(18\times 12)^4}{185\times 29{,}000\times 217.6} = 0.28 \text{ in}$$

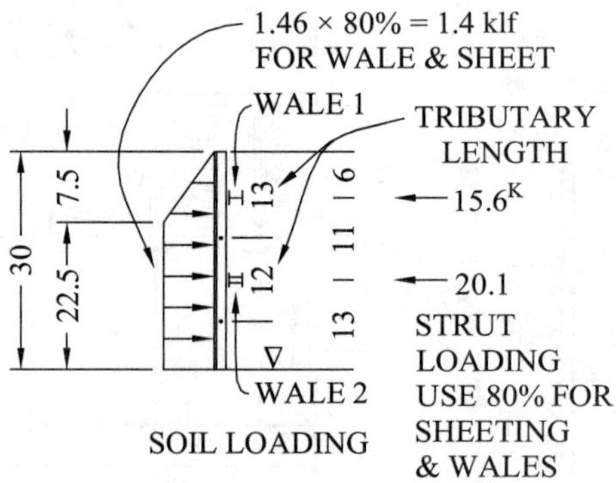

	WALE 1	WALE 2
STRUT	15.6 × 18' = 280^K	20.1 × 18' = 362^K
WALE	15.6 × 80% = 12.4 klf	20.1 × 80% = 16 klf
SHEETING	1.16 × 80% = 1.4 klf	
	LOADING PER NAVFAC—COHESIVE	

FIGURE 7.23 Strut, wale, and sheeting load from soil.

$$\text{Total bottom creep} = 0.09 \text{ in} + 0.28 \text{ in} = 0.37 \text{ in}$$

$$\text{Ground loss } V_4 = \frac{bh}{2} = \frac{0.37/12 \times 30 \times 1}{2} = 0.46 \text{ ft}^3$$

V_5 = volume from deflection of wales

Deflection for first wale, W14 × 109

$$\Delta_{max} = \frac{wl^4}{185EI} = \frac{15/12(18 \times 12)^4}{185 \times 29{,}000 \times 1240} = 0.41 \text{ in}$$

where $w = 1.76 \text{ klf} \times (6 + 11)/2 = 15 \text{ klf}$
$L = 18$ ft
$I_{x\text{-}x} = 1240 \text{ in}^4$

$$\text{Deflection area} = \frac{2bh}{3} = \frac{2 \times 18 \times 0.41/12}{3} = 0.41 \text{ ft}^2$$

$$\text{Deflection volume} = 0.41(6 + 11)/2 = 3.48 \text{ ft}^3 \text{ for 18 lin ft}$$

$$\text{Deflection volume} = 3.48 \text{ ft}^3/18 = 0.2 \text{ ft}^3/\text{ft of shoring}$$

Deflection for second wale, double W14 × 109

$$\Delta_{max} = \frac{wl^4}{185EI} = \frac{21.1/12(18 \times 12)^4}{185 \times 29{,}000 \times 2480} = 0.29 \text{ in}$$

where $w = 1.76 \text{ klf} \times (11+13)/2 = 21.1 \text{ klf}$
$L = 18$ ft
$I_{x\text{-}x} = 1240 \text{ in}^4 \times 2 = 2480 \text{ in}^4$

$$\text{Deflection area} = \frac{2bh}{3} = \frac{2 \times 18 \times 0.29/12}{3} = 0.29 \text{ ft}^2$$

$$\text{Deflection volume} = 0.29(11 + 13)/2 = 3.48 \text{ ft}^3 \text{ for 18 lin ft}$$

$$\text{Deflection volume} = 3.48 \text{ ft}^3/18 = 0.2 \text{ ft}^3/\text{ft of shoring}$$

Total V_5 = 0.2 in (top wale) + 0.2 in (bottom wale) = 0.4 ft³/lin ft of shored wall

V_6 = volume from bottom heave. There is 4 ft of soft clay at the bottom, and then the sheets are toed into dense sand. By observation the factor of safety for bottom heave is above 3.

From Table 7.11

$$V_6 = 4 \text{ ft}^3/100 \text{ ft}^2 \times 7 \text{ ft} = 0.28 \text{ ft}^3$$

F_1 = factor for soil type, Table 7.8, interpolating, cohesive soils, average blow count (6 + 4)/2 = 5 blows per foot.

$$F_1 = 1.5$$

F_2 = factor for construction procedure quality. The contractor is experienced; however, due to pipe laying procedure and speed of operation they will dig more than 2 ft below the wale prior to installing it and not keep up with placing packing at 10 ft on center.

Quality of Installation Factors for Sheet Pile, H-Pile, and Slide Rail

1.	Crew size and experience	1
2.	Equipment size appropriate for the job	1
3.	Access, haul-off, and spoil pile storage available	0.6
4.	Dewatering working and under control	1
5.	Interlock sheet piles	1
6.	Interlock corners when water is above dredge line	
7.	Template for sheet piles	1
8.	Drive sheets in one-third increments (Fig. 7.19)	1
9.	Excavate maximum of 2 ft below wale before installing	0.5
10.	All packing and struts in place before excavating below 2 ft	0.5
11.	Continuous operation, 10-h, 7-day shift during excavation and waling process	1
12.	Push slide rail panel below excavation line before next level of excavation	
13.	Slide rail panel pushed to dredge line	
	Total	8.6
	Total items	10
	% Score	0.86

$$F_2 = 0.8 + [(1 - \%\ \text{score}) \times 0.4] \tag{7.17}$$

$$F_2 = 0.8 + [(1 - 0.86) \times 0.4] = 0.86$$

where $\%\ \text{Score} = \dfrac{\text{score}}{\text{total}}$

F_3 = factor for movement restriction elements
From Table 7.12

$$F_3 = \frac{\text{rating total}}{\text{factor total}} \tag{7.18}$$

Sheeting BZ20, $s = 38.1\ \text{in}^3$

$$M = \frac{wl^2}{8} = 71.3\ \text{k} \cdot \text{ft}$$

and

$$f_b = \frac{M}{s} = \frac{71.3 \times 12}{38.1} = 22.4\ \text{ksi}$$

where $w = 1.76$ klf
$L = 18$ ft
$s = 38.1\ \text{in}^3$

Sheet piles are typically ASTM A-572 grade 50, $F_y = 50$ ksi, $F_b = 0.65$, $F_y = 32$ ksi.

$$\%\ \text{stressed} = \frac{22.4}{32} \times 100 = 70\%$$

Wale stress for W14 × 109, $s = 173$ in^3

$$M = \frac{wl^2}{8} = \frac{12.4 \times 18^2}{8} = 502 \text{ k} \cdot \text{ft} \qquad \text{wale 1}$$

$$M = \frac{wl^2}{8} = \frac{10 \times 18^2}{8} = 405 \qquad \text{wale 2}$$

where $w = 12.4$ klf at wale 1 and 10 klf at wale 2 (for each W14 × 109)
$L = 18$ ft
$S = 38.1$ in^3

$f_b = \frac{M}{s} = \frac{502 \times 12}{173} = 34.8$ ksi wale 1 (commonly 20 percent overstress allowed for temporary shoring applications)

$$f_b = \frac{M}{s} = \frac{405 \times 12}{173} = 28 \text{ ksi} \qquad \text{wale 2}$$

For 50 ksi beams $F_b = 0.66\, F_y = 33$ ksi

Stress ratio $\quad \frac{34.8}{33} \times 100 = 105\% \qquad$ wale 1

$\qquad \frac{28}{33} \times 100 = 85\% \qquad$ wale 2

From Table 7.12		Factor
Sheeting stress	70%	0.9
Wale 1 stress	105%	1.2
Wale 2 stress	85%	1.05
Preload struts	yes	1.1
Packing	stl @ 10 ft O.C.	1.1
	Total	5.35

$F_3 = 5.35/5 = 1.07$

Total ground loss volume = $V_{ss} = (2.13 + 0.625 + 0.46 + 0.4 + 0.28) \times 1.5 \times 0.86 \times 1.07 =$ **5.4 ft³/lin ft of shoring**

$V_1 = 2.13$ ft^3 $\quad V_2 = 0$ $\quad V_3 = 0.625$ ft^3 $\quad V_4 = 0.46$ ft^3 $\quad V_5 = 0.4$ ft^3 $\quad V_6 = 0.28$ ft^3

$F_1 = 1.5 \quad F_2 = 0.86 \quad F_3 = 1.07$

Step 2. Determine volume settlement distribution.

This example was taken from a project that the author worked on. Detail 1 of Figure 7.21 shows the actual measured settlement at the site. On this project there were small surcharge loads, light rail and traffic back 10 ft with no structures within 100 ft. There was observed but not measured settlement between pile driving and excavation, and the settlement from pile driving was estimated by subtracting the calculated movements from the measured ground loss at the surface. It is most likely that the large settlement within the first 4 ft is attributable to pile driving because piles in cohesive soils tend to drag the soil down with them. It is also important to note that the pile driving loss would actually be twice what is shown because during pile driving the ground subsides on both sides of the pile. The measured settlement distribution was as follows:

$$\text{Settlement from calculated ground loss} = 5.4 \text{ ft}^3$$

$$\text{Estimated settlement from pile driving} = 3.5 \text{ ft}^3$$

$$\text{Total measured settlement} = 8.9 \text{ ft}^3$$

The significant settlement occurred within a distance of 0.5 × the depth, however minimal settlement went back 1 to 2 times the depth. If the soft clay extended a large depth below the bottom, more ground loss from bottom heave would be expected and the settlement would extend 3 to 4 times the depth.

In the planning stage an approximation of volume settlement distribution in cohesive soils would be a triangle with a base 1.5 percent of the depth of the excavation at the face and the height being the extent.

$$\text{Settlement at edge} = 0.015 \times 30 \text{ ft} = 0.45 \text{ ft}$$

$$\text{Extent} = \frac{2V_{ss}}{1.5\%(\text{depth})} = \frac{2 \times 5.4 \text{ ft}^3/\text{lin ft}}{0.45 \text{ ft}} = 24 \text{ ft}$$

$$\text{Settlement volume} = \frac{bh}{2} = \frac{0.45 \text{ ft} \times 24 \text{ ft}}{2} = 5.4 \text{ ft}^3$$

where V_{ss} = calculated ground loss volume = 5.4 ft³/lin ft and depth = depth of excavation. In sands and gravels figure that the extent is approximately 0.5 percent of the depth. Calculate the settlement at the face of the excavation using calculated V_{ss}.

As one can see, there are a lot of factors that affect the amount of settlement on a shoring application. In Example 7.2 the thickness of the sheet pile and the inevitability of ground loss in cohesive soils F_3 had the greatest effect. Using a thinner wall thickness on the sheet pile would also have resulted in deflection and ground loss, most likely equivalent to the volume the thick pile used up. Ground loss is inevitable; however, by focusing on the contributing factors it is possible to control it, at least to the extent that it is cost-effective.

References

Bowels, Joseph E., *Engineering Properties of Soils and Their Measurement*, McGraw-Hill, New York,1986.

Bowels, Joseph E., *Foundation Analysis and Design*, 5th ed., McGraw-Hill, New York, 1996.

Caquot, A., and Kerisel, J., *Tables for the Calculation of Passive Pressure, Active Pressure, and Bearing Capacity of Foundations*, Gauthier-Villars, Paris, 1948.

Chouery, Farid, *Slip Surface by Variation for Smooth Wall*, Structural and Foundation Engineer, FAC Systems Inc., Seattle, Wa., 2006.

Chouery, Farid, *Variational Method in Deriving K_o*, Structural and Foundation Engineer, FAC Systems Inc., 6738 19th Ave. NW, Seattle, Wa 98117, 2006.

Hashhash, Youssef M. A., and Whittle, Andrew J., "Mechanisms of Load Transfer Arching for Braced Excavations in Clay," *Journal of Geotechnical and Geoenvironmental Engineering*, vol. 128, no. 3, March 2002.

Michalowski, R.L. and Park, N. "Arching in granular soils." Geomechanics: Testing, Modeling and Simulation. Proc. 1st Japan-U.S. Workshop on Testing, Modeling and Simulation, Boston, 2003. ASCE Geotechnical Special Publication No. 143, 2005, 255–268.

NAVFAC, *Foundation and Earth Structures*, Design Manual 7.02, pp. 7.2.61–7.2.67, Naval Facilities Engineering Command, Alexandria, Va.1971.

Occupational Safety and Health Administration, "Regulatory Review of 29 CFR 1926, Subpart P: Excavations," *Federal Register*, March 2007.

Peck, R. B., Hanson, W. E., and Thornburn, T.H. *Foundation Engineering*, John Wiley & Sons, New York, 1974.

Rehnman, S. E., and Broms, B. B., "Lateral Pressures on Basement Wall: Results from Full-Scale Tests." *Proceedings of the 5th European Conference on Soil Mechanics and Foundation Engineering*, vol. 1, 1972.

Sowers, George B., and Sowers, George F., *Introductory Soil Mechanics and Foundations*, Macmillan, New York, 1970.

Meyerhof, G.G., "The Influence of Roughness of Base and Ground Water on the Ultimate Bearing Capacity of Foundations," *Geotechnique* vol. 5, issue 3, 1955.

Terzaghi, K., and Peck, R., *Soil Mechanics in Engineering Practice*, John Wiley & Sons, New York, 1967.

Terzaghi, K., Peck, R., and Mesri, G., *Soil Mechanics in Engineering Practice*, 3rd ed., John Wiley & Sons, New York, 1996.

Wang, Cheng-Der, "Lateral Stress Caused by Horizontal and Vertical Surcharge Strip Loads on a Cross-Anisotropic Backfill," *International Journal for Numerical and Analytical Methods in Geomechanics*,vol 29, issue 14, pp. 1341–1361, 2005.

CHAPTER 8

Slope Stability and Open Cut Worker Protection Systems

8.1 Introduction

Slope stability issues involved with excavation work can be sorted into two categories: stability issues inherent with the project design and those that are involved with opening up sloped excavations for the purpose of worker protection and installing production work. "Project inherent" slope problems can be thought of as ones that should be identified in the project design process because further soils investigation and geotechnical engineering should be applied to them. The distinction is sometimes hard to determine; however, it is important for the project design engineer, the contractor's construction engineer, and the competent person to sort it out. The responsibility for, approach to solving, and solution for the problem are different for each of the two issues. The basic difference is that the inherent project slope stability problem is unique, long-term and involves the project geotechnical engineer, and the contractor's problem is short-term and involves industrywide accepted typical solutions. Within the realm of contractor-initiated stability solutions, the decision-making process has to include the ability to sort out when normal open cut solutions will work and when they are impossible to implement. This chapter will look at conditions that give rise to slope stability problems and who owns them. Project-initiated slope stability problems and solutions, aside from determining who is responsible for solving them, are the subject of another book. Here we look at recognizing potential slope stability problems, contractor-initiated solutions, safety issues, and OSHA regulations regarding open cut worker protection systems.

8.1.1 Recognizing Project Design Inherent Slope Stability Problems and Sloped Option Eliminators

As a general rule, it can be said that if standard temporary open cut or standard shoring solutions cannot be used, it is a project inherent slope stability problem that needs special consideration and clarification in the design phase and bid documents. Examples of this are as follows:

- *Cuts requiring more than* 1.5:1 *slopes* [Fig. 8.1(a)]. OSHA allows a 1.5:1 open cut option without classification of the soil by Appendix A for any soil. This does not necessarily mean that any soil will stand at that slope; it just means that at that slope OSHA figures that workers will not be in danger if the slope fails. Generally sand above the water table and clay with $q_u > 500$ psf

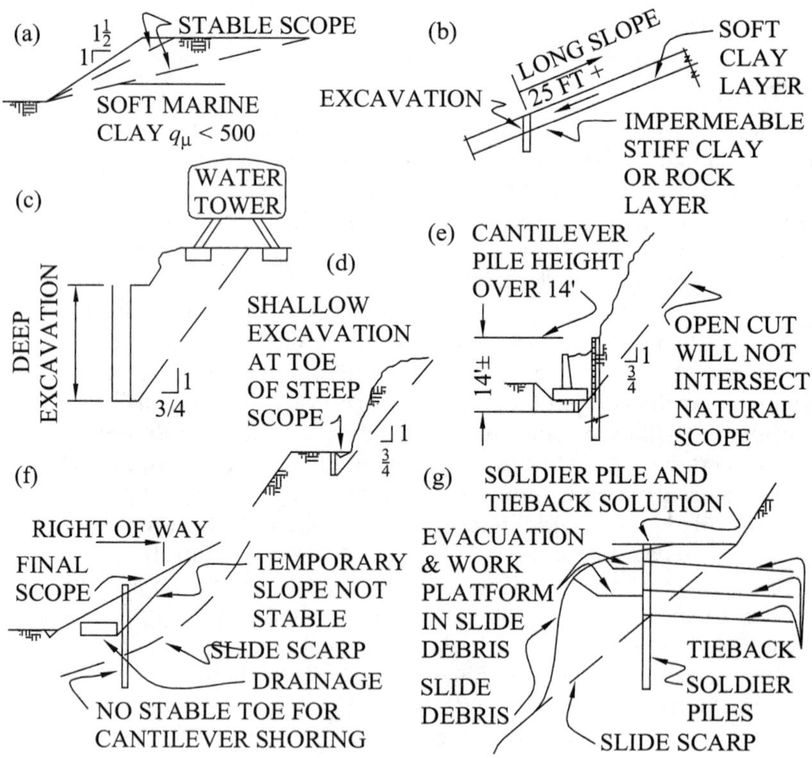

FIGURE 8.1 Project inherent slope stability problems.

will stand at that slope. Below these minimums, stable slopes can be in the range of 2:1 to 4:1. Also in soft marine clays, construction equipment loading, spoil piles, and haul truck activities can cause mud waves without even excavating. These types of soils and associated problems are project inherent and should be identified for the contractor at bid time. Expected slopes and site loading limitations should be determined by the geotechnical engineer.

- *Support systems such as sheet pile that encounter loads greater than predicted by apparent earth pressure diagrams and normal surcharge loads.* Long slopes with parallel impermeable layers and slick interfaces [Fig. 8.1(b)] can develop extreme loading at any depth when excavations cross the slope. The shoring system has to have the ability to support the force resultant of the entire layer to the top of the slope. Imagine holding back a stationary train with one car on a slope and then one with 20 cars backed up the slope. Figure 8.1(c) shows a case where a water tower is close to a deep excavation. The surcharge loading from the tower is not normally encountered. These site-specific problems should be identified in the contract.
- *Cuts required where the natural slope is greater than* ¾:1. Slopes greater than this are not allowed by OSHA except in shallow excavations open less than 12 hours. This eliminates sloping and leaves the contractor with steep slopes that generate loading on support systems much higher than normally encountered [Fig. 8.1(d)]. Cantilever shoring pile heights are limited to approximately 14 ft [Fig. 8.1(e)] due to excessive deflections and standard size shoring elements. Typically contractors and shoring rental outlets and contractors own sheet piles in section from 30 to 45 in^3. H-pile shoring is normally found in 12- to 14-in depths because deeper piles take up work space and cause excess excavation. Efficient shoring shields (lengths 16 to 24 ft) typically have maximum capacities below 1300 psf which is also less than the loading from steep slopes. The contractor and the project engineer should be aware that these conditions will call for expensive shoring solutions.
- *Excavations in slide-prone areas.* Excavations in existing landslides (sometimes thousands of years old) present unique problems. Ancient slides are not always detected in the geotechnical investigation. If the geotechnical engineer does not discover and pinpoint them, the contractor cannot be expected to root them out at bid time and plan for extra costs associated with working in them. Sometimes the excavation project is directly related to preventing, repairing, or alleviating

slides. All these cases involve working in ground that has a factor of safety for sliding of 1 or less, and the related work further reduces the safety factor until it is completed. To install drainage trenches or blankets, the slide toe has to be removed [Fig. 8.1(f)]. Safe temporary sloping required to install toe work in slide debris is always less vertical than standard sloping options and can extend to the top of the slide. In deep slides this becomes economicaly unfeasibile. Cantilever shoring in these situations can have required toe depths and cantilever heights far beyond the capacity of the pile. Slide repair approached from the top [Fig. 8.1(g)] can involve working on top of unstable slide debris. Fortunately after a slide has occurred, the forces that created it have dissipated; however, the factor of safety is only 1 at best. The safe solution in some fashion always involves dewatering and removing slide mass while working from a stable area, thereby increasing the factor of safety, prior to working in the slide. Planning and sequencing are required to accomplish this.

In addition to sorting out project inherent stability problems, it is important for the contractor to be able to sort out where open cut solutions are impossible or not feasible due to economic reasons. Some impossible situations include these:

- Right of way for sloping not available
- Staging room, equipment access, and crane pump truck reach not available
- Existing facilities and obstructions within cut slope
- Back slope steeper than safe slope [Fig. 8.1(d), (e), and (f)]

Economically unfeasible situations are as follows:

- Open cuts in roadways usually involve excessive amounts of roadway reconstruction involving fill, compaction, paving, and sidewalk reconstruction. For the project owner traffic disruption and rerouting impact are too great. Slopes greater than 1:1 involve large amounts of excavation, sometimes hauling, and backfill. The cost of these operations can dwarf the cost of shoring systems (Fig. 8.2).
- Slopes are adversely affected by environmental conditions. Wind-blown soils, sometimes referred to as loess, are highly erodible by wind and water. In long-term excavations, slopes in these soils can cause constant and expensive maintenance problems. Slopes that require constant dewatering to maintain stability are subject to pump failure over time and high dewatering costs.

FIGURE 8.2 Large excavation in slopes greater than 1:1.

- Accessibility due to distance and depth from work raises the cost of material handling and worker efficiency.

8.1.2 Slope Stability Sensitive Projects

As it relates to slope stability there are generally three classes of excavation project:

1. Some projects initiated to solve slope stability problems will involve temporary cuts or shoring.
    - Cuts for retaining walls
    - Slide repair involving removal and replacement, tiebacks, or permanent soldier pile and lagging systems
    - Installation of drainage at toe of slopes
    - Construction of drainage structures
    - Levee repairs

    Specifically within these types of projects if required excavation work will initiate slope stability problems that cannot be controlled by standard techniques, the project design engineer should specify the solution or specifically ask the contractor to propose a solution at bid time. Contractors that work on these types of projects should have knowledge and experience with the specific problems related to these projects.

2. Other projects involve excavation in sloped terrain that is required to achieve production work.

    - Cross-county pipelines
    - Belowground structures in hillsides
    - Roadway and related drainage construction

    With this type of project the specter of inherent stability problems is more elusive. To start with, the design engineer is more focused on the purpose of the project than on identifying and solving slope stability problems. The cross-country linear nature of these projects also complicates detection of stability problems. This is not to say that it is ignored—the geotechnical investigation seeks to identify these problems and provide solutions. Some of the problems are unidentified at the inception of the project, and so the cost impact is not factored in. There is a tendency to avoid the problem or classify it as a constructability problem without identifying it as one to the contractor. In many cases the problem does not reveal itself until it is encountered in the field. Due to the nature of the work, contractors have a fair amount of experience with slope stability problems. For safety and cost reasons it is important for the contractor and especially his or her field engineer and competent person to be aware of when they have stumbled into a project with an inherent stability problem.

3. Still other projects involve excavation work in flat terrain. In these projects stability issues usually arise from existing infrastructure, abnormal surcharge loading, dewatering problems, and soft soils at the base of the excavation. The possibility of having contractors with no experience or knowledge of slope stability issues is greatly increased where there is no sloped terrain. Here it is all the more important for the design engineer to identify abnormal slope stability problems.

8.2 Factors Affecting Slope Stability in Excavations

The gravity force of the soil, surcharge loads at the surface, and seepage forces cause slopes to fail. Internal friction in noncohesive soils and cohesion in cohesive soils are the forces within the soil that resist slope failure. Most soils have both cohesion and friction; see Chaps. 5 and 6 for detailed discussion of these properties and their use in soil stability problems. The *factor of safety* (FOS) is defined by the ratio of the resisting forces to failure-causing forces

$$\text{FOS} = \frac{\text{resisting forces}}{\text{causative forces}} \tag{8.1}$$

A factor of safety of 1 means that a slope is on the verge of failure, and a factor of safety from 1.3 to 1.5 is preferred but not always achieved for construction slopes. Failure or lack of resisting forces in one part of the slope requires that resisting forces in the remainder of the slope provide the required resistance, or else there will be slope failure. Another term for this is *progressive failure*.

Slope stability analysis is performed by first assuming a failure surface and then determining the forces on that surface and checking them for stability. The analysis is usually an iterative process since the worst-case failure surface is guessed at and altered until decreasing factors of safety reach a low and start to increase. Today low-cost slope stability analysis computer programs are available to take the tedium out of the analysis. As is always the case, the assumptions of c and ϕ are the critical element to accuracy. Of the two, small variations of cohesion have the greater effect on the outcome. An analysis should look at an envelope of c and ϕ values to determine the sensitivity of the results.

In construction, slope failure can be instigated by an increase in causative forces (increased slope or additional surcharge) or by a decrease in resisting forces; drying or rain at the surface, internal water from seepage, and vibrations.

In construction, determination of safe slopes is the responsibility of the contractor and her or his engineers and competent person. The fact is that most construction slope failure problems do not result in accidents. The results of failure are mostly economic impact to the contractor and poor construction quality. In construction some of the most common slope failure problems arise from the following:

- *Slope excavated too steep*. In pipeline work, steep slopes that have soil continually raveling or caving into the ditch disrupt the productivity and result in poor pipe bedding and alignment. The short-term nature of the excavation and the cost of excavating and replacing the soil usually cause the contractor to carefully evaluate and accept some level of slope degradation.
- *Soil type and layers not properly identified*. Hard layers under soils that drain easily provide a wet interface with fine soil particles that slide if the layer is sloped and can fail at the toe if it is flat. Noncohesive soils under hard layers can ravel underneath and create failure at the toe of the layer above (Fig. 8.3). Over time the result is constant slope deterioration.
- *Environmental degradation of slope surface over course of construction*. In long-term excavations the contractor is usually too optimistic about how long the slope will be exposed and the consequences of failure. Slopes that can stand at ¾:1 or 1:1 in the summer can be completely inappropriate for wet winter conditions. By the time winter arrives, these excavations usually have structures

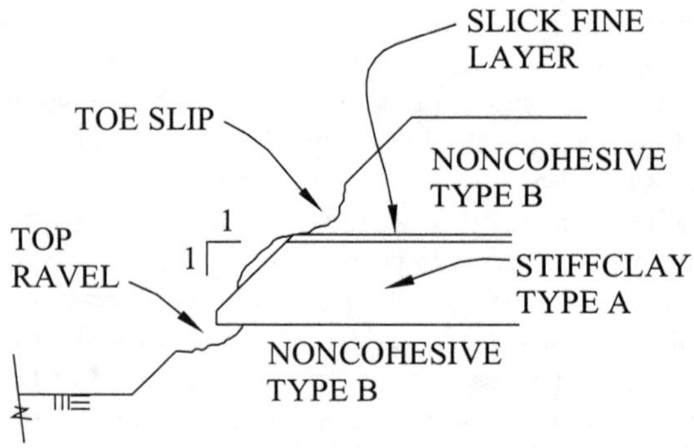

Figure 8.3 Soil layer slope degradation.

in progress and other construction activities and logistics problems at the top of the slopes so that slope repair and maintenance are impossible. Careful thought and planning should go into long-term sloped excavations.

- *Weak soil at toe and below dredge line.* With sloped excavations, since OSHA Appendix A does not require it, there is a tendency to not look below the bottom of the excavation. Soft weak soil at the bottom of an excavation or a stiff slick layer can result in a rotational failure (Fig. 8.4). The deeper the excavation, the more likely this is to happen. This is one of the reasons that OSHA requires design by a civil engineer in excavations over 20 ft deep.
- *Dewatering and seepage control.* OSHA requires for any type A or B soil that there be no water above the bottom of the excavation. In type C soil sloping is 1.5:1, often cost- and space-prohibitive. Soil layers with impermeable soils below

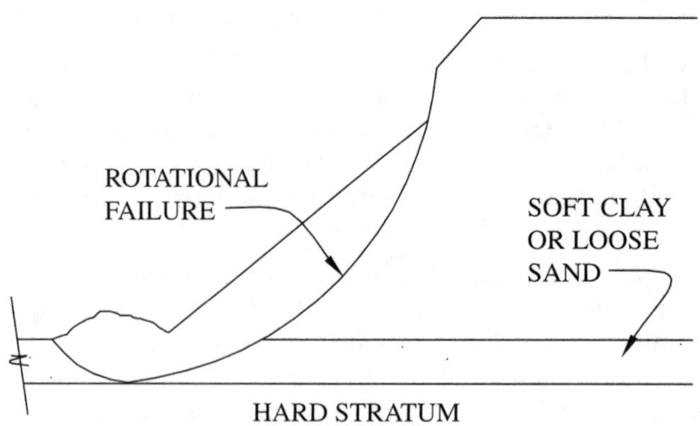

Figure 8.4 Rotational failure due to weak toe.

them can exhibit continual seeping at the seam that continues down the slope, degrading the soils below them. Often no amount of dewatering effort can prevent this. Dry seams in summer can become wet in the winter. Winter rainstorms, failed dewatering systems, or broken water mains can create water above the bottom and quick drawdown afterward. Slopes with these potential conditions should be a minimum of 1:1 regardless of the soil type and should be designed by a registered engineer.

8.3 Open Cut Sloping Worker Protection Designed under OSHA Requirements

For open cut worker protection OSHA provides a soil classification system in Appendix A and a tabulation of allowable sloping in Appendix B. This system was developed with only worker protection in mind. Many of the issues discussed above were not addressed because in the context of safety they had no bearing. The OSHA slopes contained in Appendix B are considered the standard of the industry, and anything used outside them, even when designed by an engineer, is considered risky. Geotechnical and project engineers usually specify these slopes in their specifications for temporary construction slopes. That said, clearly in many cases this tabulation is extremely conservative. The system and the slopes are generally used without further review for all construction cases. If there were flaws, aside from creating excessive costs, in the system that result in safety issues, it most certainly would have been changed by now.

8.3.1 Excerpts from OSHA 1926 Subpart P Regarding Open Cut Worker Protection with Commentary

1926.652(b)—Design of sloping and benching systems. The slopes and configurations of sloping and benching systems shall be selected and constructed by the employer or his designee and shall be in accordance with the requirements of paragraph (b)(1); or, in the alternative, paragraph (b)(2); or, in the alternative, paragraph (b)(3); or, in the alternative, paragraph (b)(4), as follows:

1926.652(b)(1)—Option (1)—Allowable configurations and slopes.

1926.652(b)(1)(i)—Excavations shall be sloped at an angle not steeper than one and one-half horizontal to one vertical (34 degrees measured from the horizontal), unless the employer uses one of the other options listed below.

1926.652(b)(1)(ii)—Slopes specified in paragraph (b)(1)(i) of this section, shall be excavated to form to configurations that are in accordance with the slopes shown for type C soil in Appendix B to this subpart.

Commentary

This is sometimes called the do-nothing option because there is no requirement to classify the soil, evaluate surcharge loads or otherwise meet the requirements of Appendix B aside from sloping at 1½:1. Users of this option should be aware that a 1½:1 slope is considered safe but does not guarantee stability. In soft noncohesive soils $q_u < 500$ psf and fine sandy soils below the water table can require much flatter slopes for stability.

1926.652(b)(2) Option (2)—Determination of slopes and configurations using Appendices A and B. Maximum allowable slopes, and allowable configurations for sloping and benching systems, shall be determined in accordance with the conditions and requirements set forth in appendices A and B to this subpart.

Commentary

This is the most often used option for sloped worker protection.

1926.652(b)(3) Option (3)—Designs using other tabulated data.

Commentary

This would be tabulated data developed by engineers for a specific region with unique soil conditions and working conditions common to the site. For example, within the site of a large wastewater treatment plant or oil refinery construction project it may be advantageous to develop sloping criteria different from those of OSHA. The advantage could be economic in the form of reduced excavation quantities or increased risk mitigation in the form of more conservative requirements.

1926.652(b)(3)(i)—Designs of sloping or benching systems shall be selected from and in accordance with tabulated data, such as tables and charts.

1926.652(b)(3)(ii)—The tabulated data shall be in written form and shall include all of the following:

1926.652(b)(3)(ii)(A)—Identification of the parameters that affect the selection of a sloping or benching system drawn from such data;

Commentary

This is a requirement for a soil classification system such as OSHA Appendix A or other systems such as the Uniform Soil Classification System (USCS). Other parameters could be proximity of existing facilities, season, etc.

1926.652(b)(3)(ii)(B)—Identification of the limits of use of the data, to include the magnitude and configuration of slopes determined to be safe;

1926.652(b)(3)(ii)(C)—Explanatory information as may be necessary to aid the user in making a correct selection of a protective system from the data.

1926.652(b)(3)(iii)—At least one copy of the tabulated data which identifies the registered professional engineer who approved the data, shall be maintained at the jobsite during construction of the protective system. After that time the data may be stored off the jobsite, but a copy of the data shall be made available to the Secretary upon request.

Commentary
Clear decision-making criteria and delineation and sloping are required for the competent person to choose the system; they are also necessary so that project design engineers, safety officials, and the courts can determine if the excavation was constructed properly.

1926.652(b)(4)—Option (4)—Design by a registered professional engineer.

1926.652(b)(4)(i)—Sloping and benching systems not utilizing Option (1) or Option (2) or Option (3) under paragraph (b) of this section shall be approved by a registered professional engineer.

1926.652(b)(4)(ii)—Designs shall be in written form and shall include at least the following:

1926.652(b)(4)(ii)(A)—The magnitude of the slopes that were determined to be safe for the particular project;

1926.652(b)(4)(ii)(B)—The configurations that were determined to be safe for the particular project;

1926.652(b)(4)(ii)(C)—The identity of the registered professional engineer approving the design.

1926.652(b)(4)(iii)—At least one copy of the design shall be maintained at the jobsite while the slope is being constructed. After that time the design need not be at the jobsite, but a copy shall be made available to the Secretary upon request.

Commentary
This is essentially the same as option 3 except that it applies to one specific excavation. The engineered option was included in the original OSHA formulation for the purpose of allowing the contractor to innovate safe and economical worker safety solutions. Acceptance by others of the solution should be based on accepted state-of-the-art engineering practice and not whichever is seen as the more stringent requirement, the OSHA Appendix B or the engineered plan. In actuality this is the most conservative option because it involves examination of the soils by a design professional. See 3.6.1 and 8.5.1 for more on site-specific engineered plans.

8.3.2 OSHA 1926 Subpart P Appendix B—Sloping and Benching with Commentary

(a) Scope and application. This appendix contains specifications for sloping and benching when used as methods of protecting employees working in excavations from cave-ins. The requirements of this appendix apply when the design of sloping and benching protective systems is to be performed in accordance with the requirements set forth in § 1926.652(b)(2).

Commentary

OSHA 1926.652(b)(2) is option 2 for open cut worker protection systems. Option 2 requires adherence to Appendix A and Appendix B.

(b) Definitions. Actual slope means the slope to which an excavation face is excavated. Distress means that the soil is in a condition where a cave-in is imminent or is likely to occur. Distress is evidenced by such phenomena as the development of fissures in the face of or adjacent to an open excavation; the subsidence of the edge of an excavation; the slumping of material from the face or the bulging or heaving of material from the bottom of an excavation; the spalling of material from the face of an excavation; and ravelling, i.e., small amounts of material such as pebbles or little clumps of material suddenly separating from the face of an excavation and trickling or rolling down into the excavation.

Commentary

A major problem for long-term excavations is that distress does not always occur until after the excavation equipment is off the site, or there are encumbrances such as structures under construction inside the excavation or at the surface that prevent additional excavation of slopes. If distress is expected to occur over the lifetime of the excavation, the actual slopes should be less steep by ½:1 than allowed in perfect conditions. This essentially turns type A slopes, ¾:1, into less steep than type B slopes, 1¼:1, and type B slopes into 1½:1 type C slopes. For long-term excavations in problem soils, it makes sense to go to design by an engineer so that intermediate options for slope maintenance can be used.

Maximum allowable slope means the steepest incline of an excavation face that is acceptable for the most favorable site conditions as protection against cave-ins, and is expressed as the ratio of horizontal distance to vertical rise (H:V).

Short term exposure means a period of time less than or equal to 24 hours that an excavation is open.

(c) Requirements—(1) Soil classification. Soil and rock deposits shall be classified in accordance with appendix A to subpart P of part 1926.

(2) Maximum allowable slope. The maximum allowable slope for a soil or rock deposit shall be determined from Table B-1 of this appendix.

(3) Actual slope. (i) The actual slope shall not be steeper than the maximum allowable slope. (ii) The actual slope shall be less steep than the maximum allowable slope, when there are signs of distress. If that situation occurs, the slope shall be cut back to an actual slope which is at least ½ horizontal to one vertical (½H:1V) less steep than the maximum allowable slope.

Commentary

See above comments regarding distress.

> (iii) When surcharge loads from stored material or equipment, operating equipment, or traffic are present, a competent person shall determine the degree to which the actual slope must be reduced below the maximum allowable slope, and shall assure that such reduction is achieved. Surcharge loads from adjacent structures shall be evaluated in accordance with § 1926.651(i).

Commentary

This requirement is most likely the least clear of all contained in Appendix B because the degree of slope reduction or method of determining it is left to the competent person with no guidance. Also the surcharge load setback, which is not addressed, is a critical element in determining the effect that it has on the slope. Technically this reads that any surcharge at any distance from the slope will require some degree of slope reduction. There is nothing within Appendix A or B to give the competent person any guidance on this issue of surcharge loads.

A complete understanding of the issue of stability of the slope first without surcharge is important in order to understand the reduction in slope stability due to surcharges. The discussion of OSHA Table B-1, Maximum Allowable Slopes, and slope stability is presented below due its order in the regulation. It is suggested that the reader skip to that section prior to reading the following on surcharge loading.

There are three different types of slope failure mechanisms associated with surcharge loads, slope stability failure, bearing capacity failure, and slope stress failure (Fig. 8.5). The following is a discussion of each problem and suggested methods of determining safe setbacks for each.

1. Reduced slope stability factor of safety on potential slip planes within the slope [Fig. 8.5(a)]. The additional weight

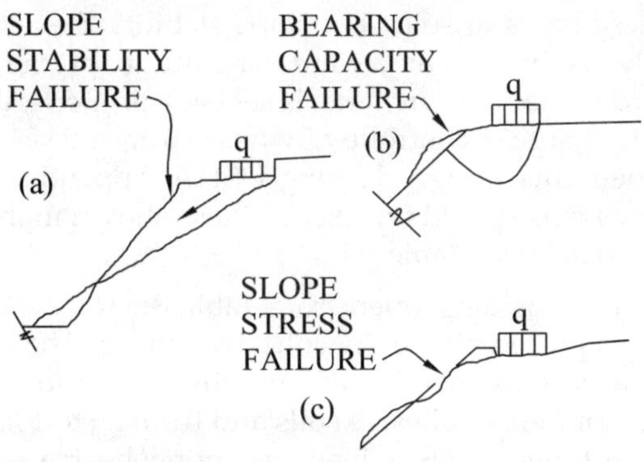

FIGURE 8.5 Surcharge load failure modes.

OSHA Soil Type	Slope (Horizontal; Vertical)	Cohesion (psf)	Maximum Surcharge (psf)	H = Depth of Excavation (ft)			
				20	15	10	5
				Factor of Safety			
A	$\frac{3}{4}$:1	1500	1500	3.2	4.1	6	10.7
B	1:1	1000	1500	2.2	2.8	4	6.6
B	1:1	500	800	1.3	1.4	1.9	2.8
C-60	$1\frac{1}{2}$:1	300	400	—	1.3	1.6	2.5
C-60	2:1	300	500	1.0	1.3	1.8	2.9
C-80	$1\frac{1}{2}$:1	200	200	—	—	1.1	2
C-80	3:1	200	200	—	1.2	1.7	3.1

TABLE 8.1 Factor of Safety for Slopes in Homogeneous Clay with Surcharge Loading

from the surcharge has to be resisted by the cohesion and internal friction of the soil.

For cohesive soils a simple way to look at this is to subtract from the cohesion value enough to carry the surcharge load and then to recalculate the factor of safety. For example, referring to Table 8.4 Factors of Safety for Slopes in Homogenous Clay, for type A soil where the minimum cohesion is 1500 psf, if the surcharge load is 1000 psf on a 20-ft-deep sloped excavation, subtract 1000 psf/20 ft = 200 psf from the cohesion, run the slope stability calculation, and look at the reduced factor of safety. If this is done for each soil type and depth to determine what the maximum surcharge load can be while still maintaining a minimum FOS = 1.3, Table 8.1 can be developed. The table shows additional slopes than those given in OSHA Table B-1 because they can become necessary in these types of soils. For slope stability the setback is not nearly as important as the magnitude of the surcharge. Provided that the surcharge is set back properly, the width of the surcharge is controlled by near surface bearing capacity. In open cuts designed using OSHA Appendix B, the surcharge loads should not exceed those shown in the table. For safe setbacks see Table 8.3.

Table 8.2 was developed from Table 8.5 for the case of surcharges placed above noncohesive slopes. The effect of the surcharge was to raise the required cohesion from 40 to 80 psf for dense cohesive soils and from 85 to 105 psf in borderline type B soils. Given that noncohesive soils usually have at least 100 psf of cohesion and that the FOS is being

OSHA Soil Type	Maximum Allowable Surcharge (psf)	Minimum Cohesion (psf)	Angle of Intenal Friction (deg)	H = Depth of Excavation (ft)	
				20	10
				Factor of Safety	
B	1500	60	38	1.3	1.8
B	1500	105	30	1.3	1.6
C	1000	40	28	1.3	1.5

TABLE 8.2 Minimum Cohesion and Angle of Internal Friction Required for FOS = 1.3 for 20-ft-Deep Slopes in Noncohesive Soil and Maximum Surcharge Allowable

achieved here, the conclusion is that surcharge loads have little effect on the sliding stability of these soils. The surcharge limitation is due to the fact that soil bearing capacity can be exceeded with high loads; see item 2 concerning bearing capacity failure.

There is the possibility of stress failure at the face of the slope, especially near the surface. This is similar to a bearing capacity failure. If the surcharge load is set too close to the slope, the resisting shear plane is cut short and the bearing capacity is reduced [Fig. 8.6(b)].

2. Based on a simple conservative bearing capacity analysis developed by Bell, and extended by Terzaghi and by Sowers, the author has developed Table 8.3 which can be used to determine a safe setback for sloped excavations. These setbacks are based on OSHA soil types. The competent person has to determine the width of the surcharge and loading. For example, a spoil pile with an average width of 8 ft and an average height of 4 ft would weigh $4 \times 8 \times 100$ lb/ft^3 = 3200 lb. The average width is approximately 5 ft, so the surcharge loading is 3200/5 = 640 psf. The example in Table 8.1 shows how to factor the setback for this load. A minimum setback of at least 2 ft should always be maintained. Also see Sec. 7.1.4 and Table 7.7 for normal surcharge loads and setbacks for shored excavations.

   The third mechanism for failure from surcharge loads is added stress at the slope surface [Fig. 8.5 (c)].

3. If the surcharge is set too close to the top hinge point, the lateral stress added to the soil can cause the slope face to ravel. By imaging a vertical plane at the hinge (Fig 8.7), the lateral force can be developed using the same Boussinesq analysis used to determine the lateral force on shoring; see

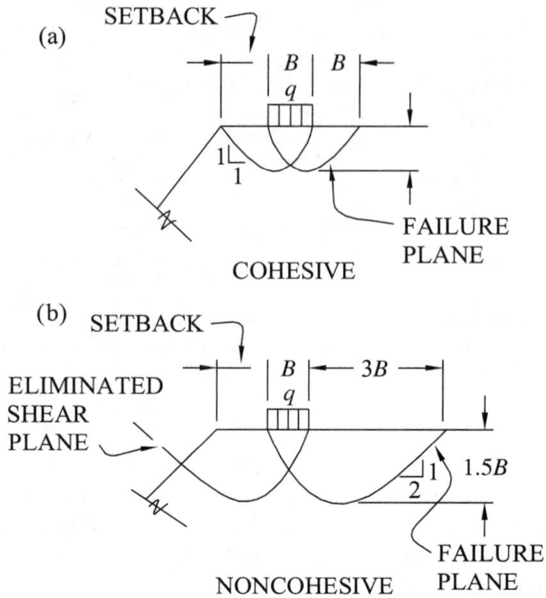

FIGURE 8.6 Bearing capacity failure planes for cohesive and noncohesive soils.

Sec. 7.1.2. If the peak lateral force from the surcharge is greater than the friction force of noncohesive soils or the shearing resistance of cohesive soils in the sloped bank, the slope surface will ravel or fall out near the top of the excavation. When the surcharge is set close to the edge, the resulting lateral forces are highest within the top 5 ft. At that level the passive resisting force from the soil is at a minimum because there is no soil weight to develop resistance from. In cohesive soils the cohesion in the soil simply needs to be larger than the peak lateral surcharge force. The setbacks given in Table 8.3 will provide enough setback to prevent this problem.

The last sentence in this requirement refers to

Stability of adjacent structures.

1926.651(i)(1)—Where the stability of adjoining buildings, walls, or other structures is endangered by excavation operations, support systems such as shoring, bracing, or underpinning shall be provided to ensure the stability of such structures for the protection of employees.

1926.651(i)(2)—Excavation below the level of the base or footing of any foundation or retaining wall that could be reasonably expected to pose a hazard to employees shall not be permitted except when:

Cohesive				Noncohesive		
Soil Type	Unconfined Compressive Strength (tsf)	Bearing Capacity (psf)	Setback* required for full bearing capacity	Angle of Internal Friction (phi)	Bearing Capacity (psf)	Setback* required for full bearing capacity
A	1.5	6000	*B*	—	—	—
B	1	4000	*B*	38	4200	4*B*
B	0.5	1500	*B*	28	1500	3*B*
C-60	0.25	750	1.5*B*	26	850	2.5*B*
C	0.1	300	3*B*	24	450	2.5*B*

Notes

1. B = width of surcharge.
2. For surcharge less than full bearing capacity, use direct proportion to determine setback. For example, using a 5-ft-wide 650 psf surcharge in type B soil with $\phi = 28°$ determine the setback.

   Full bearing capacity is 1500 psf with a setback of $3 \times 5 = 15$ ft.

   The required setback for a 650 psf surcharge is

*Minimum setback is 2 ft.

$$\text{Setback } B = \frac{650}{1500} \times 15 \text{ ft} = 6.5 \text{ ft}$$

TABLE 8.3 Required Setback from Top of Slope to Prevent Bearing Capacity Failure

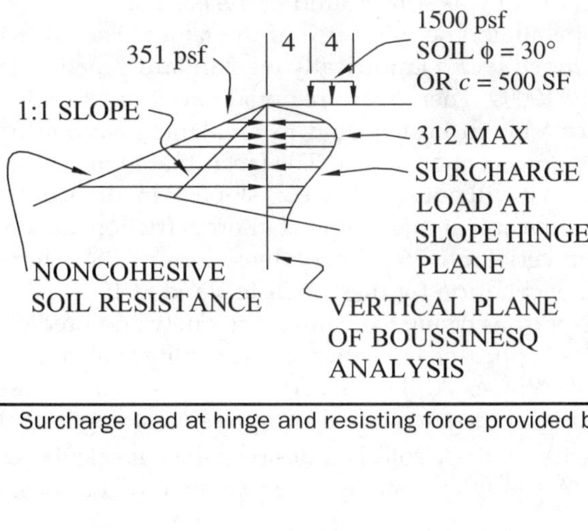

FIGURE 8.7 Surcharge load at hinge and resisting force provided by soil in the slope.

Commentary

For sloped excavations, footings with loading higher than and setbacks less than shown in Table 8.3 should be designed by an engineer.

(4) Configurations. Configurations of sloping and benching systems shall be in accordance with Table B-1.

Soil Or Rock Type	Maximum Allowable Slopes (H:V)(1) For Excavations Less Than 20 Feet Deep(3)
Stable rock	Vertical (90°)
Type A (2)	$\frac{3}{4}$:1 (53°)
Type B	1:1 (45°)
Type C	$1\frac{1}{2}$:1 (34°)

(1) Numbers shown in parentheses next to maximum allowable slopes are angles expressed in degrees from the horizontal. Angles have been rounded off.

(2) A short-term maximum allowable slope of ½H:1V (63°) is allowed in excavations in Type A soil that are 12 feet (3.67 m) or less in depth. Short-term maximum allowable slopes for excavations greater than 12 feet (3.67 m) in depth shall be 3/4 H:1V (53°).

(3) Sloping or benching for excavations greater than 20 feet deep shall be designed by a registered professional engineer.

TABLE B.1 Maximum Allowable Slopes

Commentary

OSHA Table B-1 is to some extent a result of discussions at the original 1980 draft standards workshops. Originally less steep slopes, type A soil 1:1, type B soil 1½:1, and type C soils 2:1, were proposed for excavations over 12 ft deep, with a 24-ft depth limitation. The main concern on depth was soft ground at the base of the excavation and, for the slope angle, the blind use of the tables. The table was originally developed based empirically on standard practice dating back to the early 1900s. There were regional variations to this and anecdotal information at best related to accidents resulting from excessively steep construction slopes. The fact is that there are noncohesive type B soils that will not stand at a 1:1 slope. Pure dry sands and gravels will at best stand at their angle of internal friction, which generally ranges from very loose 26° to very dense around 38°. There has to be some form of cohesion for these soils to stand at 45°.

For the persons trying to evaluate the safety and predictability of a slope, the best approach is to run a slope stability analysis. For cohesive soils in OSHA types A, B, and C this is easy to do because they are categorized by their unconfined compressive strength q_u. The major strength factor for these soils is cohesion, which is calculated by $c = \frac{q_u}{2}$. For a type A soil with a strength requirement of 1.5 tons/ft² (tsf) or

3000 psf, the cohesion is 1500 psf. The results of stability calculations for the soil types are shown in Table 8.4. A factor of safety of 1.3 for construction slopes is preferable; however 1.2 is still very safe. Safety factors greater than 1.5 do not have a lot of meaning except that the soil is safe from sliding. These calculations were based on soils with cohesion and no internal friction, even though most cohesive soils do have a friction component. Note from the table that as soon as the cohesion goes below the minimum for type C-60 soil the factor of safety drops below 1.2, with cohesion less than 400 psf. Below this range of cohesion the soil will likely move to a stable slope, FOS = 1, at excavation time or by the next day. The contractor's estimators and engineers should be able to find the cohesion in the boring logs by estimating from blow counts or unconfined compression strength test reports. The competent person should be able to estimate the shear strength by the thumb penetration test or with a pocket penetrometer.

Stability of noncohesive soils is based on the angle of internal friction and cohesion. As mentioned above, pure dry sand and gravel will stand at the slope of the angle of internal friction so that a dense sand with $\phi = 38°$ will stand at that angle with a FOS = 1, less than a 1:1 (45°) slope and loose sand with $\phi = 28°$, less than 1.5:1 (34°). Fortunately, as Table 8.3 shows, it does not take a lot of cohesion to make these soils stand safely at the OSHA prescribed slopes. For the competent person in the field there is not a lot in Appendix A to sort type B from type C noncohesive soils. It is helpful here to look at the definition of type B noncohesive soils:

> Type B Soil: (2) Granular cohesionless soils including: angular gravel (similar to crushed rock), silt, silt loam, sandy loam, and in some cases, silty clay loam and sandy clay loam.

Except for angular gravel it is almost certain that there is enough cohesion for the soils listed to stand safely at 1:1. For the competent person it is possible to visually sort for angularity, but there is no test in the appendix for cohesion in noncohesive soils. The best way for the competent person to estimate in the field the strength of noncohesive sands and gravels is to look at the grains for angularity and look at natural existing slopes. Rounded soil grains and flat slopes are usually found in streambeds. For the estimator and project engineer, the friction angle is easy to determine from blow counts in the boring logs; however, the standard method of getting the cohesion is to do a direct shear test in the laboratory. Direct shear tests report c and ϕ and are almost always performed during the soils investigation and contained in the report. Table 8.5 was developed to determine the minimum amount of cohesion required to achieve a FOS of 1.3 in OSHA soil types.

Capillary tension from moisture in the soil will almost meet the cohesion requirement; however, it cannot be relied on at the surface of the slope because it dries out. Most in situ noncohesive soils have

OSHA Soil Type	Slope (Horizontal/ Vertical)	Slope Angle (deg)	Cohesion (psf)	Soil Unit Weight (pcf)	Stability Number m	H = Depth of Excavation (ft) 20	15	10	5
						Factor of Safety			
A	$\frac{3}{4}$:1	53	1500	125	0.18	3.3	4.4	6.7	13.3
B	1:1	45	1000	125	0.17	2.4	3.1	4.7	9.4
B	1:1	45	500	125	0.17	1.2	1.6	2.4	4.7
C-60	$1\frac{1}{2}$:1	34	500	115	0.14	1.6	2.1	3.1	6.2
C-60	$1\frac{1}{2}$:1	34	400	115	0.14	1.2	1.7	2.5	5.0
C-60	2:1	26	325	115	0.12	1.2	1.6	2.4	4.7
C-80	$1\frac{1}{2}$:1	34	250	115	0.14	0.8	1.0	1.6	3.1
C-80	3:1	18	200	115	0.09	1.0	1.3	1.9	3.9

Notes

1. Results are factor of safety for cohesion.
2. Factor of safety is calculated by

$$\mathrm{SF} = \frac{c}{mH\gamma}$$

3. m = stability number from table developed after D. W. Taylor and W. Felenus. Table results based on depth factor H, solid-base at bottom excavation. For solid base below the bottom of the excavation the safety factor is reduced.

Table 8.4 Factor of Safety for Slopes in Homogeneous Clay

OSHA Soil Type	Slope (Horizontal/ Vertical)	Minimum Cohesion (psf)	Angle of Internal Friction	*H* = Depth of Excavation (ft)	
				20	10
				Factor of Safety	
B	1:1	40	38	1.3	1.8
B	1:1	85	30	1.3	1.6
C	$1\frac{1}{2}$:1	40	28	1.3	1.5

TABLE 8.5 Minimum Cohesion and Angle of Internal Friction Required for FOS = 1.3 for 20-ft-Deep Slopes in Noncohesive Soil

cohesion over 100 psf. Tables 8.4 and 8.5 were also developed so that the effect of surcharge loads could be evaluated. Surcharge loading will reduce these factors of safety; see Tables 8.1 and 8.2.

OSHA also provides no clear direction on how to measure the depth of a slope in sloped terrain. It is the conviction of the author that the OSHA standard should be self-contained, that is, the competent person should not have to look outside the text to make decisions. There is no way he should be held responsible for "OSHA department policy" or recent court decisions on the subject. That said, the best a person competent can do when there is a lack of criteria is to make reasoned decisions. Figure 8.8 shows three separate cases where slope depth is not defined by OSHA and gives a reasonably safe approach to determining the depth of the excavation. Note that even though the competent person has no way of determining the strength of the soil below the bottom of the excavation, at least in the field, it is critical to a stable excavation. The estimator and project engineer should review the soils report prior to excavation work. Not only are toe failures a safety hazard, but also they can impact the work that has been performed in the trench prior to failure.

The OSHA Appendix A definition for *stable rock* is natural solid mineral matter that can be excavated from vertical sides and remain intact while exposed.

It seems reasonable that rock that has been drilled and blasted could be included in this definition.

Virtually all rock strata have joints and bedding planes; however, the spacing could be so far apart it is not apparent. The key to stability in rock is the strength of the rock, the degree of the dip in the bedding planes, and the friction along those planes. If the bedding planes are level, the rock may still fail because it is weak and cannot hold up the rock and soil above it. The author has observed rock that had to be excavated using a rock trencher, and yet the trench walls would explode and collapse at a depth of 16 ft because the exposed rock was too weak to support the rock above. In practice in the author's experience, aside from very hard rock with practically flat bedding planes, determining

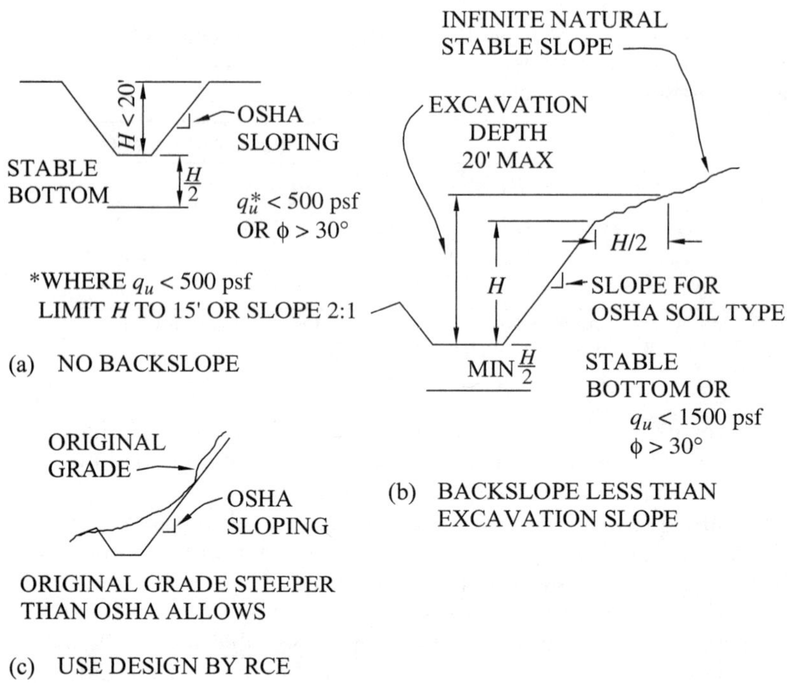

FIGURE 8.8 Determining depth of open cut excavations when using OSHA.

stable rock for vertical slopes or stable rock sloping is very complicated and needs to be done by experienced engineers. By OSHA standards if rock is not stable, the slopes automatically must be 1:1 because it is noncohesive. There is no possibility of steeper slopes unless an engineer is involved. This is for good reason because rock slope failures—even if it is one small rock falling—can easily injure workers struck by rock, unlike workers getting hit by ravel from sands and gravels.

8.3.3 OSHA Appendix B—Sloping and Benching Slope Configurations

Commentary

The following Figs. 8.9 through 8.12 for the OSHA Appendix B Slope Configurations, OSHA Figure B-1,were developed by the author. The slopes shown in these figures are exactly the same as OSHA delineates; however, there are extra notes in the drawings to add clarification.

Commentary

There are two major reasons to go through the trouble of determining type A soil:

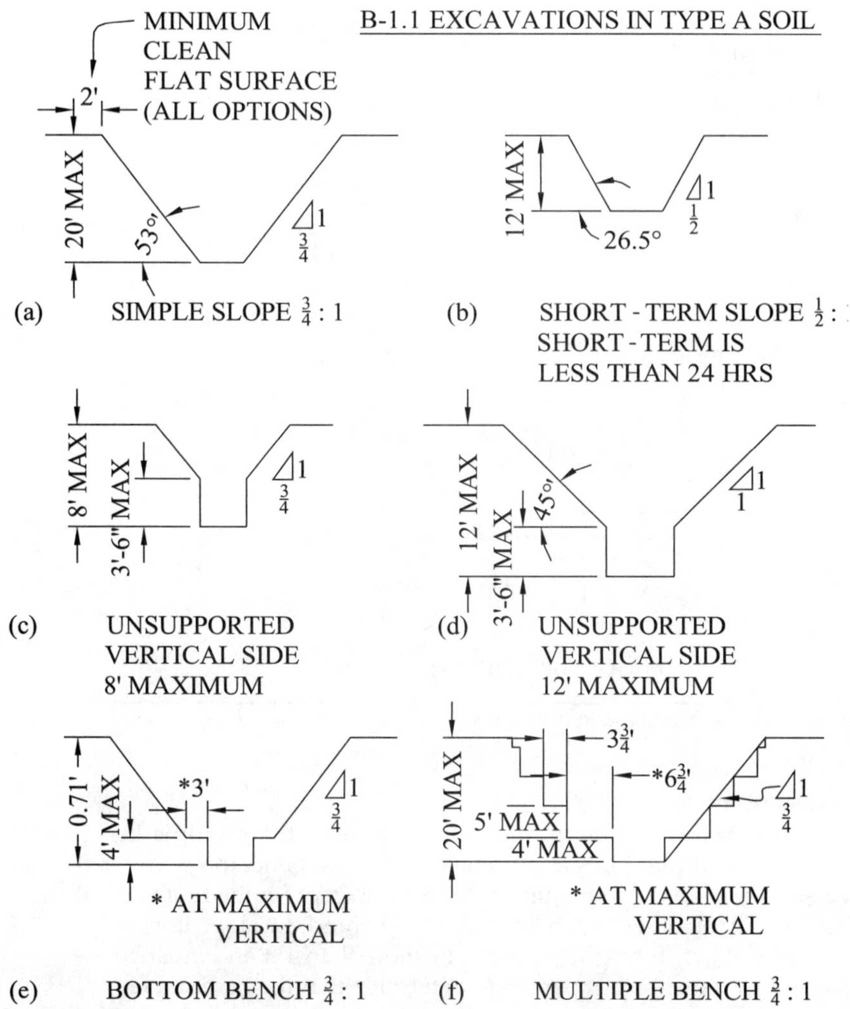

FIGURE 8.9 Excavations made in type A soil.

1. The savings in excavation quantities are substantial.
2. There are six sloping or benching options available versus three for type B and 1 for Type C.

One of the most common mistakes made with the multiple bench system [Fig. 8.9(f)] lies in not making the first benchtop wider—6-3/4 ft versus 3.75 ft—than the remaining benchtops. Except for the first vertical bench wall, all other bench lines must be behind the slope line. The first vertical bench heights may not be more than 4 ft and and the remaining bench heights cannot be more than 5 ft and must be examined by a competent person to determine that they are stable.

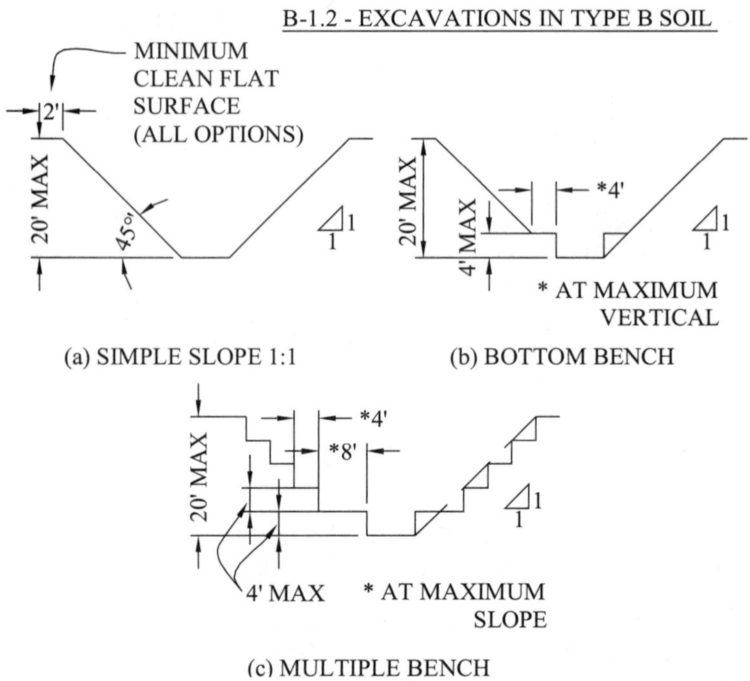

Figure 8.10 Excavations made in type B soil.

Commentary

One of the most common mistakes made with the multiple bench system [Fig. 8.10(c)] lies in not making the first benchtop wider, 8 ft versus 4 ft, than the remaining benchtops. Except for the first vertical bench wall, all other bench lines must be behind the slope line.

Vertical bench heights may not be more than 4 ft and must be examined by a competent person to determine that they are stable.

Commentary

Figure 8.11 shows a 1.5:1 slope for Type C Soil. Cohesive type C soil with unconfined compressive strength less than 0.5 tsf may not stand at 1.5:1 slopes. Slopes in these soils may be safe but may creep, showing bulge at the bottom of the slope and bottom heave. At 15 to 20 ft deep in these soils it is recommended to decrease the slope to 2:1 or greater.

Commentary

Figure 8.12 applies to all soil types with combination open and a support system below. Accidents associated with rolloff are common, and there is a high potential for serious injury. These accidents are preventable; however, it takes diligence. The top edge of the slope

B-1.3 - EXCAVATION IN TYPE C SOIL

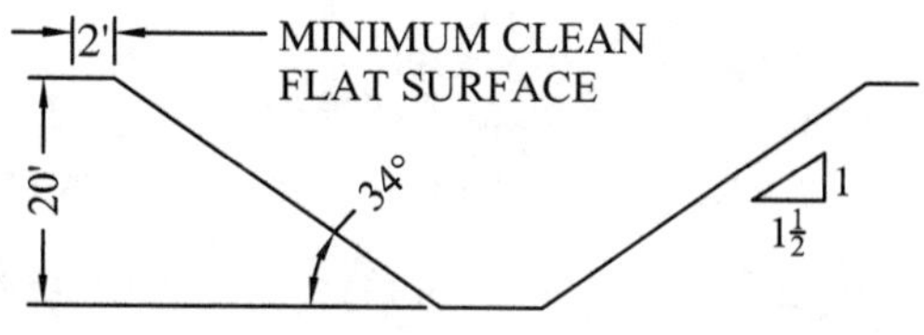

FIGURE 8.11 Excavations made in type C soil.

should be kept free from debris and loose rock, and the slopes should be inspected and cleaned if environmental degradation occurs over time.

Commentary

OSHA offers six separate drawings to explain what can be summed up about open cuts in layered systems with the rules given in Fig. 8.13.

Type C soil at or below the bottom of the excavation requires that sloping be 1½:1 to the surface regardless of soil types above.

Determining the soil type below the bottom of the excavation is not a requirement of Appendix A or B; however, it is advisable for the

EXCAVATIONS IN TYPE A, B, AND C SOILS
WITH SUPPORT SYSTEM BELOW

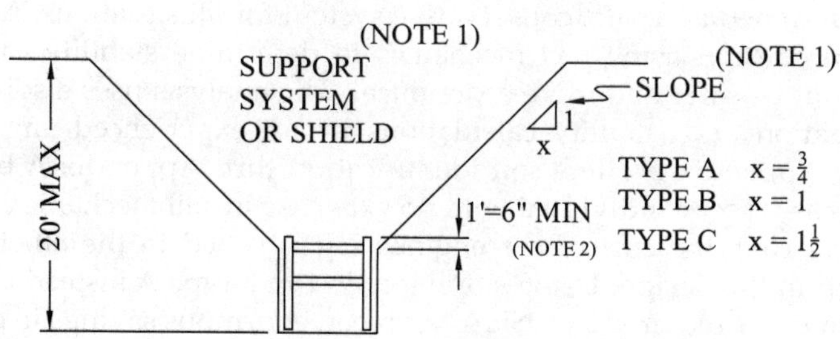

Notes

1. If shoring is a support system (active shoring), slope per soil type at top of shoring. If shoring is shield system, (passive shoring), slope per soil type at bottom of shoring
2. Follow layered soils rule above shoring level.

FIGURE 8.12 Slopes with vertically sided shored lower portion.

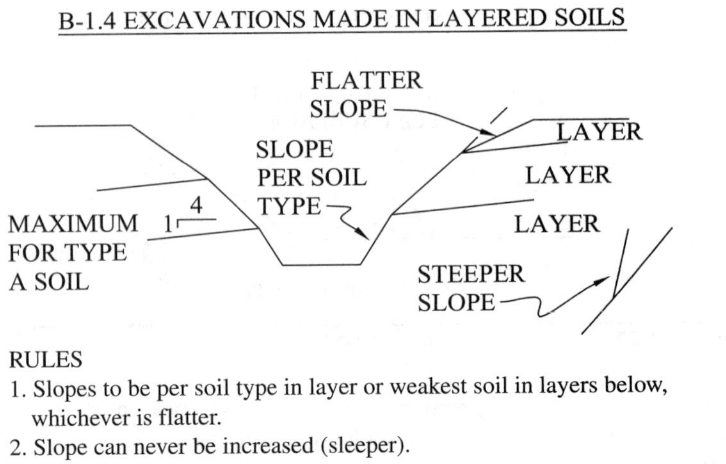

FIGURE 8.13 Excavations made in layered soils.

contractor's estimator and project manager to look for it in the soils report before formulating sloping and benching options.

8.4 Open Cut Plans by a Registered Engineer

As a result of the fact that OSHA Appendix B sloping options are easy to use and seem to cover most open cut applications, design by an engineer is often overlooked; it is seen as an expensive unnecessary alternative. In most cases this is a completely false assumption simply because design by an engineer offers an infinite number of solutions while Appendix B offers a very limited number. Appendix B was designed as a safe, conservative system for all situations. A site-specific design using soil mechanics to determine stability cannot help but be safer and more economical. The analysis uses a soils investigation and stability calculations by an experienced engineer while Appendix B offers soil identification thru Appendix A by, in most cases by an individual with no expertise in soil mechanics. As a general rule, the cost of the engineering is saved in the efficiency gained in the design. Using sloping of ½:1 in a type A instead of the required ¾:1 required by OSHA can mean enormous savings in excavation and backfill quantities on a pipeline project. In addition to cost savings, reduction of the risks of the competent person getting it wrong or failed slopes impacting work in progress is important. In the author's engineering practice there were a large number of pipeline contractors that used design by an engineer on every project because they were convinced that it made economic sense, and their insurance companies were favorably impressed with the increased safety aspect. Engineered design by outside engineers also eliminates conflict of interest for engineers within the company.

8.4.1 Engineered Design Philosophy for Open Cut Excavation Plans

In practice, design by an engineer is triggered for the following reasons:

- For economic reasons it makes sense to develop an engineered plan.
- There are obstructions such as sidewalks, trees, and existing buried facilities that make it impossible for the contractor to use the Appendix B options. Many times this is only discovered during the course of excavation work when it is too late to do anything but leave the slope steeper than allowed.
- The excavation is in place, and the slopes have deteriorated because the competent person got the classification wrong or environmental factors have rendered the original classification incorrect. By OSHA a deteriorated slope must be reduced by at least ½:1.
- Excavations are made in rock.

In Sec. 3.6.1 a complete listing of design requirements for open cut plans is presented; however, it is important to realize that circumstances are not always perfect and preplanning is not always possible. Many of the slope stability problems that came to the author's office were from a phone call that needed immediate action. Soils reports were not always available, or borings taken were not always taken where the excavations were. Many project design engineers or OSHA informed the contractor that they did not believe the excavation was in conformance with Appendix B, and the project was shut down until an engineered plan was presented. As a result of this the author established some basic rules regarding development of these plans in his office:

1. Excavations can be designed without a soils report provided that there is inspection and confirmation of the assumptions by the design engineer before workers are allowed in the excavation. It is made clear to the contractor that the final slopes will not be determined until after the excavation is complete.
2. Design slopes must be confirmed by slope stability calculations.
3. Assumptions taken from soils reports are assumptions only and must be confirmed in the field by inspection from the plan design engineer.
4. Excavation plan shall have the date it was developed and the location where it applies clearly stated on it.

These plans predominantly had slopes steeper than allowed in Appendix B for their soil types, and many times vertical slopes at the bottom were higher than allowed by OSHA. Although these plans were sometimes seen by others as a way of getting around the OSHA rules, they were usually accepted by OSHA and reviewing engineers. There was one exception to this: vertical walls greater than 5 ft high were a nonstarter. In hardpan and clays with $q_u > 2$ tsf on a short-term basis vertical walls to 6 or 7 ft would be reasonable. Calculations indicate that they would stand over 50 ft deep.

8.5 Open Cut Excavation Safety Issues

Aside from getting the soil classification right and constructing in accordance with Appendix B, there are some recurring safety issues that workers should be aware of. The key to accident prevention is physical barriers to the problem, and if that is not possible, awareness and understanding of the problem must be reinforced on an ongoing basis. Weekly safety meetings should highlight potential specific problems and in all cases develop a plan or strategy for preventing them. Here is a listing of some important issues involving open cut excavations:

- *Falling debris.* Falling rocks, dirt chunks, wood blocks, tools, etc., rolling down a slope can cause serious accidents. Pipeline excavations are not usually fenced or barricaded at the top, and the nature of pipe laying causes a lot of worker activity at the top of the slope. Workers involved in that activity should have a continuing awareness of the possibility of kicking things off the top of the slope. Spoil piles must be set back a minimum of 2 ft by OSHA requirements, and on the working side of the excavation the first 2 ft of surface before the slope hinge should be maintained clean at all times.
- *Workers falling into excavation.* At permanent excavations, guard railing is required. At this time there is a bit of a debate going on across the country about whether guard railing is required on pipeline excavations. The argument for no railing is that the work area is constantly moving and railing would interfere with the progression of the work and hence the safety of the work. If the industry takes measures and innovates to prevent accidents related to the argument, they will probably prevail; however, if they do not, there will surely be safety legislation requiring railing. In pipeline work the primary cause of workers falling of the top of the slope is loose dirt at the hinge or encroachment of the worker's space by equipment or persons from the land side. Planning and staging should focus on these problems.
- *Attractive nuisance.* All excavations should be covered or securely fenced so that the public and especially children cannot get into them. Open cuts are generally too wide to

cover them with plates, and the area surrounding the excavation can have slopes and spoil piles around it, making it hard to fence with temporary rectangular fence panels. Chain link locked with corners and slope changes closed off so that persons or animals cannot get through are a minimum requirement. If it is easy to get in without climbing over, it is not enough.

- *Slope failure on spoil piles and sand and gravel storage piles.* More than one worker has been buried in sand and gravel stockpiles. Excavation spoils should be stacked at a minimum of a 1:1 slope. Clean sands and gravels piles are on the verge of slope failure when they are dumped out of a truck or loader bucket. The classic situation for an accident occurs where pipe bedding or structure base aggregates are stockpiled at the site and then moved by a loader as needed on the job. The loader side of the pile can stand at steep angles for long periods. Generally there is no barricading because the loader is moving in and out of the area. It is surprising how often this potential hazard is ignored on a construction site. The loader operator should always knock the top of the spoil pile down at the end of the day, and workers should be warned or prevented from having access to the area during the shift.

References

Abramson, Lee W., et al., *Slope Stability and Stabilization Methods*, John Wiley & Sons, New York, 1996.

Deschamps, R. J and Leonards, G.A., *A Study of Slope Stability Analysis*, ASCE Geotechnical Special Publication No. 31 pp 222–226, 345 East 47th Street, New York, New York,10017-2398, 1992.

Franklin, John A., and Dusseult, Maurice B., *Rock Engineering Applications*, McGraw-Hill, New York, 1991.

Keaton, Jeffery R. M., Robison, Robert M., and Bott, Jacqueline D. J., *Landslide Hazard Analysis for Pipeline Design, Northeast Utah*, American Society of Civil Engineers (ASCE), New York, 1992.

Peck, R. B., Hanson, W. E., and Thornburn,T.H. *Foundation Engineering*, John Wiley & Sons, New York, 1974.

Sowers, George B., and Sowers, George F., *Introductory Soil Mechanics and Foundations*, Macmillan, New York, 1970.

Terzaghi, K., and Peck, R., *Soil Mechanics in Engineering Practice*, John Wiley & Sons, New York, 1967.

Terzaghi, K., Peck, R., and Mesri, G., *Soil Mechanics in Engineering Practice*, 3rd ed., John Wiley & Sons, New York, 1996.

US Department of Labor., *29CFR 1926, Appendix A Soil Classification*, Occupational Safety & Health Administration., 200 Constitution Avenue, NW, Washington, DC 20210, www.osha.gov. January 29 2008.

Yokel, Felix Y., and Stevanich, Ronald L., *Development of Draft Construction Safety Standards for Excavations–Volume*, National Bureau of Standards, Department of Commerce, Wasahington, April 1983.

CHAPTER 9

Shoring Systems Selected from Tabulated Data

9.1 Introduction

Excavation planning revolves around the primary assumption that a stable excavation can be achieved. Without stability nothing else can be done inside the excavation. Given enough lateral extent and a stable bottom, in theory any excavation could be stabilized by open cut methods. It may be totally out of the question economically. At the point where open cut excavations are ruled out due to spatial or economic considerations, the remaining option for excavation work started from the surface is shoring.

Selection of the shoring system is a critical function of every excavation project. The shoring system is the first thing constructed inside an excavation; construction of everything else revolves on being able to work inside the shoring. The shoring system will have an effect on the cost of construction and quality of the production work whether it is pipeline or structure.

The shoring concept should always be started in the estimating stage of the excavation project. This is truly the only way a realistic bid can be arrived at. The shoring concept may be changed completely or refined after the bid, and it may be changed again in the field during construction. This is normal and occurs for a variety of reasons. Factors that contribute to decisions about the shoring system can be categorized as follows:

- Compatibility with soil type and excavation depth
    1. Needs to support soil and surcharge loads. Structural adequacy.
    2. Bottom stability considerations. Bottom heave and water cutoff require shoring that extends below the bottom of the excavation.

3. Will the soil stand up long enough to install the shoring? Excavation needs to stand up long enough to set the shoring if trench jacks or shoring shields are to be used.

- Functional
  1. Must provide worker protection.
  2. Might need to protect existing facilities.
  3. Must allow installation of production work.
  4. Contract limitations. Sometimes certain types of shoring are disallowed by the design engineer or geotechnical engineer.
- Economic
  1. Shoring equipment availability. Owned or rented.
  2. Crew experience with the shoring system.
  3. Construction equipment access and capacity to handle shoring system being installed

Selecting the wrong shoring system can easily lead to failure to the point where the whole excavation has to be abandoned and started over or where an accident can occur.

This chapter looks closely at the shoring system selection process and provides information on how to design and install basic types of tabulated shoring systems.

9.2 Timber Shoring

Timber shoring has always been the staple of the mining and tunneling industry, and it still is to a large extent today. A unique aspect of timber shoring is that it can be lifted, cut, and installed by workers instead of requiring specialized equipment needed for steel and most aluminum shoring. In the pipeline and structure foundation excavation industry, it was used extensively until around the 1950s when the availability of heavy lifting and excavating equipment made it easier to use steel shoring. Later manufactured shoring equipment, trench jacks, and shields replaced the wood shoring techniques. In a few isolated pockets of the country, not remote, wood shoring is required partially as a result of unions trying to preserve carpenters' jobs on pipeline work and because the methods used are tried, true, and predictable. Wood is also still used today, to some extent, because it is easily adapted to different situations. It is also cheap and easy to leave in place. In soft clays and runny noncohesive soils, it is possible to advance vertical timbers, 6 to 8 in wide, behind horizontal wales with sledge hammers or light pounding equipment as the excavation is being advanced downward. Working from scaffolds to start the shoring, 20-ft-deep excavations can be shored using only the workforce. These excavations can be stepped in and extended even deeper. The advent of steel road plates, usually 1 in thick by 8 ft wide, has replaced timber lagging in pile and lagging shoring schemes. The rental cost on steel road plate usually eclipses the cost of timber after 3 months, so shoring that will be in the ground less than 3 months is usually steel and not abandoned in place.

Most likely today the place where timber shoring applications will be encountered is in special circumstances, such as places along an excavation where trench jacks or shields cannot be used due to obstructions like crossing lines or concrete structures that extend into the pipeline excavation zone. Timber is also often used to advance small crawl holes down to existing lines or tunnel under footings and pipelines. Lagging is the other major use for wood. For shoring purposes wood is readily available at the nearest lumber company and should not be overlooked when an unplanned shoring need arises.

9.2.1 Timber Shoring Design Using Tabulated Data

Even though timber shoring is not used extensively today, it is important to develop a basic understanding of what can be done with it. In an emergency or unplanned need for shoring, lumber is readily available at the nearest lumber company, and quite often adequate materials are sitting on the ground at the construction site.

OSHA provides tabulated data for timber shoring in 1926 Subpart P Appendix C to section 1926.652 (c)(1). Tables for use in type A-25, B-45, and C-80 soil using oak timber and surfaced lumber are provided for trench shoring systems consisting of vertical sheeting, wales, and cross-struts. From the tables, member sizes and spacing can be obtained. The tables are specific to this type of system although there is the option for equivalent timber members or equivalent members of different materials. It is possible to use the tables to develop an entire steel sheet, wale, and strut system; however, it is more likely that an equivalent steel member would be substituted for one of the timber members. The timber tables provide an opportunity for a competent person to piece together an abnormal shoring configuration using tabulated data and should not be overlooked This can mean that when an abnormal situation arises, the operation does not have to be shut down until an engineer can come out and figure out a solution; the competent person can do it and justify it through tabulated data. Remember that the tables provide for substitution of materials and not for different member spacing than what is tabulated, and that the competent person has to be able to read the tables and know how to determine equivalent member sizes (see Sec. 9.2.2). The only productive way to acquire a good understanding of the tables is to step through a few examples. OSHA does a good job with that, and this book will not go any deeper into it, except to further discuss the notes and terminology being used.

Understanding the following distinctions is basic to achieving safe timber shoring design:

- Due to size differences there are two sets of tables, one set for mixed oak timber or equivalent with an allowable bending stress of 850 psi and the other for dimensional lumber with an equivalent allowable bending strength of 1500 psi. The geometric properties of any shape are directly related to the dimensions. The term *timber* is generally used to describe unsurfaced lumber that measures the full size of the description. Finished wood is called dimensional lumber and has a nominal dimension that is ½ in less than the description for 2 to 6 in and ¾ in less than the dimension for 7 in and larger. A 6 × 8 board measures 5½ in × 7¼ in, and a 6 × 8 timber measures 6 in × 8 in. Dimensional lumber will measure different if it is not surfaced on all sides. It comes in S4S (surfaced four sides), and S2S. The second set of OSHA tables refers to lumber that is S4S. Lumber sizes are not as strong as timber sizes. A 6 × 6 timber is stronger than 6 × 6 lumber of the same material.
- Lumber species and quality have a large effect on strength. There are several different grades of each type of lumber. The best is select structural and the worst is No. 3. The most common species of lumber used in shoring is Douglas fir and mixed oak. Most of the fir lumber that comes off the shelf at lumber companies is No. 2 and better. Select structural and No.1 are special-order materials. Lumber used for building structures is green (wet) and is considered to have a natural moisture content of 19 percent or less. The tables specify that oak timber have a minimum bending strength of 850 psi, F_b = 850 psi, and Douglas fir F_b = 1500 psi. Design strength values (Table 9.2) are published for green lumber. When it dries to less than 19 percent as is usually the case with used lumber, the strength values are still correct. If the lumber is being used in wet conditions such as under water or used for long periods of time, several weeks, in wet ground, then the strength becomes degraded; see wet service factors in Table 9.2. Green Douglas fir weighs approximately 32 pounds per cubic foot (pcf) and green oak weighs approximately 45 pcf. By grading rules used lumber is as strong as new lumber unless it is damaged, cracked, split, or worn to different dimensions. All used lumber should be visually inspected by a competent person with no hesitancy to cast aside poor-quality material.
- The tables are specific to cross-members, wales, and timber lagging in trenches. Except to allow a 250-lb load on cross-braces, combined axial loads and bending loads are not

addressed; this means that in a four-sided excavation an independent timber system is required for each direction. In other words, members in one direction should not be used to brace the shoring in the opposite direction because they are put into a combination of axial and bending forces.

- Built-up wood sizes are not equivalent to sawn wood sizes. Two 4 × 4 timbers are not as strong as a 4 × 8, etc.
- Tight sheeting holds back water. With the water table at the surface, noncohesive soils load the shoring at approximately 80 lb/ft of depth and soft clays, $c < 500$, at about 85 to 90 psf; and over time, overnight in soft marine clays and days in normal clays, the loading can increase significantly. In soft clays there is also the possibility of bottom heave that can have an effect on stability of the shoring. Even though the tables can be used to a depth of 20 ft in C-80 soil, it is recommended that in soft clays the shoring system be designed by an engineer.
- When using the tables, the competent person should also review OSHA 1926.652(d) Materials and Equipment and 1926.652(e) Installation of Removal and Support. The tables allow vertical spacing of wales at 4 to 5 ft while the notes to the tables limit the distance from the last wale to the bottom of the excavation to 2.5 ft if the sheeting is not embedded and to 3 ft if it is. The reason for this is that the bending moment for a cantilever member is 2 times that of a supported member. It is not clear if the wording *bottom of the excavation* means bottom at any given time during the excavation process or just the final bottom. Calculations indicate that in type B and C soils after the bottom is approximately 10 ft below the surface, the sheeting will become overstressed and at 20 ft dangerously overstressed if the wale is not placed within 2.5 ft of the bottom. This means that a temporary wale would have to be placed halfway until the 4- to 5-ft wale is in place.

Notes to Timber Shoring Tables

Figure 9.1 is intended to illustrate the following text found in OSHA Appendix C (g) Notes for All Tables.

1. Member sizes at spacings other than indicated are to be determined as specified in 1926.652(c), Design of Protective Systems.
2. When conditions are saturated or submerged, use tight sheeting. *Tight sheeting* refers to the use of specially edged timber planks (e.g., tongue and groove) at least 3 in thick, steel sheet piling, or similar construction that when driven or placed in position,

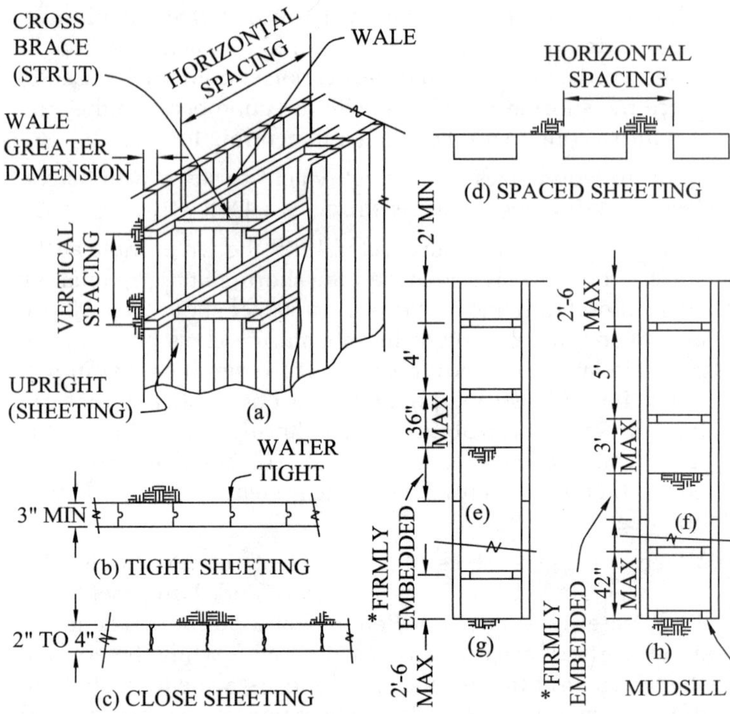

FIGURE 9.1 Illustration of Appendix C (g) Notes for All Tables. (a) Elements of timber trench shoring system; (b), (c), and (d) sheeting configurations; (e) and (f) table alternatives for wale spacing and toe-in; (g) and (h) alternatives for no toe-in.

provide a tight wall to resist the lateral pressure of water and to prevent the loss of backfill material. *Close sheeting* refers to the placement of planks side by side, allowing as little space as possible between them.

3. All spacing indicated is measured center to center.
4. Wales are to be installed with the greater dimension horizontal.
5. If the vertical distance from the center of the lowest cross-brace to the bottom of the trench exceeds 2½ ft, uprights shall be firmly embedded or a mudsill shall be used. Where uprights are embedded, the vertical distance from the center of the lowest cross-brace to the bottom of the trench shall not exceed 36 in. When mudsills are used, the vertical distance shall not exceed 42 in. Mudsills are wales that are installed at the toe of the trench side.
6. Trench jacks may be used in lieu of or in combination with timber cross-braces.
7. Placement of cross-braces. When the vertical spacing of cross-braces is 4 ft, place the top cross-brace no more than 2 ft below

	OSHA Soil Type				
	A-25	B-45		C-80	
Depth (ft)	$q_u = 1.5$ tsf	$\phi = 32$ deg	$q_u = 0.75$ tsf	$\phi = 28$ deg	$q_u = 0.7$ tsf
10	0.25	2	1	3	3
15	0.25	2.5	2	4	3
20	0.25	4	3	5	3

Table 9.1 Timber Sheeting Toe-in Required for 3-ft Clear Span to Bottom of Excavation (ft)

the top of the trench. When the vertical spacing of cross-braces is 5 ft, place the top cross-brace no more than 2.5 ft below the top of the trench

In Appendix C (g) note 5 the term *firmly embed* is used to refer to how far the sheeting toe needs to be driven below the bottom of the excavation; however, it is not defined. Table 9.1 shows the result of calculations and values derived by the author for required embedment. It is important to note that in type C soils at 20 ft deep if the cohesion of the soil is less than 600 psf, there is no toe-in depth that will brace the toe; however, the toe-in will help to prevent bottom heave. In this type of soil a temporary brace should be used at 2 ft below the bottom wale, and then a mudsill should be installed. In noncohesive soil where the water table inside the trench is at the bottom of the excavation, the toe-in resistance is also questionable and temporary bracing should be used at 2 ft above.

9.2.2 Timber Shoring Design by a Registered Engineer

Learning timber design is definitely a case of teaching a person to fish instead of giving that person fish, and in this case OSHA has limited the given fish to what can be caught from a single pond in the form of a sheeting-wale-strut system. Timber shoring design requires only basic engineering principles and is easy to learn. By using the basic principles any timber system or single member can be designed by the competent person, estimator, or field engineer. This is not to suggest that the final design should not be approved by a registered engineer, just that the preliminary work can be performed by anyone interested in the subject. The procedure presented here focuses on and is limited to shoring design; however, the principles apply to all wood design applications. The design approach comprises four steps:

1. Draw the configuration.
2. Determine the soil loading and distribution of forces on the members.

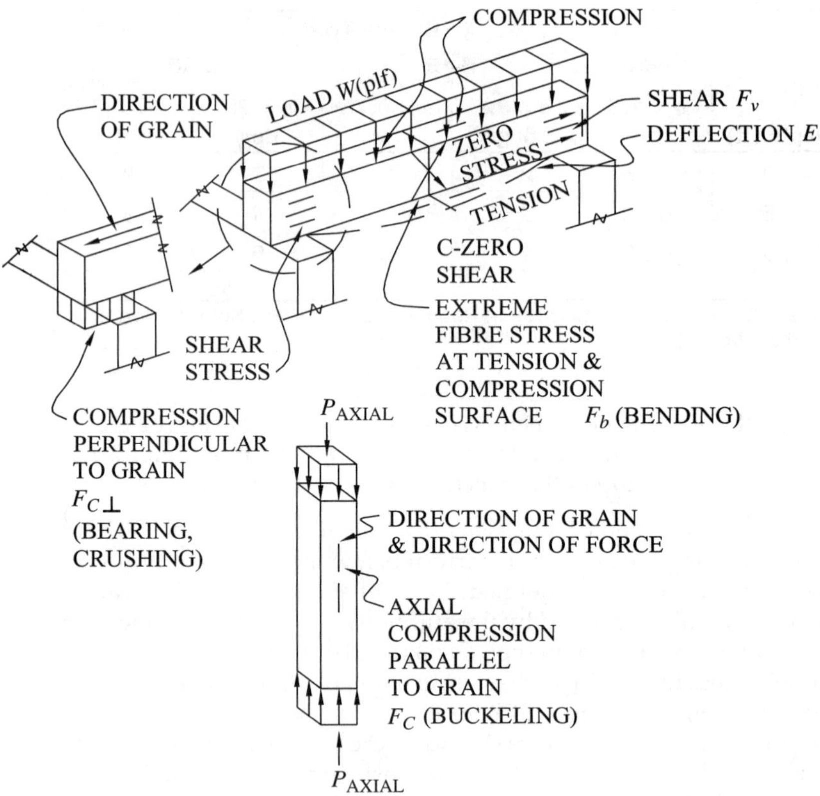

Figure 9.2 Stresses on wood beam and column.

3. Determine the bending, shear, and axial forces on the members.
4. Size the members, using allowable stress design.

Allowable Stress Design Concept and Formulas

Figure 9.2 illustrates the location and direction of stresses in wood design. In shoring design, sheeting and wales are sized for bending and checked for shear stress. The deeper the member, the greater the resistance to bending; for instance, a 4 × 8 is stronger than a 6 × 6 even though the 6 × 6 has greater area. Deflection is usually checked only if there is a minimum deflection specified. Cross-members or struts are designed to support end loading that causes buckling. Balanced area is the significant factor in column or strut design. A 6 × 6 makes a better strut than a 4 × 8.

In wood design all beams are assumed to be simple even though they may be continuous over more than one support. Due to this the analysis usually winds down to distributed loads or point loads on simple beams and cantilever beams that can be calculated by using a few simple formulas. The most common are as follows:

$M = \frac{w\ell^2}{8}$	maximum moment for simple beam with distributed load w	(9.1)
$V = \frac{w\ell}{2}$	maximum shear for simple beam with distributed load w	(9.2)
$M = \frac{w\ell^2}{2}$	maximum moment for cantilever beam with distributed load w	(9.3)
$V = w\ell$	maximum shear for cantilever beam with distributed load w	(9.4)
$M = \frac{P\ell}{4}$	maximum moment for simple beam with point load P at center	(9.5)
$V = \frac{P}{2}$	maximum shear for simple beam with point load P at center	(9.6)

where M = bending moment (ft·lb)
V = shear (lb) or (K = 1000 lb)
w = distributed load (lb/ft)
ℓ = length (ft)
P = force (lb or k = 1000 lb)

Figure 9.3 shows loading diagrams, shear diagrams, and moment diagrams for these simple and cantilever beams. The purpose of shear and moment diagrams is to find where the maximum shear and moment are on the beam. With simple beams the moment is always greatest near the middle of the span, and the shear is greatest at the supports. With cantilever beams the moment and shear are always greatest at the support. A large assortment of beam diagrams and formulas are available in the AISC manual of steel construction.

Stress is the result of a force applied to an area. Geometric properties of the section shape, area, moment of inertia, section modulus, and radius of gyration are used in conjunction with shear, moments, and reactions to determine stress. The stress formulas that apply to wood are as follows:

$f_b = \frac{M}{S}$	bending stress	(9.7)
$f_v = \frac{3V}{2bd}$	shear stress	(9.8)
$f_a = \frac{P}{A}$	axial stress parallel to grain	(9.9)

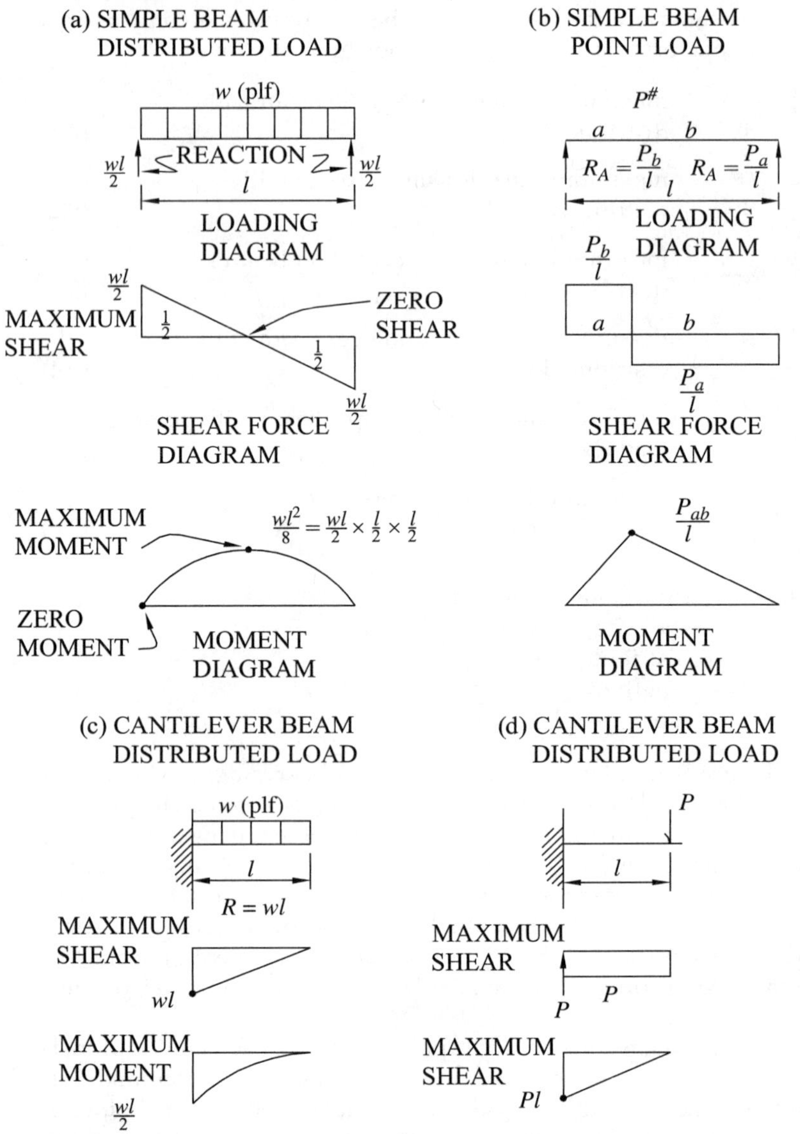

FIGURE 9.3 Simple and cantilever beam diagrams and formulas for shoring design.

$$f_{\perp} = \frac{P}{A} \qquad \text{axial stress perpendicular to grain} \qquad (9.10)$$

where S = section modulus, in^3 $\quad S = \dfrac{bd^2}{6}$

b = beam width, in

d = beam depth, in

A = area = bd, in^2

Computed stress from the actual loading condition is referred to with a lowercase f and allowable stress with an uppercase F, both subscripted with the type of stress referred to; for example, f_b refers to computed bending stress and F_b refers to allowable bending stress. The design process entails sizing structural members so that $f_{cumputed} \leq F_{allowable}$.

Allowable Stress Design Values

Allowable stress values for timber were first developed by a technical design advisory committee starting in 1944 and accepted as a national standard in 1991. They are published in the *National Design Specification for Wood Construction* (NDS). In 1997 the advisory committee developed a set of adjustment factors for use with the stress tables to account for variations in materials and types of uses of wood products. These factors were not in use when the OSHA timber shoring tables and the construction use design values were first developed. Table 9.2 gives allowable stress values and adjustment factors most applicable to shoring design.

In construction, timber of mostly Douglas fir and oak are used extensively, mostly for bridge falsework, concrete formwork, and shoring. Since these uses are temporary, the quality and source of the materials vary, they are often reused, and grade stamps are lost. The construction industry has adopted their own simplified allowable stress values with their root based on the modulus of elasticity E (Table 9.3). The table does not refer to lumber grades; however, it does require that the values be factored by the ratio $E_{lumber}/1{,}600{,}000$. The author suggests that for construction applications the easiest way to determine allowable stress values is to use Table 9.3. Table 9.2 is presented in order to see just how much design values between Douglas fir and oak vary and as a means for comparison. Also the construction standard values are for lumber with $E \geq 1{,}600{,}000$ psi; to use other species of lumber, the allowable stresses should be factored. For instance, if mixed oak is to be used for a bending member, the F_b of Table 9.3 should be factored by

$$F_b = \frac{E_{oak}}{1{,}600{,}000} \times F_{b\,table\,9.3}$$

One factor, among others, that is not listed in Table 9.2 is the frequently used load duration factor C_D, because it applies to all species. The ASD Table 2.3.2 found in that document specifies that for 2 months' duration (snow load) $C_D = 1.15$ and for 7 days (construction loads) $C_D = 1.25$. The factor table also excludes use of the factor on the modulus of elasticity E and $F_{c\perp}$. This factor is similar to the shoring industry devised "shoring use factor = 1.33" and should not be used in addition to C_D. From the comparisons below it is evident that the shoring use factor is incorporated into the table.

Species and Commercial Grade	Size Classification	Bending F_b	Tension Parallel to Grain F_t	Shear Parallel to Grain F_v	Compression Perpendicular to Grain $F_{g,\perp}$	Compression Parallel to Grain F_c	Modulus of Elasticity E	Grading Rules
Douglas fir-larch	S4S							
No. 1 & better	2–4 in thick	1200	800	95	625	1550	1,800,000	WWPA
No. 2 & better	2 in and wider	900	575	95	625	1350	1,600,000	WWPA
Mixed oak	S4S							
No. 1 & better	2–4 in thick	825	500	85	800	825	1,000,000	NELMA
No. 2 & better	2–4 in thick	800	475	85	800	625	900,000	NELMA
Adjustment Factors[†]								
	Repetitive member	$C_r = 1.15$						
	Size	$C_f = 1.1$	$C_f = 1.1$			$C_f = 1.1$		
	Flat use	$C_{fu} = 1.15$						
	Shear stress			$C_h = 1$				
	Wet service	$C_m = 0.85$	$C_m = 1$	$C_m = 0.97$	$C_m = 0.67$	$C_m = 0.8$		
Douglas fir-larch	Timber							
No. 1	< 5 × 5 in beams and stringers	1350	675	85	625	925	1,600,000	WWPA
No. 2 & better	< 5 × 5 in beams and stringers	875	425	85	625	600	1,300,000	WWPA

No. 1	< 5 × 5 in posts and timbers	1200	825	85	625	1000	1,600,000	WWPA
No. 2 & better	< 5 × 5 in posts and timbers	700	475	85	625	475	1,300,000	WWPA
Mixed oak	Timber							
No. 1	< 5 × 5 in beams and stringers	1150	550	80	800	825	1,000,000	NELMA
No. 2 & better	< 5 × 5 in beams and stringers	725	375	80	800	450	800,000	NELMA
No. 1	< 5 × 5 in posts and timbers	1000	675	80	800	775	1,000,000	NELMA
No. 2 & better	< 5 × 5 in posts and timbers	575	400	80	800	350	800,000	NELMA
Adjustment Factors[†]								
	Repetitive member	$C_r = 1$						
	Size	$C_f = 1$	$C_f = 1$			$C_f = 1$		
	Flat use	NA						
	Shear stress			$C_h = 1$				
	Wet service	$C_m = 0.85$	$C_m = 1$	$C_m = 1$	$C_m = 0.67$	$C_m = 0.91$		

* Tables are extracted from *National Design Specification for Wood Construction,* 1997 edition.

† These adjustment factors are selected as most often applying to shoring aplications. Adjustment factors vary with use and should be reviewed in NDS Tables 4A and 4D.

Table 9.2 Design Values for Visually Graded Lumber (Allowable Stress), psi*

These design stresses are based on Douglas fir larch group II or equivalent lumber or timber with a modulus of elasticity $E \geq 1{,}600{,}000$.

Compression perpendicular to grain $F_{\perp}$ 450 psi

Compression parallel to grain $F_c = \dfrac{480{,}000}{(L/D)^2}$ psi ≤ 1600 psi

Where L = unsupported length, in
d = least dimension, in

Flexural stress
$F_b = 1800$ psi for depth > 8 in
$F_b = 1500$ psi for depth ≤ 8 in

Horizontal shear
$F_v = 140$ psi

Axial tension
$F_t = 1200$ psi

Modulus of elasticity
$E = 1{,}600{,}000$ psi

Timber connections shall be designed in accordance with NDS for stress graded lumber and its fastenings except that reductions in allowable loads required therein for high-moisture condition of the lumber and service conditions shall not apply.

Notes:

1. Timber is considered non-surfaced lumber that is full nominal dimension. Lumber is considered surfaced four sides, S4S, and is ½ in less than nominal dimension for boards to 7 in wide and ¾ in less than nominal for boards 8 in wide and greater.
2. The designer should carefully evaluate the grade and condition of the proposed lumber especially if it is used material. These short form design stresses have been in use since before 1997 when NDS came out with adjustment factors for the effects of knots, slope of grain, splits, checks, size, moisture content, duration of load, and repetitive use. When buying new lumber a No. 1 grade should be requested.

TABLE 9.3 Timber Design Values for Construction Applications

As a comparison, if a mixed oak No. 2 and better, 3 × 8 timber were to be used for sheeting, the allowable stress from Table 9.2 would be

$$F_b = 800 \times C_r \times C_F \times C_{fu} \times C_D = 800 \times 1.15 \times 1.1 \times 1.15 \times 1.25$$
$$= 1455 \text{ psi}$$

Table 9.3 allows $F_b = 1500 \times 900{,}000/1{,}600{,}000 = 850$ psi
And a Douglas fir No. 2, 6 × 8 wale

$$F_b = 825 \times C_D = 825 \times 1.25 = 1031 \text{ psi}$$

Table 9.3 allows $F_b = 1500 \times 1{,}300{,}000/1{,}600{,}000 = 1219$ psi
For a Douglas fir No. 1, 6 × 8 wale, the calculation yields $F_b = 1350 \times 1.25 = 1687$ psi and Table 9.3 allows 1500 psi. In the case of wales it is obviously important to get the grade right as No. 1 lumber is 40 percent stronger than No. 2.

Further similar calculations show that Table 9.3 is conservative compared to NDS requirements except in the case of horizontal shear F_V. In Table 9.3, $F_V = 140$ psi and for Table 9.1 $F_V = 80 \times 1.25 = 100$ psi to $F_V = 95 \times 1.25 = 119$ psi. The significance of this is hard to evaluate; however, having designed a lot of bridge falsework, the author can state that in most cases calculations using S4S 2 × 4 and 4 × 4 lumber, by far the most used in bridge falsework, would fail if F_V were limited to 119 psi. In the field there is no evidence of failure due to allowing $F_V = 140$ psi.

Connections

Although there is no requirement given on how to accomplish it, OSHA requires in 1926.652(e)(1)(i) that

> Members of support systems shall be securely connected together to prevent sliding, falling, kickouts, or other predictable failure.

Duplex nails and 8d and 16d and 20d spikes are the common method of fastening timber. Table 9.4 gives NDS tabulated values for these fasteners to be used in conjunction with adjustment factors. The 1.25 duration factor could be used on the tabulated values; however, considering that most likely 1 out of 5 nails probably splits the wood or misses the mark, it is advisable to be conservative. Failed connections in shoring generally result in bad falls from workers standing on failing connections. NDS also provides tables for bolts, lag bolts, and screws.

In the OSHA timber tables there is a limit of 240 lb at the middle of a cross-member. It is very likely that a 240-lb worker would climb the cross-struts at the wall and cause a 240-lb shear force at that connection. From the table, 4d to 16d box nails toe nailed would provide 336 lb of shear and should be considered as a minimum nailing requirement. Wales are usually deeper than wide and have a tendency to roll, especially when they are in the horizontal direction, so there should be a 16d toe nail every 2 ft on center staggered (Fig. 9.4).

Box Nails							
Nail				Douglas Fir-Larch (N) Alignment		Red Oak Alignment	
Penny-weight	Diameter Box	Nail Length	Penetration	Perp.	Toe	Perp.	Toe
	(in)	(in)	(in)	T	T	T	T
8d	0.113	$2\frac{1}{2}$	1	71	59	94	78
16d	0.128	$3\frac{1}{2}$	$1\frac{1}{4}$	101	84	135	112
20d	0.148	4	$1\frac{1}{2}$	115	95	154	128
Common Nails							
Nail				Douglas Fir-Larch (N) Alignment		Red Oak Alignment	
Penny-weight	Diameter Common	Nail Length	Penetration	Perp.	Toe	Perp.	Toe
	(in)	(in)	(in)	T	T	T	T
8d	0.131	$2\frac{1}{2}$	$\frac{3}{4}$	87	72	127	105
16d	0.162	$3\frac{1}{2}$	$1\frac{1}{4}$	138	115	184	153
20d	0.192	4	$1\frac{1}{2}$	166	138	222	184
Spikes							
Nail				Douglas Fir-Larch (N) Alignment		Red Oak Alignment	
Penny-weight	Diameter Spike	Nail Length	Penetration	Perp.	Toe	Perp.	Toe
	(in)	(in)	(in)	T	T	T	T
16d	0.207	$3\frac{1}{2}$	1	162	134	243	202
20d	0.225	4	$1\frac{1}{4}$	193	160	268	222

Applicable Adjustment Factors			
Load duration	C_D	1.25	Not included in table C_D impact factor does not apply to connections.
Wet service	C_M	0.7	Not included in table
Penetration depth	C_d	$C_d = \frac{P}{12D} \leq 1$	Included in table
Toe nail	C_{tn}	0.83	Included in table

Table 9.4 Nail and Spike Single-Shear Design Values, lb

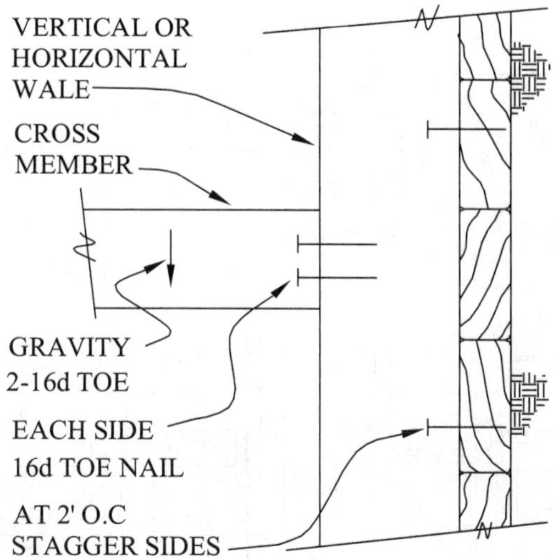

FIGURE 9.4 Wale and strut nailing recommendation.

Example 9.1
A 60-in-OD concrete pipeline project has to cross under an 8-ft × 12-ft concrete box culvert. A bobcat-dug, hand shored excavation is planned using timber shoring (Fig. 9.5). Determine the size and spacing of the shoring members.

Step 1. Determine soil loading.

From Fig. 9.5 Section B-1, the top 10 ft is stiff clay weighing 120 pcf, and the bottom 10 ft is soft marine clay weighing 90 pcf with the boring log showing 6 blows per foot and $q_u = 0.7$ tsf. Since $q_u = 2c = 1400$ psf, $c = 700$ psf. At the bottom of the excavation $q_u = 0.5$ tsf. Dewatering is to the bottom of the excavation. Using the apparent pressure diagram from Peck, Hanson, and Thornburn in Fig. 6.9 gives

$$\gamma H = 120 \times 10 + 90 \times 10 = 2100 \text{ lb}$$

$$\frac{\gamma H}{c_b} = \frac{2100}{500} = 4.2 > 4$$

$$\therefore \quad \text{Use } P_a = \gamma H - 4mc \qquad \text{where } m = 0.4$$

$$P_a = 2100 - 4 \times 0.4 \times 700 = 980 \text{ psf}$$

Since the apparent pressure diagram is for strut loading, sheeting and wale loading will be approximately 20 percent less than calculated for struts. This soil is fairly fluid so 85 percent of the 980 psf load will be used for sheeting and wales and 100 percent for the struts. Close sheeting, not tight, should be used to prevent water table from backing up behind sheeting.

Sheeting loading	980 psf × 85% + 100 psf surcharge = 935 psf
Wale loading	980 psf × 85% + 100 psf surcharge = 935 psf
Strut loading	980 psf soil + 100 psf surcharge = 1080 psf

Bottom heave should also be investigated.

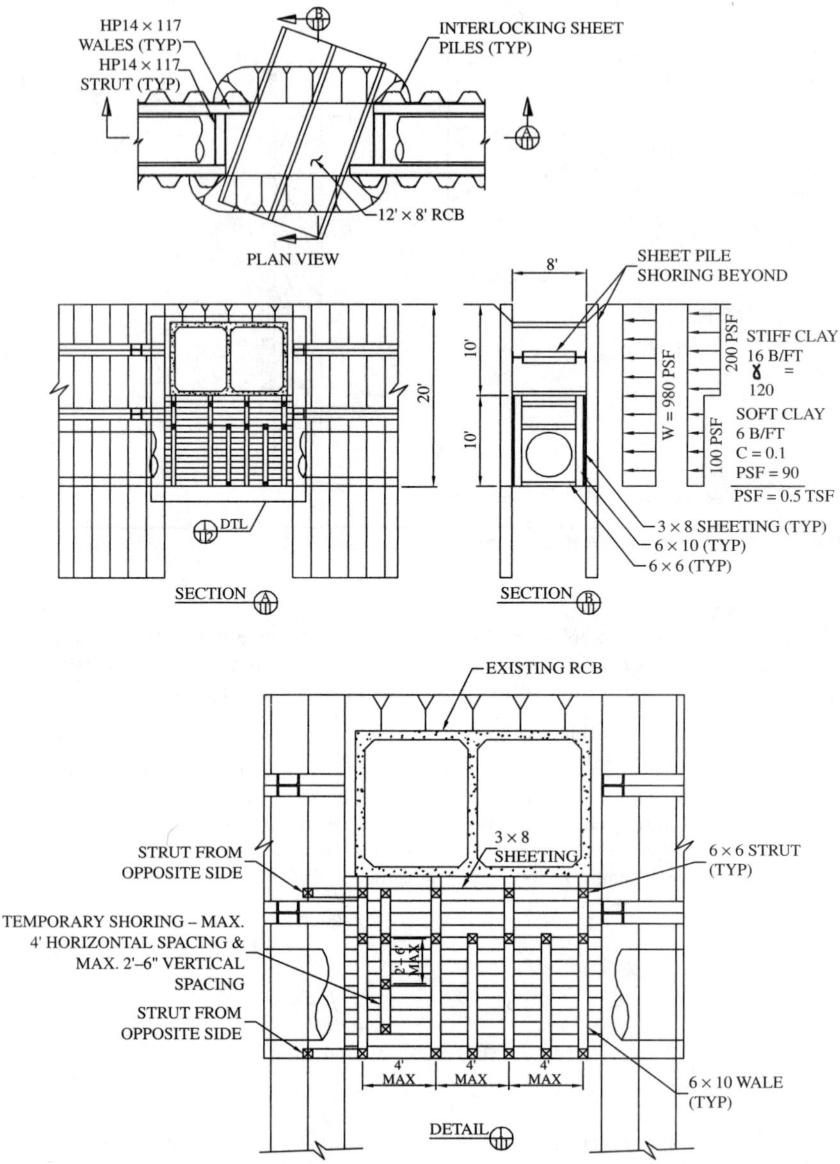

Figure 9.5 Example 9.1 timber shoring plan at 8 × 12 RCB.

Step 2. Sheeting design

$$\text{Maximum wale spacing} = 4 \text{ ft}$$

$$\text{Soil loading } w = 935 \text{ psf}$$

$$M = \frac{w\ell^2}{8} = \frac{935 \times 4^2}{8} = 1870 \text{ ft·lb}$$

From Table 9.3,

$$F_b = 1500 \text{ psi} \qquad \text{Douglas fir } E = 1{,}600{,}000 \text{ psi}$$

Since

$$\frac{M}{S} = F_b \Rightarrow S_{req} = \frac{M}{F_b} = \frac{1870 \times 12}{1500} = 15 \text{ in}^3/\text{ft of sheeting width}$$

use 3-in timber.

$$S = \frac{3^2 \times 12}{6} = 18 \text{ in}^3 < 15 \text{ in}^3 \qquad \text{OK}$$

To make timber easier to handle, use 3 × 8 or 3 × 6. Check shear.

$$V = \frac{935 \times 4}{2} = 1870 \text{ lb}$$

$$f_v = \frac{3V}{bd} = \frac{3 \times 1870}{2 \times 3 \times 12} = 78 \text{ psi} < 140 \qquad \text{OK}$$

Step 3 Design wale. Ultimately a 6-ft clear span will be required to clear the 5-ft-OD pipe that is going to be installed. The installation procedure will be to excavate an approximately 2- to 3-ft-deep lift, temporally shore and strut, and afterward install full-length wales and struts so that the 6-ft span can be achieved. Preliminary calculations indicate that an 8 × 10 wale will span 4 ft if spaced at 4 ft on center and 6 ft if spaced 2 ft on center.

$$\text{Tributary sheeting load for wale spaced at 4 ft OC} = 4 \times 935 = 3740 \text{ plf}$$

$$\text{Wale span strut to strut} = 4 \text{ ft OC}$$

$$W = 3740 \text{ plf}$$

$$M = \frac{3740 \times 4^2}{8} = 7480 \text{ ft·lb}$$

Section modulus for 6 × 10 is

$$S = \frac{10^2 \times 8}{6} = 133 \text{ in}^3$$

$$f_b = \frac{7480 \times 12}{133} = 897 \text{ psi} < 1800 \qquad \text{OK (Table 9.3)}$$

$$\text{Shear} = 3740 \times 4/2 = 7480 \text{ lb}$$

$$f_v = \frac{3}{2} \times \frac{7480}{8 \times 10} = 140 \text{ psi} \le 140 \text{ psi} \qquad \text{OK}$$

At 6-ft span spaced at 2 ft on center,

$$\text{Tributary sheeting load for wale spaced at 2 ft OC} = 2 \times 935 = 1870 \text{ plf}$$

$$\text{Wale span strut to strut} = 6 \text{ ft OC}$$

$$W = 1870 \text{ plf}$$

$$M = \frac{1870 \times 6^2}{8} = 8415 \text{ ft·lb}$$

$$\text{Section modulus for } 6 \times 10 \text{ is } S = \frac{10^2 \times 8}{6} = 133 \text{ in}^3$$

$$f_b = \frac{8415 \times 12}{133} = 760 \text{ psi} < 1800 \qquad \text{OK}$$

$$\text{Shear} = 1870 \times 6/2 = 5610 \text{ lb}$$

$$f_v = \frac{3}{2} \times \frac{5610}{8 \times 10} = 105 \text{ psi} \le 140 \text{ psi} \qquad \text{OK}$$

Step 4. Design struts.

$$\text{Strut tributary area} = (4 \times 2) + (2 \times 3) = 14 \text{ ft}^2$$

$$\text{Strut loading } P = 1080 \text{ psf} \times 14 \text{ ft}^2 = 15{,}120 \text{ lb}$$

$$\text{Strut length} = 8 \text{ ft}$$

Try 4 × 6:

$$F_c = f_c = \frac{15{,}120}{4 \times 6} \; 630 \text{ psi}$$

From Table 9.3,

$$F_c = \frac{480{,}000}{(L/d)^2} \text{ psi} \le 1600 \text{ psi}$$

$$F_c = \frac{480{,}000}{(8 \times 12/4)^2} = 833 \text{ psi}$$

Check allowable bearing.

$$F_{c\perp} = 450 \text{ psi} < 630 \text{ psi} \qquad \textit{No Good,}$$

Try 6 × 6:

$$f_{c\perp} = \frac{15{,}120}{6 \times 6} = 420 \text{ psi} \qquad \text{OK}$$

Try 4 × 8:

$$f_{c\perp} = \frac{15{,}120}{4 \times 8} = 472 \text{ psi} \qquad \text{NG}$$

Use 6 × 6 timber struts.

9.2.3 Equivalent Section

Equivalent members can be substituted for given members based on section geometric properties and allowable strength requirements for the new materials. For instance, given a section modulus requirement for a certain type of material, any other type of material can be substituted by using $M = SF_b$ if the allowable bending strength of both materials is known, using

$$S_2 F_{b2} = S_1 F_{b1}$$

and

$$S_2 = \frac{S_1 F_{b1}}{F_{b2}} \tag{9.11}$$

where S_1 = section modulus of original material
F_{b1} = allowable bending stress of original material
S_2 = to be solved for section modulus of new section
F_{b2} = allowable bending stress for new material

In Example 9.1 the 8 × 10 timber wale spaced at 2 ft on center could be substituted with a steel member spaced at 4 ft OC that could handle the 6-ft span. The tributary beam loading and hence moment are twice those of the 2-ft spacing,

$$M = \frac{1870 \times 2 \times 6^2}{8} = 15{,}830 \text{ ft·lb}$$

and

$$S_{req} = \frac{M}{F_b} = \frac{15{,}830 \times 12}{1500} = 123 \text{ in}^3 \qquad \text{for wood}$$

$$S_2 = \frac{S_1 F_{b1}}{F_{b2}} = \frac{123 \times 1500}{44{,}000} = 4.2 \text{ in}^3 \qquad \text{for steel}$$

where $S_1 = 123 \text{ in}^3$ section modulus of wood
$F_{b1} = 1500$ psi
S_2 = to be solved for section modulus of steel section
$F_{b2} = 50{,}000 \text{ psi} \times 0.66 \times 1.33 = 44{,}000$ psi
Allowable bending strength of 50 ksi steel

A W8 × 10 has a section modulus of 7.81 in^3. Considering that an 8 × 10 timber weighs approximately 18 lb/ft and 180 lb for a 10-ft beam, the equivalent 8 × 10 steel beam would weigh 100 lb and fewer beams would be required. This beam would need to be checked for shear and point load bearing at the struts. The timber strut load would be increased and would need to be checked. A stable steel-to-wood strut connection would have to be designed.

When substitutions are made, the new member should always be checked for all affected design stresses.

9.2.4 Soil Arching and Timber Lagging

Timber lagging is used in pile and lagging systems to prevent soil raveling or sloughing between the piles (Fig. 9.6). In this system there is an assumption that there is soil arching between the pile, (see Sec. 6.4 on soil arching). Although it is counterintuitive, soil arching is not depth-dependent, as it works at any depth. Arching capacity does vary with soil type and strength. The more fluid the soil gets, the shorter the reliable arch distance. Soft clays can shear and squeeze out, and noncohesive soils can ravel out in the area of the arch void, the area between the arch and the lagging. Table 9.5 was developed based on the arching assumption from experience and empirical rules and can be used where it is certain that arching will occur. This table has been used extensively in shoring design since it was developed by the Federal Highway Administration in 1976. Use of this table is not included in OSHA Appendix C tabulated data for timber and is only an option for designs by a registered engineer.

When the table is used, the soil should be in contact with the piles for the arching assumption to be valid. In the case of drilled and set in piles, if the pile hole fill is not in place or runs out during lagging installation, there is no place for the soil to arch to. Lagging

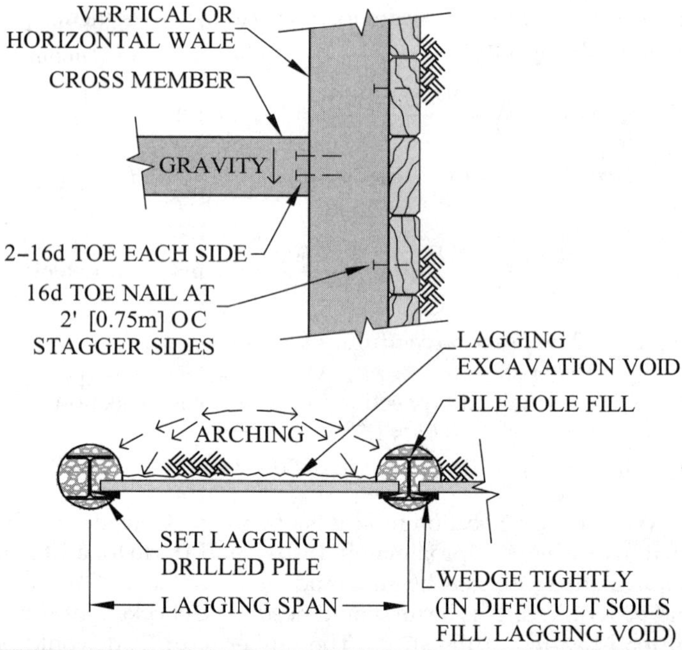

FIGURE 9.6 Pile and lagging system.

installation usually requires hand excavation to make a space for the timbers to fit behind the vertical beams. The excavation should be neat so that there is very little void behind the sheeting and the pile. Wood wedges should be placed between the beam flange and the lagging to get soil contact. In noncohesive soils this prevents the soil from raveling down and leaving a large void behind the sheeting. If the soil is cohesive, the soil can slough into the void and induce settlement and ground movement or voids in other places. If the soil mass from either of these conditions shears into a large void, the soil arch is broken and the lagging can fail rapidly. In loose sand and gravel or soft plastic soil, it is necessary to fill the voids behind lagging by dumping sand or grout in from above or using a grout pump to fill the voids. Seepage and rainstorms can create large voids behind sheeting. It is important to inspect lagging periodically to make sure voids are not developing. When work is done in very soft clays above and below the water table, and in plastic silts and noncohesive soils below the water table, the appropriateness of lagging is questionable. When the ratio of overburden stress to undrained shear strength $\gamma H/c \leq 5$, workers installing lagging should be aware that soil can move during hand excavation and squeeze between openings in the lagging. When lagged walls are opened for pipe intrusions or tunneling

Soil Description Classification	Unified Soil System	Depth (ft)	Recommended Thickness of Lagging (Timber) for Clear Spans of: (ft)					
			5	6	7	8	9	10
Competent Soils								
Silts or fine sand and silt above water table	ML, SM-ML	0 to 25	2"	3"	3"	3"	4"	4"
Sands and gravels (medium dense to dense)	GW, GP, GM, GC, SW, SP, SM							
Clays (stiff to very stiff); nonfissured	CL, CH	25 to 60	3"	3"	3"	4"	4"	5"
Clays, medium consistency and $\gamma H/c < 5$	CL, CH							
Difficult Soils								
Sands and silty sands (loose)	SW, SP, SM	0 to 25	3"	3"	3"	4"	4"	5"
Clayey sands (medium dense to dense) below water table	SC							
Clays, heavily overconsolidated, fissured	CL, CH	25 to 60	3"	3"	4"	4"	5"	5"
Cohesionless silt or fine sand and silt below water table	ML; SM-ML							
Potentially Dangerous Soils (appropriateness of lagging is questionable)								
Soft clays $\gamma H/c > 5$	CL, CH	0 to 15	3"	3"	4"	5"	-	-
Slightly plastic silts below water table	ML	15 to 25	3"	4"	5"	6"	-	-
Clayey sands (loose) below water table	SC	25 to 35	3"	5"	6"	-	-	-

* Adapted and revised from April 1976 Federal Highway Administration Report No. FHWA-RD-75-130.

Notes:

This table was developed by empirical method, s and on the basis of construction grade lumber $F_b = 1200$ psi and $E = 1{,}500{,}000$ psi. Expected deflections are < 1 in. Equivalent steel plate thickness:

3- and 4-in-thick timber	$F_b = 1200$ psi	1-in-thick plate	$F_y = 50$ ksi
5- and 6-in-thick timber	$F_b = 1200$ psi	1-in-thick plate	$F_y = 50$ ksi

Table 9.5 Recommended Thickness of Wood Lagging When Soil Arching Will Be Developed (for Locations without Surcharge Loadings)*

machine entrance, there is the potential danger of soil collapse and worker inundation.

Equivalent Uniform Pressure and Surcharge Loading

Because Table 9.5 is developed empirically it is important to use it conservatively and within the limitations it was developed from. The table was developed from experience with deep excavations and has little correlation to surcharge loading. The table heading states that it applies in locations without surcharge loadings. At 20 ft deep most surcharge loads are reduced to less than 100 psf; however, railroads, cranes, and large spoil piles can easily create larger surcharge loads. By back calculating the distributed load that will produce maximum allowable bending at the cutoff points for different lagging thicknesses, it is possible to develop a theoretical distributed loading on the lagging. The concept of an equivalent distributed load and reserve lagging capacity can be used to deal with lagging materials with different allowable bending strengths and surcharge loads. Table 9.6 was developed by the author for this purpose. This table gives a percentage of the calculated lateral soil load as the distributed lagging load. Because the lagging loading does not increase with depth, the percent lagging load decreases at deeper elevations unless very soft cohesive soils are present. If the equivalent lagging distributed load is used at depths of 25 and 60 ft, the lagging thickness could be calculated from $M = SF_b$, and the results would be very close to Table 9.5 and there would be no reserve capacity in the lagging. It is important to remember that the lagging thickness given in Table 9.5 is still the minimum lagging thickness required.

9.2.5 Equivalent Steel Lagging Section for Timber Lagging

Steel plate is often substituted for timber lagging because it can be cheaper to rent and can be handled and pushed down with the excavator that is digging the hole. Steel plate sections that are equivalent to the lagging sizes in the table can be derived; however, the method of installing the steel plate, pushing in and slicing off soil, will cause the plate to have additional bending forces. Typically 1-in-thick and 1½-in-thick plate is used and works well. Other than saving installation cost, the big advantage is that voids behind the plate are practically eliminated. Using the largest timber size in Table 9.5, 6 in thick, the equivalent plate thickness is, from Eq. (9.11),

$$S_2 = \frac{S_1 F_{b1}}{F_{b2}} = \frac{72 \times 1200}{44{,}000} = 1.96 \text{ in}^3$$

where $S_1 = 72 \text{ in}^3$, section modulus of 6×12 timber
$F_{b1} = 1200$ psi, allowable bending stress for wood
S_2 = to be solved for section modulus of steel section
$F_{b2} = 50{,}000 \text{ psi} \times 0.66 \times 1.33 = 44{,}000$ psi
Allowable bending strength of 50 ksi steel

Equivalent Lagging Distributed Load from Soil, psf			
		Lagging load as % of lateral soil loading (%)	
		Depth (ft)	
Table 8-4 Category		0–25	25–60
Competent soils	Noncohesive	23	11
	Cohesive	21	10
Difficult soils	Noncohesive	20	11
	Cohesive	17	10
Potentially dangerous soils		**0–25**	**25–35**
	Cohesive	38	38

Allowable Timber Lagging Loads at Span, psf							
	Allowable	Lagging Span (ft)					
Lagging Thickness (in)	Bending Moment (ft·lb)	5	6	7	8	9	10
2	800	256	178	131	100	79	64
3	1800	576	400	294	225	178	144
4	3200	1024	711	522	400	316	256
5	5000	1600	1111	816	625	494	400
6	7200	2304	1600	1176	900	711	576

* Table based on timber sizes, $F_b = 1200$ psi, $E = 1{,}500{,}000$

Notes:

1. This table to be used in conjunction with Table 8.4. Lagging thickness at depths shown are minimum requirement.
2. This table can be used to determine lagging reserve strength that can be used to resist additional lagging loads such as surcharge by the following formula: Lagging reserve LR = Allowable lagging load at span − Lateral soil load × equivalent lagging load %
   Example: Lagging between soldier piles spaced at 8 ft OC. Soil loading at 15 ft deep in competent noncohesive soil is 540 psf. What is lagging reserve capacity that can be used for surcharge loads? Table 8.4 requires use of 3-in lagging.
   LR = 225 psf − 0.23 × 540 psf = 100 psf allowable surcharge
   If larger surcharge is anticipated, use 4-in-thick lagging to get
   LR = 400 psf − 0.23 × 540 psf = 275 psf allowable surcharge

TABLE 9.6 Equivalent Lagging Distributed Load and Lagging Reserve Strength*

A 1-in-thick plate has $S = (1^2 \times 12)/6 = 2$ in^3 and a 1½-in plate has $S =$ 4.5 in^3. Due to increased bending stress from advancing the plate, the 1-in plate should be used for 3- and 4-in timber and 1½-in plate for 5- and 6-in timber.

9.2.6 Timber Shoring Safety Issues

In addition to points made in Sec. 9.2.1, other factors that affect safety when shoring with timber include these:

- Timber installation to some extent requires workers to work inside unshored trenches. Even when work is done within the confines of the shored area of the trench, placement of new members requires some sort of momentary incursion into the unshored area.

  Careful thought and planning should go into the installation procedure, keeping in mind that workers have to be protected at all times from cave-in. Temporary intermediate waling and strutting may be required to make this possible. If this is not possible, some other method of shoring is required.

- Timber is the weakest and least predictable in terms of consistent material properties. Failure is usually instantaneous with very little warning and can be progressive due to the close proximity of repetitive members. Excessive deflection is usually the only clue.

  Make sure that soil and surcharge assumptions are correct. Do not let water back up behind lagging unless it is designed for C-80 soil with tight interlocked sheeting.

- Timber shoring members are often used to climb on when one is entering and exiting the trench. Wood cracks and splits around nails easily. Connection failure due to standing on members results in falls and failed shoring elements.

  A ladder should be used. Secure strut connections with a minimum of 4–16d toe nails (Fig. 9.4).

- In soft clays when the overburden stress to undrained shear strength $\gamma H/c > 5$, the soil is potentially dangerous. When $\gamma H/c > \cong 7$, the soil is on the verge of imminent failure. Using an average unit soil weight of 120 psf, these points are reached at the following depths:

Depth (ft)	Cohesion (psf)	
	$\frac{\gamma H}{c} = 5$	$\frac{\gamma H}{c} = 7$
10	240	170
15	360	260
20	480	340
25	600	430
30	720	515

When workers are required to work on shoring in areas where these conditions exist, workers should be made aware of it. Excavations below soil lagging should be limited to 2 ft. An engineered plan should be developed for any lagging removal for pipe penetrations or tunnel heads.

9.3 Aluminum Hydraulic Shoring

Aluminum hydraulic shores have been used successfully in the United States since the early 1950s. At that time they were a major technological advancement over timber shoring due to the fact that they could be installed and removed from outside the trench, so that the worker did not have to risk entering an unshored excavation to install the shoring. Another unique principle of the design that often goes unnoticed is that once the cylinders are pumped out, they are locked into place because in order to rotate back they would have to get wider before they can get smaller (Fig. 9.7). This changes the jack from a set of pin-connected hinges with no stability to an internally braced structure; cylinder buckling stability and stability from rotation are enhanced.

The extent of this added stability is debatable because unlike steel and concrete, earth is compressible and can therefore allow some movement into the trench wall. Added buckling stability is more prominent than resistance of uplift, causing folding. The fact that trench jacks were manufactured, reusable, and hinged so that they could swing into a trench and be pumped out dramatically increased productivity. With the equipment on the trench bank, an 8-ft trench jack can be set in less than 1 min with a two-person crew, where it can take 5 min with a two-person crew working inside an unsafe trench to install the timber shoring.

The original design has changed very little since the early days, and today there are several manufacturers of hydraulic trench jacks. They all look essentially the same with the parts being almost interchangeable. The basic elements of the trench jack are the rail, pin,

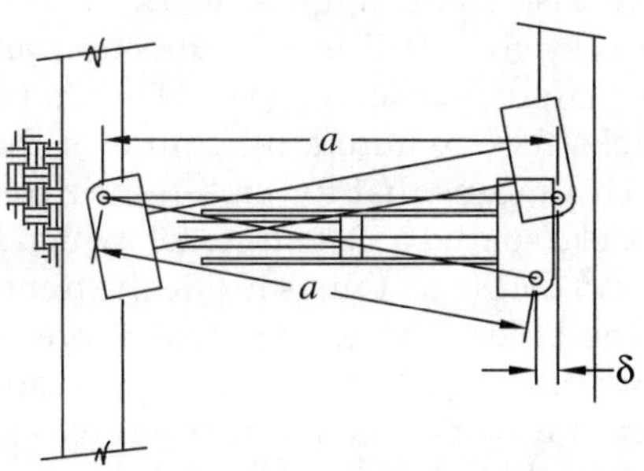

Figure 9.7 Trench jack locking principle.

cylinder block, hydraulic cylinder, and hydraulic hoses. The rail lengths are 1.5, 3.5, 5, 7, 8, 12, and 16 ft, and the cylinder ranges can vary from 10 to 36 in and the full extension varies from 17 to 144 in. All parts are manufactured from aluminum, and the shore weight varies from 21 lb for a 1.5-ft rail to 235 lb for a 16-ft rail. See Fig. 9.8.

9.3.1 Basic Theory of How Trench Jacks Work

Trench jacks work on the theory of soil arching; Article 6.4, and Fig. 6.12 is repeated here for reference. Arching theory has some important implications regarding use and stability of trench jacks that users should understand and be aware of:

- The soil arches between the jacks in the vertical and horizontal directions. If any jack cylinder location is not touching the soil, the arch goes to the next available cylinder [Fig. 9.9(a)], either a maximum of 8 ft away horizontally or 4 ft away vertically. A 16-ft arch is less stable than an 8-ft arch, and more importantly the load transmitted to the jack cylinder is 1.5 times the intended load and can exceed the allowable strength of the cylinder. If a cylinder fails, there can be progressive failure of the entire system or local failure of the wall at the arch void. It is important to make sure all cylinders are engaged with the soil.
- The soil arches from the jack cylinders to the end corners and bottom corners of the trench [Fig. 9.9(b)]. This is why the ends of an unshored trench will stand up when the longer walls will collapse. There is some hesitancy to allow workers between the last trench jack and the end of the trench; however, due to arching as a general rule if the width of the trench is less than two-thirds the jack spacing and the distance from the last jack to the end of the trench is one-half the jack spacing, then it should be safe to work in that area.
- Arching works on round and rectangular holes; however, it does not make sense to extend this theory to eliminate trench jacks from a small rectangular work pit completely because they are easy to install in one direction and are still more effective than corner arching. In OSHA's tabulated data for trench jacks they require a minimum of three jacks equally spaced. This means that for an 8-ft-long trench there would be three jacks spaced at 2 ft on center, which leaves little room to work and construct things inside the trench. The three-jack theory was based on the idea that if one jack failed, there would still be two jacks to work between and that there is no protection if a worker is not between two jacks. This theory disregards arching to the corners of the excavation. Perhaps a better way to look at this is that arching to corners is a

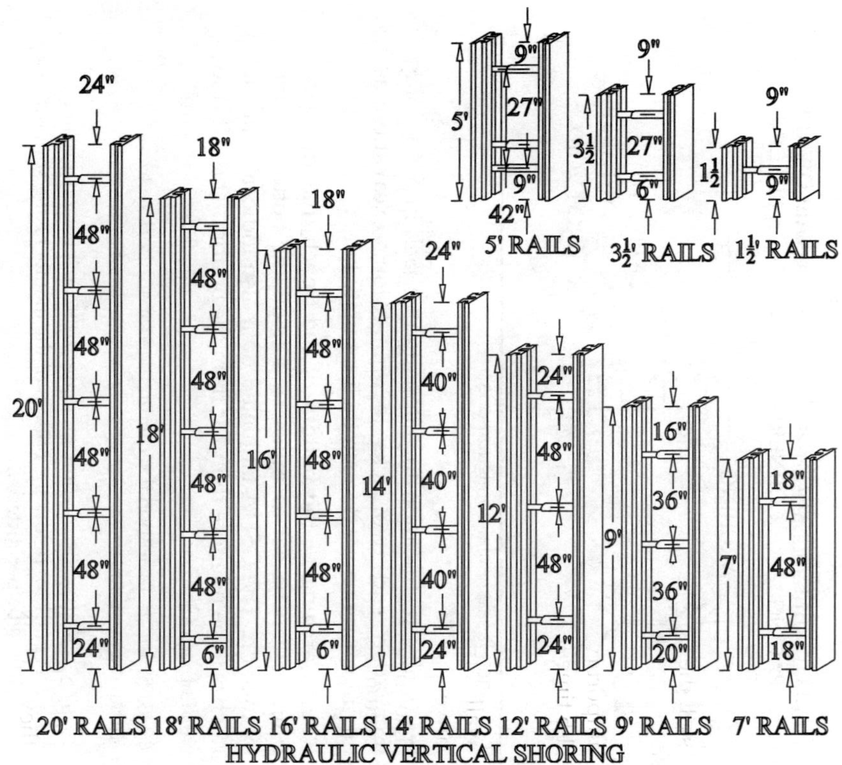

Figure 9.8 Typical manufactured trench jack sizes.

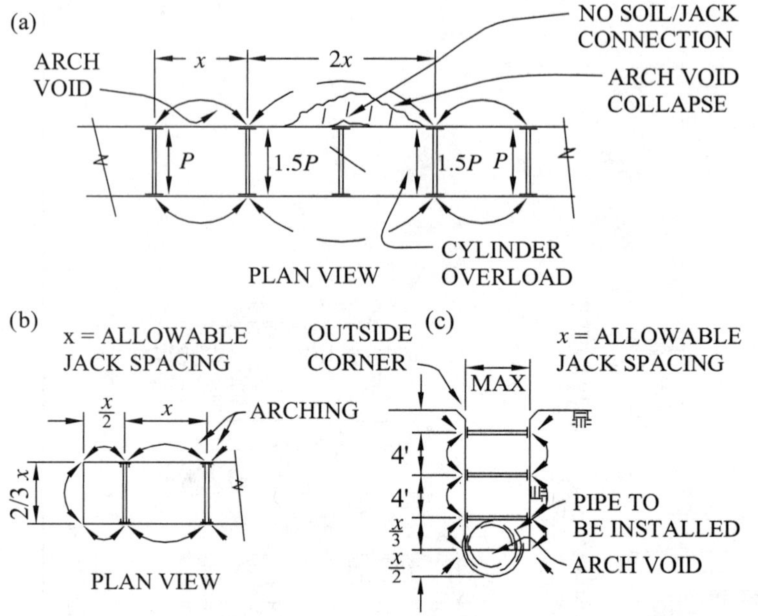

FIGURE 9.9 Trench jack special configurations.

reasonable justification for allowing one jack in a 4-ft-long excavation and two jacks in an 8-ft-long excavation as manufacturers have done in their tabulated data.

- Due to arching at the bottom of the trench, the first jack cylinder can be set as much as 4 ft off the bottom of the excavation. The OSHA tabulated data also indicate that the rail has to be no more than 2 ft from the bottom. The 2-ft rule stems from OSHA:

  1926.652(e)(2)—Additional requirements for support systems for trench excavations.

  1926.652(e)(2)(i)—Excavation of material to a level no greater than 2 feet (.61 m) below the bottom of the members of a support system shall be permitted, …

  OSHA holds that this rule is an additional requirement that is effective regardless of what the tabulated data say. They also claim that design by a registered civil engineer cannot supersede this rule. In the case of trench jacks this is an unjustified and costly limitation. The rule presents a problem for rails that do not extend at least 24 in below the cylinder. Standard trench jacks 9 ft long and under have rails that do not normally extend 24 in below the cylinder. Under the OSHA tabulated data, the solution is to set the cylinder less than 4 ft from the bottom and maintain the minimum 2-ft rail requirement. The problem is that the closer the first cylinder is to the bottom, the less working room is available to install pipe and for workers to move under the jacks. Manufacturers have chosen to ignore the 2-ft blade rule in their tabulated data by only stating the 4-ft rule; however, users should be aware that they are technically in violation of OSHA requirements if they do not adhere to the 2-ft rule. This is more of a technicality than it is a safety issue because to the author's knowledge there have been no accidents related to this. Justification for ignoring the 2-ft rule is that the soil arches to the jack cylinder and not the rail. Where no sheeting is used behind the jacks, all the soil wall between the trench jacks is exposed, as much as 8 ft, so why not allow exposure of the bottom 4 ft? There are two other possible solutions to this problem—changing the OSHA rule or manufacturing 9-ft and under trench jacks with a 24-in rail extension. The first option is pretty much off the table until sometime in the future when OSHA does another regulation review and combines this problem with other major changes in the rules. The second option is not as simple as it sounds because innovation and usage requirements were the reason that the jacks with short rail legs were developed; adding more rail lengths to an already standardized line of trench jacks adds extra equipment purchasing, inventory management, and user knowledge requirements to the system.

- In the early 1970s cast-in-place pipe was innovated with a round-bottom casting machine and round-bottom excavator bucket. The round bottom provides a more uniform bedding thickness for pipes and also reduces the excavation and backfill quantities. Today round-bottom trenches for all types of large pipe are common. Round-bottom trenches were not envisioned when tabulated data were first developed for trench jacks, and they are still not addressed in the data today. Soil arching theory provides a reasonable explanation for why round-bottom trenches stand up so well. Essentially removing the soil in the bottom of the excavation is the same as removing the arch void and does not weaken the arching capacity of the trench bottom. This configuration is also quite similar to an inverted tunnel arch, the difference being that gravity is working in the opposite direction. If the cylinder location has to be a maximum of 4 ft off the bottom, it completely defeats the purpose of the round-bottom trench because the jack cylinder would interfere with a round pipe that was being fit into the trench. Figure 9.9(c) shows a round-bottom trench jack shoring configuration based on this theory that has worked well in the past. Users of this detail should again be aware that

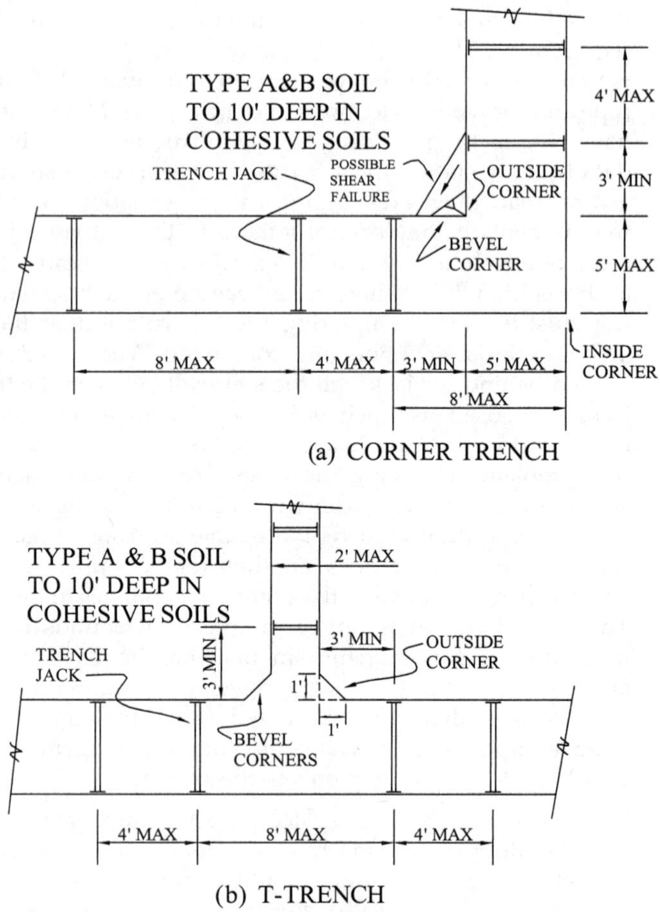

FIGURE 9.10 (a) Inside corner and (b) T-trench shoring.

technically even under design by a registered civil engineer it could be considered in violation of OSHA standards.

- Soil arches to inside corners and causes stress concentrations at outside corners. The top edge of the trench and corners of intersecting and corner trenches are examples of outside corners [Fig. 9.10(a) and (b)]. The problem exists with all materials. The solution is beveling the edges the same as is done with wood, steel, and concrete structures. In the case of soil, the outside corners of excavations are prone to falling off in large chunks and injuring workers below. It is best for the excavator operator to knock the corners off before they fall off. Placing trench jacks at these corners is also problematic because the stress at the jack can also cause the corners to break apart and cause the jack to destabilize. Trench jack

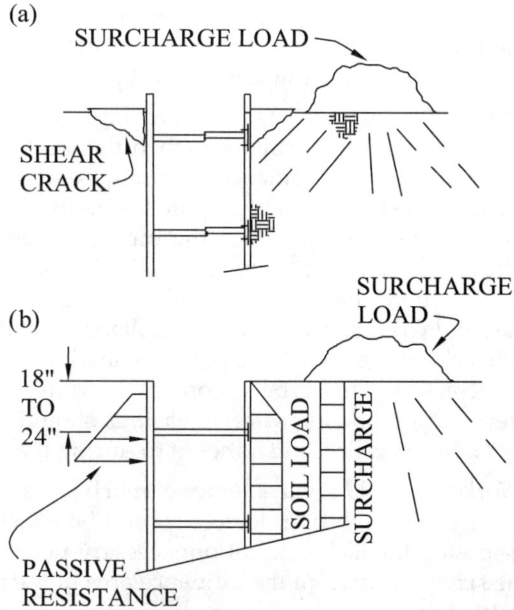

FIGURE 9.11 (a) Trench top edge failure and (b) trench top edge loads.

shoring in the vicinity of T and inside trench corners should be avoided due to this shearing problem. If the trench jack is set 3 ft back in cohesive type A and B soils, there is a large enough shear plane to resist a shear failure. The configuration shown in Fig. 9.10(a) and (b) has worked successfully for excavations less than 10 ft deep; otherwise trench box or open cut should be considered for these trench conditions. There is very little cohesion in noncohesive soils. Trench jack use at outside corners in these soils should be avoided.

- At the top of the trench OSHA tabulated data limit the distance between the cylinder and the surface to 18 in and give no minimum. Most manufacturers' data set a 24-in maximum and a 12-in minimum from the surface. One way to look at this issue is that the closer the top cylinder gets to the surface, the less passive soil resistance is available from the soil and the more likely the surface corner will shear off and cause that portion of the jack to fail [Fig. 9.11(a) and (b)]. In type B soils with a top cylinder depth less than 2 ft from the surface, calculations indicate that there is not enough passive soil pressure to resist the soil and surcharge load. Since most soils that will stand up at all have enough strength to stand 2 to 3 ft without support, most of the work that the top cylinder is doing is resisting surcharge loads. If the cylinder location is too close to the surface, the surcharge loads will pass beneath it. For these reasons setting the top jack cylinder between 18 and 24 in below the surface is the best practice.

- Once trench jacks are in place and arching is set up, the chance that the soil bank will collapse when the jack is removed is greatly increased. The practice of setting trench jacks to allow workers inside the trench to prepare the bottom, removing the jacks to allow room to set in the pipe or other production product, and then replacing the jacks is common. Several serious accidents have occurred due to concrete trucks, boom trucks, and workers falling into the collapsed trench. After the trench jacks are replaced, the potential loading on the jacks is also increased. This type of operation is reasonably safe; however, the following safety measures should be reviewed prior to the operation and adhered to during the work.
  1. Workers should always remove trench jacks from outside the trench. All lifting devices should be attached prior to releasing the jack, and all unnecessary personnel should be safely away from the influence area of the jack prior to releasing it.
  2. Verify that a trench wall collapse will not disrupt existing facilities or structures. If they are at risk, this type of reshoring operation should not be allowed.
  3. All lifting equipment used in setting the production product into the unshored area should be stationed a safe distance away or adjacent to a shored area. Working on the surface from corners of a rectangular unshored excavation is safer than working from the middle.
  4. Set jacks back into their prior location when replacing the jacks for workers to enter the reshored excavation. If original jacks were set under maximum loading conditions, cylinders loaded at 18,000 lb add extra trench jacks to cut the jack spacing to 80 percent of the original spacing.

Removal of trench jacks during regular operations is usually performed in conjunction with backfill operations. All workers working with and around trench jacks should understand that due to arching, removal of the jack is similar to removing a column from under an arch, and that trench wall collapse is more likely when the jack is being removed.

9.3.2 Trench Jack Engineering

There are several manufacturers of trench jacks in the United States. For 2-in cylinder trench jacks, the design and engineering principles are the same for all manufacturers. This is not to say that all trench

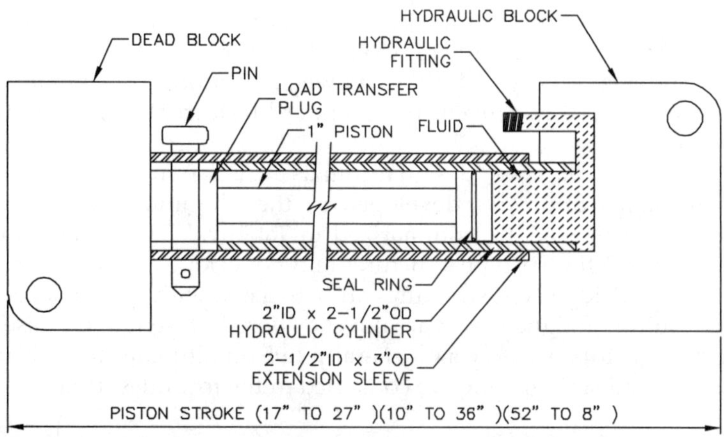

Figure 9.12 Hydraulic trench jack cylinder.

jacks are created equal. The manufacturers' material acquisition, handling, and manufacturing processes are all different. Innovations such as finger guards and faster hydraulic pumps, quality control, customer service, and presentation of tabulated data all help to differentiate the manufacturers. However, because some parts are interchangeable and because of the advent of replacement parts also coming from different sources, it can be hard to identify the original manufacturer. Since manufacturers hesitate to give out their design and manufacturing information for competitors and new entrants into the market to use, it is hard to find a set of engineering calculations for trench jacks. OSHA only requires that the tabulated data have a registered engineer's stamp, and they do not review the engineering calculations. Engineers who review shoring submittals are asked to take the tabulated data on faith that the manufacturer's engineer knows what she or he is doing. The following engineering calculations are intended to provide a more detailed understanding of the structural aspects of trench jacks.

The design concept around trench jack engineering (Fig. 9.12) is to first look at the safe working load, or bursting strength, of the hydraulic cylinder and then check to be sure that all the components will support the safe working load. The piston is axially loaded and can fail in buckling. The cylinder barrel, in addition to confining the fluid, has a bending moment applied to it due to the bending movement of the piston arm. The oversleeve receives a bending load from the buckled piston. Neither the barrel nor the oversleeve receives an axial load. There is a ⅜-in pin at the dead block end holding the oversleeve and the dead end piston block in place, and there are ⅜-in pins holding the dead block and hydraulic block to the rail. The pins act strictly to hold the jack together for handling and installation and are under no pressure when the jack is in operation; they are designed for durability only. There are two types of trench jack blades, light-duty and heavy-duty (Fig. 9.16). As well as holding the cylinders in place they serve to prevent the cylinder block from punching into the soil.

The allowable stress design formulas are taken from *Specifications for Aluminum Structures*, developed by the Aluminum Association, Inc. Formulas used are summarized in Table 9.7. The design is for building and similar type structures and not bridge and similar type structures. Bridge-type structures are associated with cyclical loading and resulting fatigue which are not considered present in the case of shoring. Factors of safety are generally 1.95 on ultimate strength and 1.65 on yield strength and are contained in the formulas used.

Trench Jack Cylinder Engineering

The 2-in hydraulic cylinders used for trench jacks are manufactured with an extension range from 17 to 27 in up to a range of 52 to 88 in with piston strokes from 10 to 36 in. The hydraulic cylinder is 2-in-ID by ¼-in wall 6061-T6 aluminum tubing with a tensile stress $F_u = 42{,}000$ psi and yield stress $F_y = 35{,}000$ psi. Laboratory testing shows that failure in the short range is from bursting and in the long range from buckling.

The most widely used and accepted formula, Eq. (9.1), for bursting and bulging is Barlow's:

$$p = \frac{2St}{D}$$

where p = bursting or bulging pressure, psi
$S = F_u$ = material ultimate strength, psi, used for bursting
$S = F_y$ = material yield strength, psi, used for bulging
t = wall thickness, in
D = outside diameter, in

Calculate the ultimate bursting and bulging pressure and load for a 6061-T6 aluminum hydraulic cylinder:

$$p_{bursting} = \frac{2 \times 42{,}000 \times 0.25}{2.5} = 8400 \text{ psi}$$

$$p_{bulging} = \frac{2 \times 35{,}000 \times 0.25}{2.5} = 7000 \text{ psi}$$

where p = bursting or bulging pressure, psi
$S = F_{tu} = 42{,}000$ psi for bursting
$S = F_{ty} = 35{,}000$ psi for bulging
$t = 0.25$ in
$D = 2.5$ in

Type of Stress	Type of Member or Component	Spec No.	Allowable Stress (ksi)
Tension, axial, net section	Any tension member	1	19
Tension in beams, extreme fiber, net section	Rectangular tubes, structural shapes bent about strong axis	2	19
	Round or oval tubes	3	24
	Shapes bent about weak axis, rectangular bars, plates	4	28
Bearing	On rivets and bolts	5	34
	On flat surfaces and pins in slotted holes	6	23

Allowable Stress for Building and Similar Type Structures 6061-T6, -T651, -T6510, -T6511 Extrusion up through 1 in, Sheet & Plate, Standard Structural Shapes, Rolled Rod and Bar, Drawn Tube, Pipe, 6351-T5 Extrusions

Not—these formulas do not apply within 1 inch of weld

Type of Stress	Type of Member or Component	Spec. No.	Allowable Stress (ksi) Slenderness ∏ S_1	Slenderness Limit S_1	Allowable Stress (ksi), Slenderness between S_1 and S_2	Slenderness Limit S_2	Allowable Stress (ksi) Slenderness √ S_1
Compression in columns, axial gross section	All columns	7	19	$\frac{L}{r} = 9.5$	$20.2 - 0.126\frac{L}{r}$	$\frac{L}{r} = 9.5$	$\frac{51{,}000}{(L/r)^2}$

Table 9.7 Allowable Stress for Aluminum Building and Similar Type Structures

Type of Stress	Type of Member or Component	Spec. No.	Allowable Stress (ksi) Slenderness ∏ S_1	Slenderness Limit S_1	Allowable Stress (ksi), Slenderness between S_1 and S_2	Slenderness Limit S_2	Allowable Stress (ksi) Slenderness √ S_1
Compression in components of columns, gross section	Outstanding flanges and legs	8	19	$\frac{b}{t}=5.2$	$23.1-0.70\frac{b}{t}$	$\frac{b}{t}=12$	$\frac{1970}{(b/t)^2}$
	Flat plates with both edges supported	9	19	$\frac{b}{t}=16$	$23.1-0.25\frac{b}{t}$	$\frac{b}{t}=33$	$\frac{490}{(b/t)}$
	Curved plates supported on both edges, walls of round or oval tubes	10	19	$\frac{R}{t}=16$	$20.2-0.80\sqrt{\frac{R}{t}}$	$\frac{R}{t}=141$	$\frac{3200}{(R/t)(1+\sqrt{R/t/35})^2}$
Compression in beams, extreme fiber, gross section	Single web beams bent about strong axis	11	21	$\frac{L_b}{r_y}=23$	$23.9-0.124\frac{L_b}{r_y}$	$\frac{L_b}{r_y}=79$	$\frac{87{,}000}{(L_b/r_y)^2}$
	Round or oval tubes	12	25	$\frac{R_b}{t}=28$	$39.3-2.7\sqrt{\frac{R_b}{t}}$	$\frac{R_b}{t}=81$	$\frac{3200}{(R/t)(1+\sqrt{R/t/35})^2}$
	Solid rectangular beams	13	28	$\frac{d}{t}\sqrt{\frac{L_b}{d}}=13$	$40.5-0.93\frac{d}{t}\sqrt{\frac{L_b}{d}}$	$\frac{d}{t}\sqrt{\frac{L_b}{d}}=29$	$\frac{11{,}400}{(d/t)^2(L_b/d)}$
	Rectangular tubes and box sections	14	21	$\frac{L_bS_c}{I_y}=146$	$23.9-0.24\sqrt{\frac{L_bS_c}{I_y}}$	$\frac{L_bS_c}{I_y}=1700$	$\frac{24{,}000}{(L_bS_c/I_y)}$

Compression in components of beams, (component under uniform compression), gross section	Outstanding flanges	15	21	$\frac{b}{t}=6.8$	$27.3-0.93\frac{b}{t}$	$\frac{b}{t}=10$	$\frac{182}{(b/t)}$
	Flat plates with both edges supported	16	25	$\frac{b}{t}=22$	$27.3-0.29\frac{b}{t}$	$\frac{b}{t}=33$	$\frac{580}{b/t}$
Compression in components of beams, (component under uniform compression), gross section	Flat plates with compression edge free, tension edge supported	17	28	$\frac{b}{t}=8.9$	$40.5-1.41\frac{b}{t}$	$\frac{b}{t}=19$	$\frac{4900}{(b/t)^2}$
	Flat plates with both edges supported	18	28	$\frac{h}{t}=46$	$40.5-0.27\frac{h}{t}$	$\frac{h}{t}=75$	$\frac{1520}{(h/t)}$
	Flat plates with horrizontal stiffener, both edges supported	19	28	$\frac{h}{t}=107$	$40.5-0.117\frac{h}{t}$	$\frac{h}{t}=173$	$\frac{3500}{(h/t)}$
Shear in webs, gross section	Unstiffened flat webs	20	12	$\frac{h}{t}=36$	$15.6-0.099\frac{h}{t}$	$\frac{h}{t}=65$	$\frac{39{,}000}{(h/t)^2}$
	Stiffened flat webs	21	12	$\frac{h}{t}=36$	12	$\frac{a_r}{t}=66$	$\frac{53{,}000}{(a_r/t)^2}$

Table 9.7 Allowable Stress for Aluminum Building and Similar Type Structures (*Continued*)

Calculate the ultimate load for bursting and bulging:

$$\text{Load } P = \text{pressure} \times \text{area}$$

$$\text{Bursting force} = 8400 \times 3.14 = 26{,}000 \text{ lb}$$

$$\text{Bulging force} = 7000 \times 3.14 = 22{,}000 \text{ lb}$$

where Area $A = \dfrac{2^2\pi}{4} = 3.14 \text{ in}^2$

$$p_{bursting} = 8400 \text{ psi}$$
$$p_{bulging} = 7000 \text{ psi}$$

Using a 1.5 factor of safety on bursting force, the safe working strength of a 2-in jack is 26,000/1.5 = 17,300 lb. If a 1.3 factor for shoring loading is put on this, the safe working load is 22,500 lb. OSHA allows 2-in aluminum hydraulic cylinders to have a safe working load at full extension of 18,000 lb, and manufacturers do not state the maximum cylinder load used when calculating their tabulated data.

Several manufacturers have had their cylinders tested. The results indicate a bulging rupture. The average rupture force for seamless drawn tubing was found to be 33,000 lb. With a 1.5 factor of safety on the seamless tubing, the safe working load will be 22,000 lb.

Piston Buckling

Buckling can occur in the piston rod, but it is prevented from going into complete failure by the 3-in-OD × 3/16-in wall oversleeve (Fig. 9.13). There are two types of piston rod being used, 1¾-in-OD × 1¼-in-ID drawn tube from 6061-T6 tubing, $F_y = 35{,}000$ psi, $F_u = 42{,}000$ psi, and a 1¼-in solid rod from 2014-T6, $F_y = 55{,}000$ psi, and $F_u = 65$ ksi. With the tube piston, in addition to overall buckling there can be local buckling, so both conditions should be checked. In both cases the ends of the rod are fixed by the piston end block and head into the oversleeve and cylinder barrel, allowing an effective length factor K of 0.65 (Table 9.8). The longest piston rod is 48 in long, and the effective length is $0.65L = 31$ in. This allows an overall trench jack extension width of 7 ft 4 in. Aluminum allowable strength formulas are from Table 9.7.

From Table 9.7 specification 7, find overall buckling for the 1¾-in-OD × ¼-in wall tubing.

$$\text{Slenderness } s = \frac{L}{r} = 58$$

where L = 31-in effective length
r = 0.5377-in radius of gyration

$$\therefore \quad s \text{ is between } s_1 \text{ and } s_2$$

$$\text{Allowable stress } F_{cy} = 20.2 - 0.126\frac{L}{r} = 12.9 \text{ ksi}$$

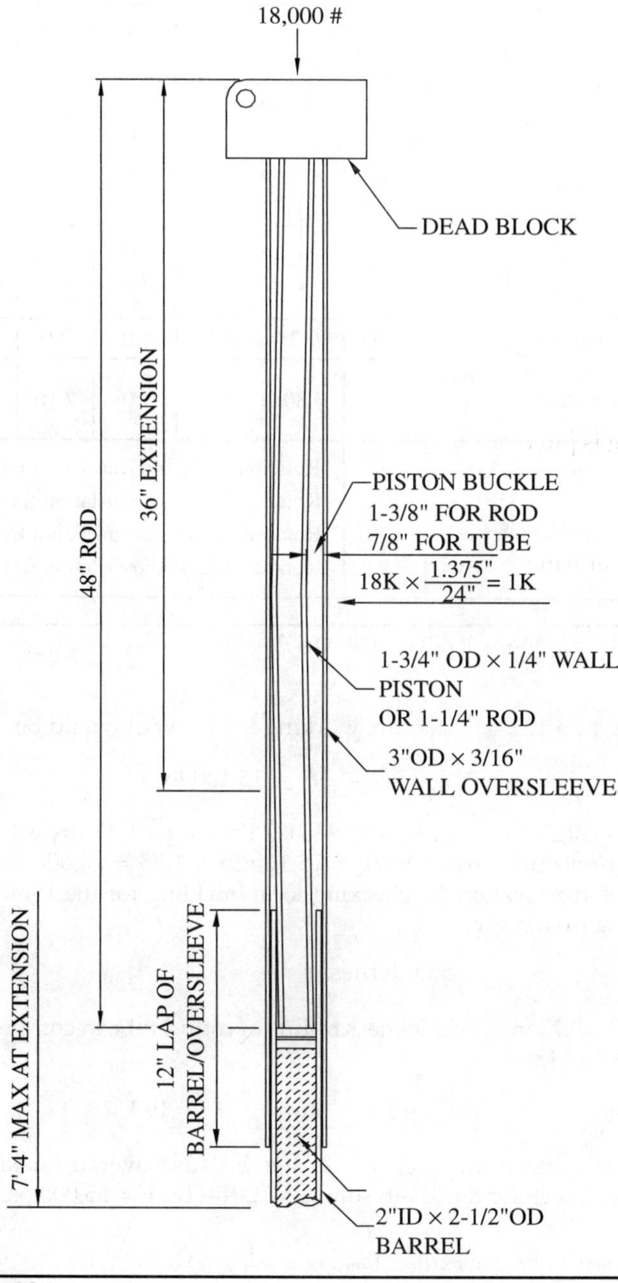
18,000 #
DEAD BLOCK
36" EXTENSION
48" ROD
PISTON BUCKLE
1-3/8" FOR ROD
7/8" FOR TUBE
18K × 1.375"/24" = 1K
1-3/4" OD × 1/4" WALL
PISTON
OR 1-1/4" ROD
3"OD × 3/16"
WALL OVERSLEEVE
7'-4" MAX AT EXTENSION
12" LAP OF
BARREL/OVERSLEEVE
2"ID × 2-1/2"OD
BARREL

FIGURE 9.13

	(a)	(b)	(c)	(d)	(e)	(f)
Buckled shape of column is shown by dashed line	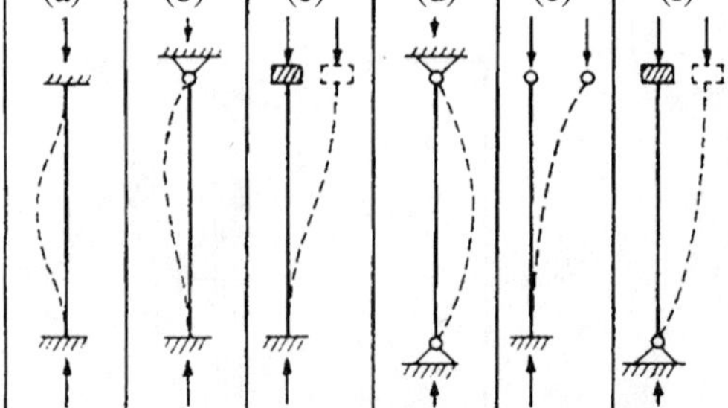					
Theoretical *K* value	0.5	0.7	1.0	1.0	2.0	2.0
Recommended design value when ideal conditions are approximated	0.65	0.80	1.2	1.0	2.10	2.0
End condition code	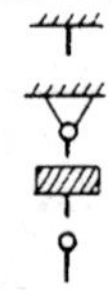	Rotation fixed and translation fixed Rotation free and translation fixed Rotation fixed and translation free Rotation free and translation free				

TABLE 9.8 Effective length of columns, AISC Table C-C2.1

Based on 12.9 ksi, the safe working load (SWL) would be

$$\text{SWL} = P = AF_{cy} = 15{,}100 \text{ lb}$$

where $A = 1.781$ in^2 and $F_{cy} = 12.9$ ksi. Allowing a 1.33 stress increase for shoring application, SWL = 15,100 lb × 1.33 = 20,000 lb. From Table 9.8 specification 10, checking local buckling for the 1¾-in-OD × ¼-in wall tubing gives

$$\text{Slenderness } s = \frac{R}{t} = 3$$

where $R = 0.75$ in (midthickness radius of round tube in compression) and $T = 0.25$ in.

$$\therefore \quad s < s_1 = 16 \qquad \text{and} \qquad F_{cy} = 19 \text{ ksi}$$

There is no need to check further because overall buckling F_{cu} controls. Checking the 1¼-in solid rod, 2014-T6, $F_y = 55{,}000$ psi gives

$$\text{Slenderness s} = \frac{L}{r} = 99$$

where $L = 31$-in effective length and $r = 0.313$ in radius of gyration.

$$\therefore \quad s > s_2 \qquad \text{and} \qquad F_{cy} = \frac{54{,}000}{(L/r)^2} = 5.5 \text{ ksi}$$

Using the same formulation as above including the 1.33 factor, the safe working load for the 1¼-in sold piston is 9000 lb.

The buckling failure calculated above is controlled by the 3-ft × 3/16-in wall oversleeve; however, if the jack is to be operated at the safe working load for bulging, 18,000 lb that is allowed by OSHA, and it is considered important to not allow any buckling in the pistons, then their safe working length can be calculated by using the same formulas. The working stress for the 1¾-in × ¼-in wall tubing is

$$\frac{P}{A} = \frac{18.000}{1.1781} = 15.27 \text{ ksi}$$

$$F_{cy} = 20.2 - 0.126\frac{L}{r} = 15.27 \text{ ksi}$$

where $r = 0.5377$ and

$$\text{Effective length } KL = \frac{(15.27 - 20.2) \times r}{-0.126} = 21 \text{ in}$$

$$L = \frac{21}{K} = 32 \text{ in}$$

where $K = 0.65$. Use the same procedure for the 1¼-in rod, $L = 33$ in. This result allows a cylinder stroke of approximately 24 in and safe jack extension of 5 ft 4 in.

What can be concluded from this is that *trench jacks with an extension over 5 ft require an oversleeve because the piston will buckle at longer extensions.*

Barrel and Oversleeve Bending or Buckling

When the piston buckles and the buckling is resisted by the oversleeve, a perpendicular force and resulting moment are delivered to the oversleeve/piston beam, as shown in Fig. 9.14.

Because the dead block and hydraulic block are pinned and the soil can compress slightly, the ends are assumed pinned. In the previous calculation the piston head and block were internal to the oversleeve/piston so it was considered fixed at the ends. From Fig. 9.13 the force transmitted to the oversleeve is $P = 18 \text{ k} \times (1.375 \text{ in}/24 \text{ in}) = 1.0$ k, and from Fig. 9.14 the maximum moment is 1.45 k·ft and 0.75 k·ft at the juncture of the 3- and 2.5-in tube. The bending stress for the 2.5-in hydraulic cylinder is

$$f_b = \frac{M}{S} = \frac{0.75 \times 12}{0.9057} = 9.9 \text{ ksi}$$

where $M = 0.75$ k·ft and

$S = 0.9057\text{-in}^3$ section modulus for 2.5-in × ¼-in wall tube

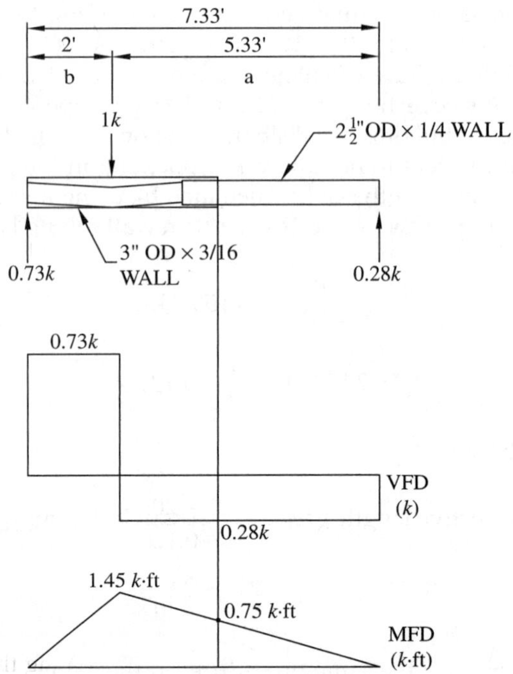

FIGURE 9.14 Piston buckling shear and moment diagram.

And for the oversleeve

$$f_b = \frac{M}{S} = \frac{1.45 \times 12}{1.0969} = 15.86 \text{ ksi}$$

where $M = 1.45$ k·ft and

$S = 1.0969\text{-in}^3$ section modulus for 3-in × 3/16-in wall tube

From Table 9.8 specification 12 for compression in bending of the 3-in × 3/16-in wall tube

$$s = \frac{R_b}{t} = \frac{1.4}{0.1875} = 7.4$$

where $R_b = 1.4$ in

$t = 3/16$ in (midthickness radius of round tube in bending)

$$\therefore \quad s < s_1 \qquad \text{and} \qquad F_{bc} = 25 \text{ ksi}$$

And for tension

$$F_{bt} = 24 \text{ ksi} \qquad \text{(Table 9.7 specification 3)}$$

These results leave 24 − 15.9 = 8.1 ksi for additional bending forces such as an external point load from a worker climbing on it or

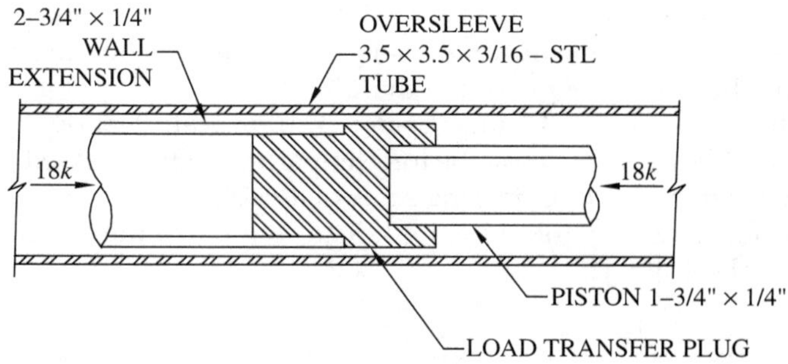

Figure 9.15 Cylinder extension and steel oversleeve.

eccentricities caused by the jack blade not setting perpendicular to the cylinder/oversleeve. If a 250-lb worker stood on the middle of the fully extended jack, the moment would be

$$M = \frac{PL}{4} = \frac{250 \times 7.33}{4 \times 1000} = 0.458 \text{ k·ft}$$

and $$F_b = 0.5 \text{ ksi} \cong 4.8 \text{ ksi} < 8.1 \text{ ksi} \qquad \text{OK}$$

By adding 8 in for the thickness of two end blocks and the blades, the safe working extension becomes 8 ft, and by using the full capacity of the 3-in oversleeve the safe extended length of the jack is 9 ft 4 in. After that length the jack needs a stronger oversleeve.

Cylinder Extension

To use trench jacks in trenches wider than 8 ft and up to 15 ft, the jack is extended by adding a transfer plug and another 2.5-in × ¼-in wall inner sleeve inside a 3.5-in × 3.5-in × 3/16-in wall steel tube oversleeve (Fig. 9.15).

In this configuration the 2.75-in × ¼-in wall tube is in compression and fixed at both ends. The tube is 56 in long, and the effective length is 0.65 × 56 = 36.4 in.

Check buckling in 2.75-in × ¼-in wall extension:

$$\text{Slenderness s} = \frac{L}{r} = 45.4$$

where L = 36.4-in effective length
r = 0.8003-in radius of gyration

$$\therefore \quad \text{s is between } s_1 \text{ and } s_2$$

and

$$\text{Allowable stress } F_{cy} = 20.2 - 0.126\frac{L}{r} = 14.5 \text{ ksi}$$

$$f_c = \frac{P}{A} = 10.2 \text{ ksi} \qquad \text{OK}$$

where $P = 18{,}000$ lb and $A = 1.7671$ in². Through back calculation it can be determined that the maximum length of the extension can be 96 in, making the overall length of the shore 15 ft.

Check bending in 3.5-in × 3.5-in × 3/16-in wall steel oversleeve. Due to the larger oversleeve the force imparted from a buckled piston calculates to be 1.4 k. Using the same analysis as shown in Fig. 9.14 except for a 12-ft span and a 1.4-k load, the maximum moment is found to be 4.14 k·ft.

$$f_b = \frac{M}{S} = 20.27 \text{ ksi}$$

where $M = 4.14$ k·ft and $S = 2.45$ in². Tube steel is ASTM A-500 grade B, $F_y = 46$ ksi. From AISC

$$F_b = 0.66\, F_y = 30.4 \text{ ksi} \qquad \text{OK}$$

At an extension of 15 ft the moment is 5.5 k·ft and the bending stress is 27.3 ksi. OSHA allows an extension using 3.5-in × 3.5-in × 3/16-in tube steel up to 12 ft, and some manufacturers allow up to 15 ft.

Trench Jack Rails

Rails are an extruded shape made from 6061-T6 aluminum. Manufacturers use two different sizes, standard and heavy-duty (Fig. 9.16). Both blades are 8 in wide and approximately 3/16 in thick for the standard and ¼ in thick for the heavy-duty. The difference in rail sizes is strictly related to durability and handling strength and not to structural requirements. As a general rule, the light-duty rail is used on shores to 7 ft long and the heavy-duty from 8 to 20 ft long. OSHA requires a minimum blade section modulus of 0.4 in³, and manufacturers in their tabulated data require that their particular brand of rails be used. In all tabulated data, OSHA or manufacturers', there is no requirement to use the heavy-duty rail. Each manufacturer owns their own dye that is used for extrusion at the aluminum casting plant. There are slight variations, but they all have a minimum 3½-in space between the legs that allows the cylinder blocks to fit into them. With the correct drilling pattern through the rail they are completely interchangeable with any manufacturers' 2-in hydraulic cylinder.

The main purpose of the blade is to keep the cylinder block from punching into the soil. Secondary purposes are to connect a series of usually 4-ft spaced hydraulic cylinders together into a unit so that they can be handled and lifted into the trench during installation, and to attach sheeting to the system.

For the purpose of preventing punching into the soil, the rail has to have some cantilever strength beyond the center of the 4-in-long block (Fig. 9.17). The following analysis determines the cantilever length and the surface area provided by the blade and then compares

VERTICAL RAIL SPECIFICATION SHEET

SECTION PROPERTIES	STANDARD RAIL	HEAVY DUTY RAIL
MATERIAL	ALUMINUM	ALUMINUM
ALLOY	6061–T6	6061–T6
AREA	2.45 in^2	3.47 in^2
WEIGHT	2.94 plf	4.17 plf
SECTION–MODULUS – TOP (LEG SIDE)	0.44 in^3	1.25 in^3
SECTION–MODULUS – BOTTOM (BLADE SIDE)	1.29 in^3	2.38 in^3
EQUIVALENT TIMBER SIZE * (#2 DOULAS FIR)	3x10 (FLAT)	4x10 (FLAT)

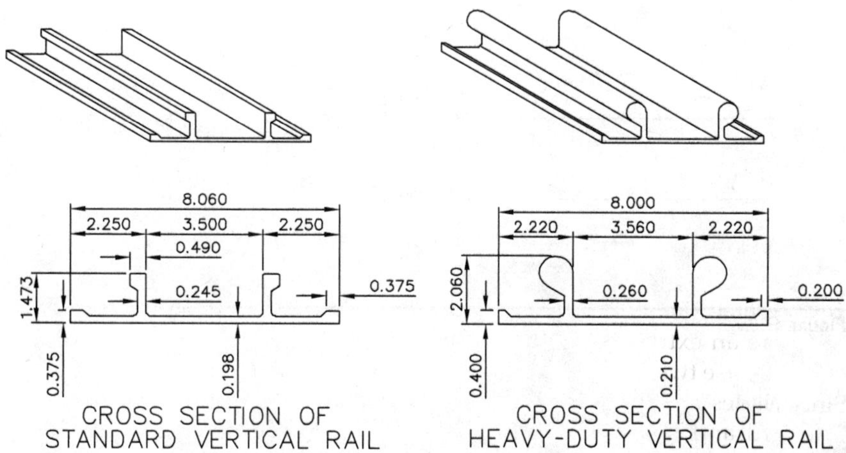

FIGURE 9.16 Trench jack rails.

it to the bearing capacity of the soil. A normal assumption used when dealing with braced or cantilevered piles in lateral bearing is that the soil shears at a 45° angle from the edge of the pile until it reaches one-half the width of the pile face (Fig. 9.17), section A. This gives a bearing width of 2 times the pile face width. The same assumption is made for the 8-in-wide trench jack rail.

Determine the cantilever length for standard and heavy-duty rail. From Table 9.7, compression in components of beams, formula 17

$$s = \frac{b}{t} = \frac{1.275}{0.245} = 5.2 \text{ standard duty} \quad s = \frac{b}{t} = \frac{1.850}{0.260} = 7.11 \text{ heavy duty}$$

where $b = 1.275$ in standard-duty rail
$b = 1.850$ in heavy-duty rail
$t = 0.245$ in standard-duty rail
$t = 0.260$ in heavy-duty rail

$$\therefore \quad s < 8.9 \quad \text{and} \quad F_{bc} = 28 \text{ ksi}$$

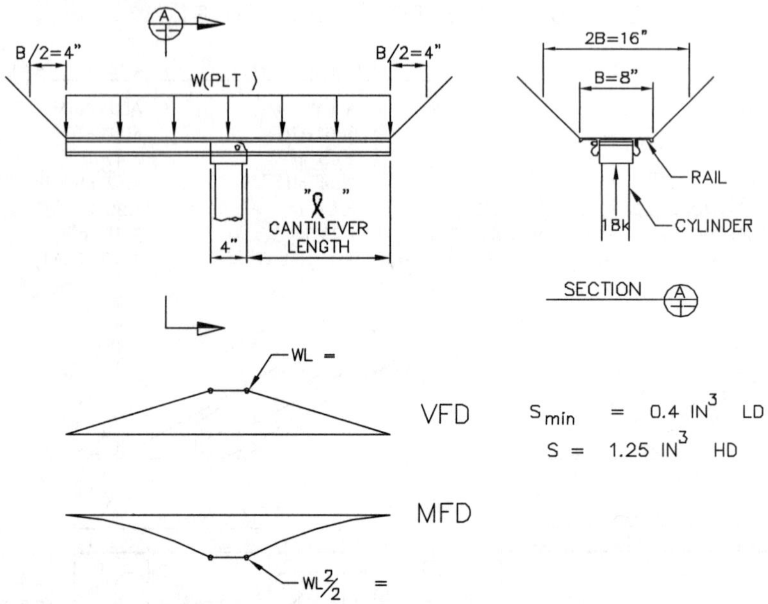

FIGURE 9.17

Since $M = sF_b$,

$$M_{std} = 28 \text{ ksi} \times 0.4 \text{ in}^3 = 11.2 \text{ k·in}$$

$$M_{hvy} = 28 \text{ ksi} \times 1.25 \text{ in}^3 = 35 \text{ k·in}$$

Also from the cantilever moment

$$M = \frac{wL^2}{2}$$

$$V = wL$$

where L = cantilever length and w = distributed load. And by limiting the cylinder load to 18 k, one-half of the load coming from above the cylinder and one-half from below,

$$V = 9 \text{ k} = w(L + 2 \text{ in})$$

$$w = \frac{9 \text{ k}}{L + 2}$$

Substituting for w gives

$$M_{std} = \left(\frac{9}{L+2}\right) \times \frac{L^2}{2} = 11.2 \text{ k·in}$$

and $$M_{hvy} = \left(\frac{9}{L+2}\right) \times \frac{L^2}{2} = 35 \text{ k·in}$$

By trial and error

$$L_{std} = 3.8 \text{ in} \qquad \text{and} \qquad L_{hvy} = 9.4 \text{ in}$$

This provides a bearing area of

[3.8-in leg + 3.8-in leg + 4-in block + 8-in (45° bearing width assumption)] × [8-in rail width + 8-in(bearing width assumption)] = 313 in² = 2 ft²   standard-duty rail

and

[9.4-in leg + 9.4-in leg + 4-in block + 8-in (45° bearing width assumption)] × [8-in rail width + 8-in (bearing width assumption)] = 493 in² = 3.4 ft²   heavy-duty rail

Bearing capacity in soft clays is going to be the worst condition for the rail punching into the soil. In OSHA C-60 soil the unconfined compressive strength q_u is less than 0.5 tsf and cohesion $c = q_u/2 =$ 500 psf. Using $p_p = \gamma H + 2c$ for passive bearing capacity and assuming a depth of 21 ft which is where the last jack cylinder would be for an allowable depth of 25 ft, the available passive pressure is

$$p_p = 120 \times 21 + 2 \times 500 = 3520 \text{ psf}$$

Based on the bearing capacity of high-end OSHA C-60 soil or low-end OSHA B soil, the rail can support a total force before punching into the soil of

$$3250 \text{ psf} \times 2.0 \text{ ft}^2 = 6500 \text{ lb} \qquad \text{standard-duty rail}$$

$$3250 \text{ psf} \times 3.4 \text{ ft}^2 = 11{,}050 \text{ lb} \qquad \text{heavy-duty rail}$$

For high-end B soil and low-end A soil $q_u = 1.5$ tsf and $c = 1500$ psf; the rail can support a total force before punching into the soil of

$$p_p = 120 \times 21 + 2 \times 1500 = 5520 \text{ psf}$$

and

$$5520 \text{ psf} \times 2.0 \text{ ft}^2 = 11{,}040 \text{ lb} \qquad \text{standard-duty rail}$$

$$5520 \text{ psf} \times 3.4 \text{ ft}^2 = 18{,}768 \text{ lb} \qquad \text{heavy-duty rail}$$

In loose ($\phi = 28$ degrees, $K_p = 2.7$) to medium dense sands and gravels ($\phi = 32$ degrees, $K_p = 3.25$), the passive pressure using the Rankine formula at 21-ft depth is 6500 to 7500 psf, giving a bearing force from 13,000 to 15,000 lb for the standard rail and 22,000 to 25,000 lb for the heavy-duty rail.

The results of theses calculations indicate that in most cases if the trench jacks are loaded through arching to their full 18,000-lb cylinder capacity, the soil can experience a shear failure around the rail/cylinder. The rail will curl away from the soil bank and allow soil movement until the soil arch tightens to take up some of the load or it shears into the trench. A long history of trench jack use confirms that this effect is unobservable or is not happening. Most likely soil arching is not allowing enough movement of the soil to develop the predicted loads from the soils. From an engineering standpoint, these numbers are real, and there is little doubt that if high enough loads are delivered to the cylinders, there will be bearing capacity failure at the cylinder location. One factor in the calculations that makes them conservative is the 1.65 factor of safety on the allowable stress for aluminum. Using $F_y = 35$ ksi would yield a longer cantilever length and hence slightly higher bearing areas. The way to eliminate the possibility of this type of failure is to design the jack spacing to limit the loads to those calculated above. Adding a 1.33 increase to those forces for shoring application would also be reasonable.

For the competent person and the design engineer making decisions regarding use of standard-duty and heavy-duty rails, the following rules of thumb should be considered:

- In OSHA type B and C noncohesive soils, standard-duty rails should not be loaded to more than 15,000 lb per cylinder.
- Limit standard-duty vertical rail use to excavations less than 10 ft deep.
- In OSHA type B and C cohesive soils use heavy-duty rails only.
- In OSHA type C cohesive soils consider using other shoring alternatives over 16 ft deep.
- Surcharge loads are greatest in the top 10 ft of the excavation. If higher than normal surcharge loads are anticipated, use heavy-duty rails.

Steel sheeting such as plate or sheet piles can be used to extend the bearing area of the jack rail. Plywood has very little bending strength and would not significantly increase the bearing area of the jack rail.

Sheeting

When soil arches to trench jack cylinders, there is still a section of soil between the arch and the trench wall that is not confined by the arch. The void volume does not experience lateral loading, but there is still vertical gravity loading. In clay soils, cohesion holds the void in place until drying at the trench wall face destroys the clay particle bond. This can take hours to days. In noncohesive soils, capillary action between the soil particles holds the soil in place until the drying effect takes over. Capillary action is weaker and drying takes place more rapidly. Sands and gravels or dried clay pieces falling off the face of the trench wall is called raveling.

As clay tension cracks form at the surface or in the void area between the jacks, the soil separates and falls off in sheets. This process is referred to as *sloughing* ("sluffing").

In soft clays the weight of the soil above can cause the lower bank to bulge at the trench wall. This is also referred to as sloughing, but it is a different mechanism for trench wall failure. It indicates that the soil is on the borderline between C-60 and C-80 soil. In this condition trench jacks should have sheeting and be closely spaced. In this condition soil is almost fluid and will transmit the full anticipated load to the jacks.

Fissured clay soil has a discontinuity of cohesive bond at the fissure lines. In the arch void, large chunks of fissured soil can break off and fall into the trench. This type of movement could be put into the category of sloughing; however, the size of the falling chunks can be large, and the weight can be enough to bend ¾-in plywood.

Sheeting is not considered part of a trench jack although there are two cases where it is required to be used with trench jacks:

1. In all cases where sloughing or raveling occurs. If there is a gap between the sheeting due to jack spacing, the spacing must be decreased until the sloughing or raveling stops.
2. In C-60 soil over 10 ft deep.

Sheeting is not considered a structural member and is only intended to prevent soil and rock from falling on the workers below. Given this functional requirement, it seems that anything that fulfills it would be adequate. If sand and pebbles are raveling off the wall, geofabric or thin plywood would serve the purpose whereas if a 3-ft^3, 375-lb, chunk of fissured clay was trying to make its way out of the trench wall, 1⅛-in wood or plywood might not be enough to stop it. OSHA has set minimal requirements for sheeting, 1¼-in plywood or ¾-in-thick, 14 -ply, arctic white birch (Finland form, they must have had a representative there when the specifications were being developed), and manufacturers have followed this lead, allowing steel plate, ¾-in-thick 13-ply plywood Omniform and a variety of aluminum and steel sheeting.

The basic structural aspect desired in all these materials is bending strength in a 2-ft cantilever span. Note that OSHA does not spell out the width or length of the sheeting although it is normally assumed that they are 4 × 8 plywood sheets. Table 9.9 shows the basic bending strength properties of the OSHA specified materials. The least bending moment comes from the 1⅛-in plywood

$$M = F_b KS = 1100 \times 0.840 = 924 \text{ in·lb/ft} \quad \text{or} \quad 77 \text{ ft·lb/ft}$$

Material	Grade Stress Level	Effective Section Modulus KS	Allowable Bending F_b
1-1/8"-2.4.1 int APA Plywood	S-2	0.840 in^3/ft	1100 psi
Finland Form ¾" All-Birch	S-1	0.4826	3600 psi

TABLE 9.9 Bending Properties for OSHA Sheeting

As a general rule, anything that possesses the minimum bending strength and area of the material required by OSHA should be seen as equivalent.

If the sheeting is too stiff, the arching capacity of the soil is diminished, and the shoring system becomes a braced sheeted wall with active soil pressure acting over the entire area of the sheeting. Steel plates 1 in thick or 3 × 12 timber would be an example of this. It is still reasonable to use trench jacks to brace these systems, but they should be designed by a civil engineer.

In some parts of the United States sheeting is normally bolted onto the trench jack before it is sent out to the job site, and in other parts it is placed into the trench and then the jacks are set against it. Durability is also an issue for the suppliers of trench jacks, and on-the-job handling strength of sheeting is important. The weight of a ¾-in plywood sheeted trench jack can cause the corners of the sheeting to bend as it is hoisted off the ground.

Hydraulic System

The hydraulic system consists of a 5-gal hand pump, 10,000 psi hydraulic hoses and fittings, and a remote release tool. The fluid is ethylene glycol–based and biodegradable. Hydraulic hoses are connected between vertical flights of cylinders. Larger and faster electric hydraulic pumps are also available.

Manufacturers require that the jacks be loaded to between 750 and 1500 psi, resulting in a 2350- to 4700-lb force on the trench wall amounting to an approximate 500 to 1000 psf force on the soil as it extends up and down the trench jack rail. This rigid unyielding contact point becomes the pillar that the soil arches to. As the soil load increases, the pressure inside the jack can increase to 7000 psi or 22,000 lb before it will theoretically bulge.

9.3.3 Development of Trench Jack Tabulated Data

Based on OSHA soil types A-25, B-45, and C-60, trench jack spacing can be calculated to limit the jack load to 18,000 lb. The other given is that the vertical spacing is limited to 4 ft and the horizontal spacing is limited to 8 ft. These distances have worked reasonably well throughout the historical use of trench jacks and in tunnel mining applications and it seems that no manufacturers' engineer wants to challenge them. One reasoning for the 4-ft vertical spacing being one-half the horizontal spacing is that gravity works in a vertical direction and not in a horizontal direction. The loading increases as the depth increases while the loading does not increase as the length of the trench increases. To further facilitate the development of tabulated data, OSHA has categorized trench depths in 5-ft increments to 20 ft deep, and manufacturers have gone to 25 ft deep. When OSHA developed their data, they concluded that trench jacks could not be used in the extreme case of C-80 soil because it is fluid and will not stand up long enough to install the jacks; therefore they did not develop a table for trench jacks in C soils. Manufacturers concluded that C soil ranging from C-46 to C-60 would stand up long enough to install the jacks and developed tables for it.

Figure 9.18 shows the basic assumptions for development of tabulated data for trench jacks. The loading diagram is assumed to be rectangular, and the formula for soil loading, Eq. (9.2), is

$$w = \text{OSTF} \times H + \text{surcharge}$$

where w = lateral soil pressure (psf)
OSTF = OSHA soil type factor
OSTF = 25 lb/ft of depth for type A soil
= 45 lb/ft of depth for type B soil
= 60 lb/ft of depth for type C soil
H = full depth of trench (ft)
Surcharge = 72 psf represents 2 ft of soil or lateral effect of 20,000 lb equipment

The trench jack cylinder load [Eq. (9.3)] is the influence area, vertical jack spacing (4 ft) × horizontal jack spacing × the soil plus surcharge load w:

$$JL = 4wS$$

where JL = jack load, lb
w = lateral soil pressure, psf
S = horizontal jack spacing, ft

Substituting w and solving for jack spacing give, from Eq. (9.4),

$$S = \frac{JL}{4 \times (\text{OSTF} \times H + 72)}$$

Setting the jack load to 18,000 lb gives the results shown in Table 9.10(a). Table 9.10(b) and (c) shows the jack spacing allowed by OSHA and by trench jack manufacturer's tabulated data.

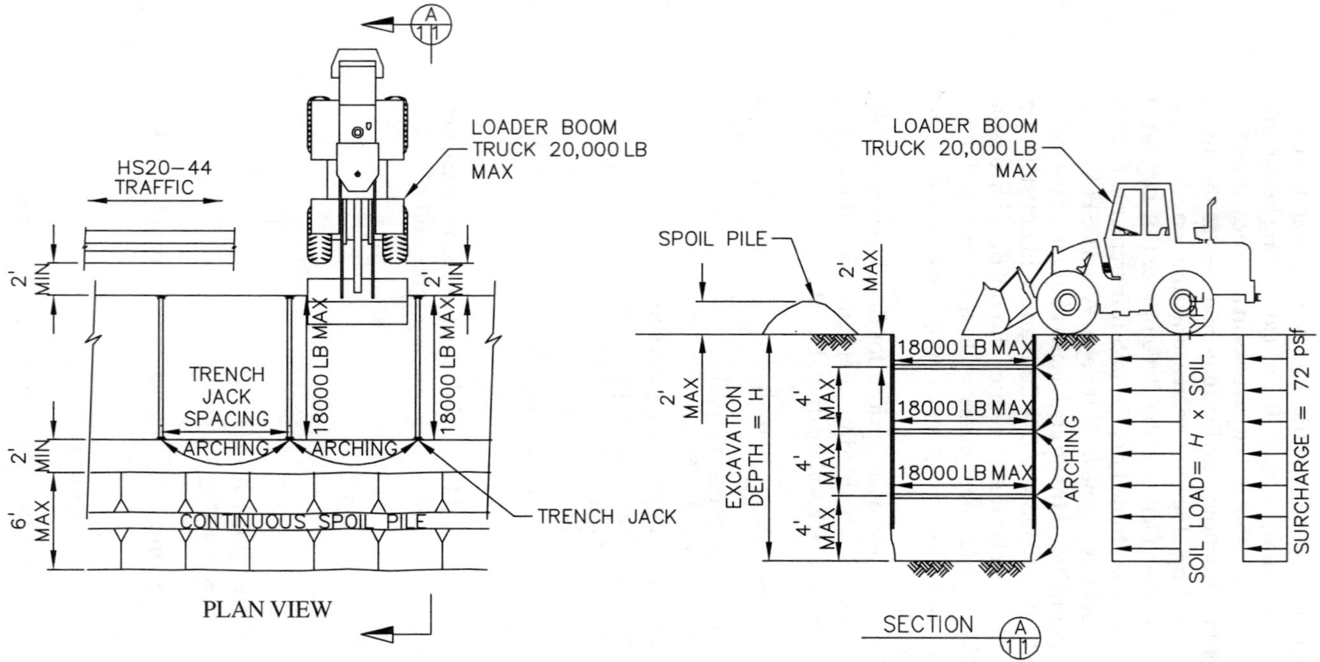

SOIL TYPE (OSHA)	PRESSURE PER FOOT OF DEPTH
A	25
B	45
C	C-60

FIGURE 9.18 Basic trench jack loading assumption.

(a)

Calculated Horizontal Jack Spacing *S* (ft)			
Trench Depth	OSHA Soil Type		
H (ft)	A-25	B-45	C-60
Up–10	14	8.6	6.7
11–15	10	6	4.6
16–20	8	4.6	3
21–25	6.5	3.75	3.5

(b)

OSHA Horizontal Jack Spacing *S* (ft)			
Trench Depth	OSHA Soil Type		
H (ft)	A-25	B-45	C-60
Up–10	8	8	0
11–15	8	6.5	0
16–20	7	5.5	0
21–25	0	0	0

(c)

Manufacturers' Horizontal Jack Spacing *S* (ft)			
Trench Depth	OSHA Soil Type		
H (ft)	A-25	B-45	C-60
Up–10	8	8	6
11–15	8	7	5
16–20	8	6	4
21–25	8	5	3

(d)

Calculated Horizontal Jack Load *P* (lb)			
Trench Depth	OSHA Soil Type		
H (ft)	A-25	B-45	C-60
Up–10	18,000	18,000	18,000
11–15	18,000	18,000	18,000
16–20	18,000	18,000	18,000
21–25	18,000	18,000	18,000

(e)

OSHA Horizontal Jack Load *P* (lb)			
Trench Depth	OSHA Soil Type		
H (ft)	A-25	B-45	C-60
Up–10	10,300	16,700	0
11–15	14,300	19,420	0
16–20	16,000	21,380	0
21–25	0	0	0

(f)

Manufacturers' Horizontal Jack Load *P** (lb)			
Trench Depth	OSHA Soil Type		
H (ft)	A-25	B-45	C-60
Up–10	10,300	16,700	16,130
11–15	14,300	20,900	19,440
16–20	16,000	23,300	20,350
21–25	20,000*	22,500*	18,000*

*Assumes no surcharge effect after 20 ft deep.

Table 9.10 Trench Jack Spacing and Cylinder Load

By using jack spacing the jack load can be determined from

$$JL = S \times 4 \times (\text{OSTF} \times H + 72)$$

Table 9.10(d), (e), and(f) shows the results of this calculation. These results indicate that OSHA has allowed the jack load to get to 21,300 lb and manufacturers have allowed it to get to 23,300 lb. With lab-tested cylinder bulging at 33,000 lb this still leaves a factor of safety for OSHA of 1.54 and 1.41 for manufacturers.

Maintaining trench jack spacing is important, but how that is done is not so important. Two 8-ft jacks set one above the other are just as good as a 16-ft shore as long as the distance between the cylinders is not more than 4 ft vertically and at the proper horizontal spacing. As long as spacing requirements are maintained, jacks can be staggered, set out of plumb, or even set horizontally.

9.3.4 Safe Handling and Use of Trench Jacks

By removing the shoring installer from the unshored trench and making shoring equipment more available and easy to install, trench jacks have no doubt had a huge impact on excavation safety. Utilizing trench jacks for shoring still has safety hazards that users should understand and protect workers from. These things happen rarely, but it is still important that workers be informed of the risks they are taking before they do so. The following are hazards and safety procedures associated with the use of trench jacks:

- *Injury to back and muscles from lifting heavy objects*. An 8-ft-long 52 × 88 extension trench jack weighs approximately 120 lb. A two-person crew can safely lift, set, and remove it from the trench. Anything longer or heavier should be lifted and set with equipment such as a backhoe or boom truck.
- *Overhead lifting hazard*. When jacks are being hoisted by sling from a tractor bucket or boom truck, the swinging jack presents a hazard to workers guiding it. Loose plywood and rocks can also fall off onto workers. Workers should stand clear and guide with a lead rope.
- *Finger and hand protection*. Trench jacks have moving parts at the connection between the cylinder and the rail. When the jack swings open, fingers can be crushed under the cylinder block; and when it is swung closed, fingers can easily be sheared off if they are between the block and the rail leg. When the hydraulic hose is being connected to the block fitting and when the jack is being lifted by hand, shearing and crushing are most likely to happen. Awareness through safety instruction and hand placement at a safe distance, 12 in, from the blocks are safe practices. Some manufacturers have optional finger guards, but it is still possible to get fingers

under the block and wrists cut and banged when the jack folds or unfolds. A method of eliminating the problem would be to lock the jacks into the open position by inserting a second pin in the block. Using this method they could be locked open while they are on the job and completely eliminate the possibility of finger and hand damage.

- *Bank collapse with worker standing on it.* When the jack is being set, it is still possible for the trench wall to collapse from the additional weight and activity going on around it. Trench jack installation should closely follow the excavation activity.

  During jack removal the arch column is being literally removed with the load still on it. Pipe bedding and initial backfill cut the trench depth, adding some stability prior to removing the jack. If backfill operations are closely following jack removal, the length of unshored collapsible trench wall becomes short. Soil arching back to the backfilled area is likely, and trench wall failure becomes less likely. Remote backfill operation such as excavator wheel or vibraplate or remote operated compactors must always be used for compaction outside the shored area.

  When trench jacks are being removed to allow pipe installation and then reset, there is a greater likelihood of trench wall collapse. Equipment and personnel in close proximity are at risk of losing the ground under their feet. Keep equipment and personnel, except those needed to remove the jack, a safe distance away. This type of operation is not uncommon and most often works safely, but if there is any evidence of trench wall collapse, the operation should be discontinued and a different method of getting production materials into the trench or a different shoring system should be used. Several bad accidents have occurred in conjunction with this type of operation.

- *Get the surcharge loads right.* Equipment over 20,000 lb and large spoil piles over 2 ft high quickly add extra surcharges, especially in the top 10 ft, that can easily overload the trench jack. If one cylinder fails, a progressive failure to the bottom of the trench and then down the length of the trench is possible. A boom truck or backhoe outrigger placed next to a trench jack can trigger this. The way to adjust for additional surcharge load is to move the load away from the trench, spread the load with timber pad or steel plate, or decrease the trench jack spacing. Centering the surcharge load on the jack places most of the load on that jack. The alternative, centering the load between the jacks, distributes the load evenly between the jacks, but it increases the possibility of the arch void to fall out or arch shear failure at the jack. One alternative may not be any better than the other.

- *Trench jack fold up failure [Fig. 9.19(a)].* If all the jacks were unfolded into the trench from one side of the trench it is possible to get a bank failure that can lift the rotating jack leg. This type of failure is not common; however the author has spoken with more than one worker that has fortunately from outside the trench witnessed this type of failure. No workers were inside the trench. The story goes that 40 ft of trench

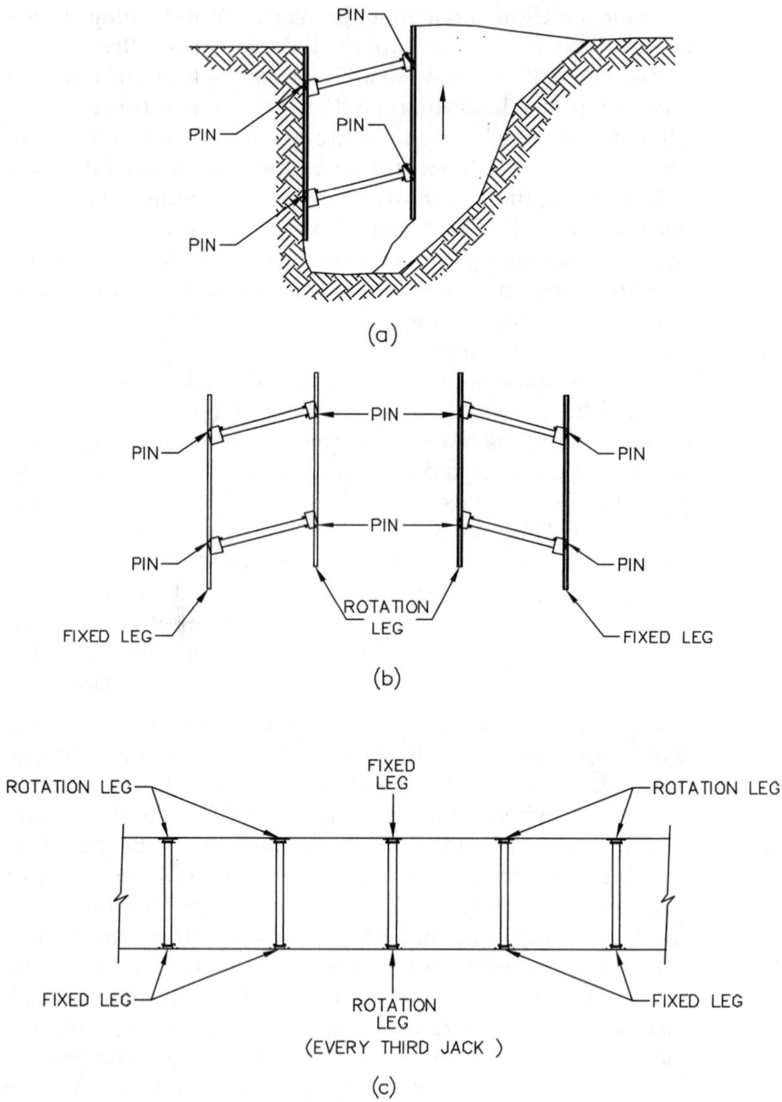

Figure 9.19 (a) Trench jack fold-up failure, (b) leg rotation, and (c) jack rotation to prevent fold-up failure.

folded up the jacks and collapsed. The solution is to rotate the every other jack so that the rotation leg is on the other side of the trench. The problem is that the installers have to move to the other side of the trench to set and pressurize the jack. Two soil conditions that this would be most likely to happen are in medium dense to loose noncohesive soils and soft clays with high surcharge loads.

- *Loose trench jacks in the trench.* Jacks that are not pressurized in the trench are not setting up arching and preventing trench collapse. In this condition the jacks can also fall down on workers below them. Jacks should not leak at all. Pressure can change slightly up or down due to temperature changes or increase due to loading, but it should never loosen up in the ditch. If jacks are left overnight, they should be checked before entering the trench in the morning. Simply tap them with a hammer or bar of metal, and they will sound loose if they are. Remove and replace jacks that bleed off. If the trench wall has voids where the cylinder hits the wall, use wood blocking to extend the connection to the soil [Fig.9.20(a)].

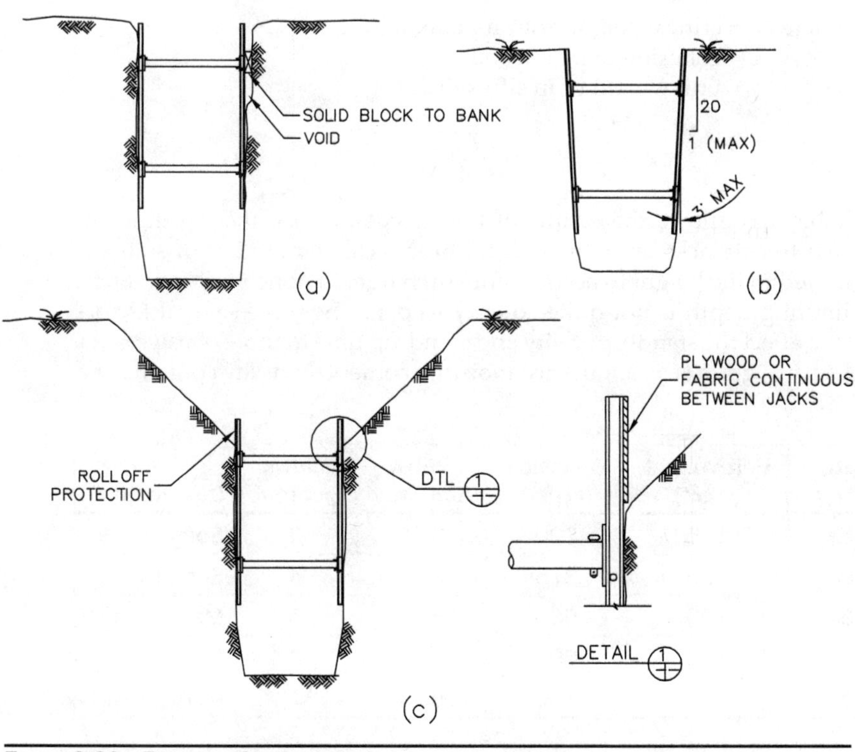

FIGURE 9.20 Trench safety issues.

- *Nonvertical trench walls*. When trench walls are not vertical, an inverted A shape, the trench jack is not stable [Fig. 9.20(b)]. Assuming a coefficient of friction of 0.1 between the soil and the aluminum rail and applying a factor of safety of 1.5 calculations indicate that the slope of the trench wall should not exceed 3 degrees or the jack will lift up and fail to provide an arching point.

  In trenches that are sloped above, extending the jack 18 in above the hinge point does not provide rolloff protection for workers below because the jack is spaced. Place fabric or boards behind the jack rail to stop objects at the surface and bank ravel from falling on workers [Fig. 9.20(c)].

9.3.5 Trench Jack Use Criteria

The limiting soil condition for trench jack use is whether the soil will stand up long enough at trench depth to install the jack. In cohesive soils this can be predicted, from Eq. (6.29), by calculating the critical height of the soil

$$H_c = \frac{4c}{\gamma}$$

where H_c = critical height prior to bank failure
c = cohesion of in situ soil
γ = unit weight of in situ soil

or
$$c = \frac{\gamma H_c}{4}$$

Table 9.11 shows the results of this calculation for the depth levels that trench jacks are tabulated to. In this calculation 2 ft of soil was added to the height to account for surcharge. In noncohesive soils the limiting depth is not quite so easy to peg. There are several factors that affect the stand-up ability and stand-up time for noncohesive soils. Density, grain size, angularity, moisture content, and clay content are a

Depth (ft)	Unit Weight γ (pcf)	Cohesion *C* (psf)	OSHA Soil Type	Blows per foot	Description
5	125	220	C-60	3	Soft
10	125	375	C-60	4	Soft
15	125	530	B	5	Medium stiff
20	125	690	B	6	Medium stiff
25	125	850	B	7	Medium stiff

TABLE 9.11 Minimum Soil Strength for Trench Jacks in Cohesive Soils

few. Some of this can be picked up in the description section of the bore log. Generally clean sands and gravels, GW, GP, SW, and SP in decreasing order of stability, are poor candidates for trench jack use. In increasing order of stability are SC, SM, GC, and GM when the blow count is above 10 for trenches less than 10 ft deep and 15 for trenches to 16 ft deep. If the particles are rounded, from streambeds, alluvial fans, etc., the chance of stand-up is highly diminished. In shales the slope of the bedding plain can have a large effect; the soil can fall out on one side of the trench and stand on the other side. Thin layers of soft clay or gravels, less than 6 in, will usually stay in place until the trench jacks are set; however, all exposed layers should be looked at closely because if the upper layer falls out before the bottom layer is reached, the trench is going to be useless.

Existing parallel utility trenches create problems for trench jack use. Not only is the possibility of trench wall collapse before jack insertion increased, but also the danger of serious damage to the existing utility is present. Existing trenches usually carry collected water in their bedding material that seeps into surrounding soil, creating a weak layer at the bottom, and so water may be present in the new excavation even when it was not expected. Problems with existing trenches can be controlled to a certain extent by keeping the jack installation close to the excavation operation. If the allowable jack spacing is 8 ft, then the excavation should not be more than 8 ft ahead of the last jack. Trench-crossing utility trenches can have their bedding material spilling into the new trench, and cross lines can interfere with jack spacing and sheeting requirements. Single jacks and timber sheeting are usually used in these situations.

The decision about whether to use trench jacks or open cut usually hinges on right of way, existing utilities, and availability of contractor-owned excavation equipment. In new subdivision construction where most sewer pipe is light enough to be laid by hand and there is plenty of room to open up trenches and store excavated dirt, open cut excavations allow each operation, dig, lay, and backfill to work separately at maximum speed and efficiency. The trench jack operation is more conducive to the opposite type of working condition. Open a short trench length, less than 100 ft, place shores, lay pipe and backfill, with all working crews together in a tight space, none working at their optimum productivity. Figure 9.21.shows the results of a cost survey between the use of trench jack and open cut worker protection. The efficiency of the optimized open cut operations at a depth of approximately 9 ft is overcome when the cost of the increased quantity of dirt that has to be excavated and recompacted rapidly exceeds the cost of using longer trench jacks and a less efficient crew. One of the key factors that leads contractors to use open cut beyond this depth is that if there is any question about whether the trench will stand up long enough to install the trench jacks, that doubt is eliminated with open cut. Trench walls that fail before getting the shoring in can really

OPEN CUT			SHORE @ 6' O.C	
ACTIVITY	ACTIVITY EFFICIENCY		ACTIVITY	ACTIVITY EFFICIENCY
EXCAVATION TRENCH	100%	STEP ①	EXCAVATION TRENCH	100%
EXCAVATE OPEN CUT	100%	STEP ②	SHORE @ 6' O.C	100%
BED PIPE	100%	STEP ③	BED PIPE	90%
LAY PIPE	100%	STEP ④	LAY PIPE	50%
INITIAL BACKFILL	100%	STEP ⑤	INITIAL BACKFILL	50%
BACKFILL OPEN CUT	100%	STEP ⑥	REMOVE TRENCH JACKS	100%
BACKFILL TRENCH	100%	STEP ⑦	BACKFILL TRENCH	100%

A. E. - ACTIVITY EFFICIENCY

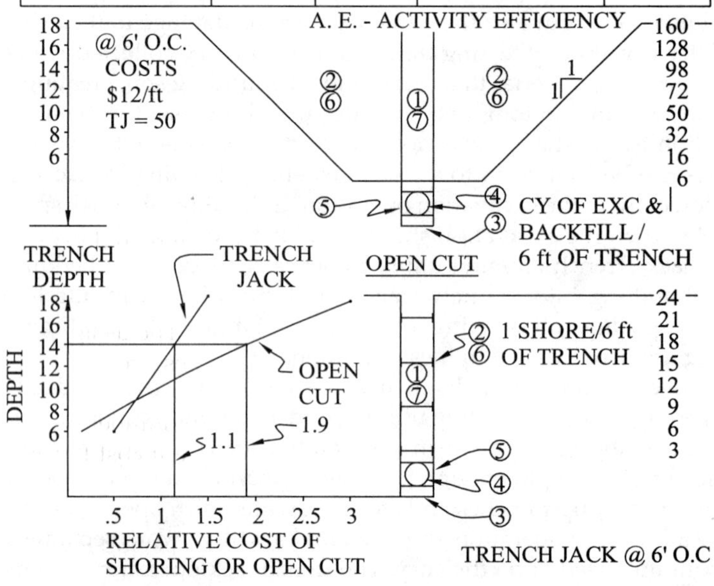

FIGURE 9.21 Relative cost of using trench jacks versus open cut.

foul up the speed of the operation, and trench jacks are no longer an option at that point. The other powerful deciding factor is optimized equipment and workforce utilization. With open cut the dig crew and lay crew do not even have to be on the same project at the same time, and there can be days between one operation and the next.

In urban areas when any of the following conditions are present, trench jacks are the most favored and least expensive option:

- *Existing streets.* Open cut causes increased roadbed reconstruction costs and eliminates traffic access.

- *Full traffic to be allowed on streets at end of shift.* Trench jacked trenches can be plated over. Open cut trenches are usually too wide for plating.
- *Crossing utilities.* In open cut, crossing utilities have to be supported and protected. Shoring boxes cannot be pulled through crossing utilities.
- *Existing facilities that must be protected.*Trench jacks provide positive support, trench boxes do not.

Trench jacks cannot be used under the following conditions:

- If the trench wall will not stand up long enough to get the jack in.
- In pipeline operations where the outside diameter is greater than 36 in, bedding plus outside diameter must be less than 48 in.
- Trenches over 15 ft wide unless special oversleeves are designed.

9.3.6 Trench Jack Shoring Design by a Registered Engineer

The major advantage of having a trench jack shoring plan developed by an engineer instead of working from tabulated data is that the engineer uses design parameters from the soil, cohesion *c* and angle of internal friction *f*, and not from OSHA soil types. The design parameter method is more accurate; some see it as less conservative, making it possible to space the jacks farther apart. In large projects with repetitive use of trench jacks, spacing and the number of times the jack must be set and removed can be a significant cost factor. In theory, spacing of more than 8 ft on center and cylinder spacing more than 4 ft from the bottom should be achievable. The 8-ft spacing is almost universally adhered to because increased spacing increases the arch void to the point where the impact of the void falling out on the workers is as dangerous as a trench wall collapse. In very stiff cohesive soils, a reasonable argument can be made for increasing the jack spacing beyond 8 ft. Increased vertical spacing beyond 4 ft at the bottom is quite often requested by contractors to make it possible to lay larger-diameter pipe. Engineers should have some discretion about this distance; however, OSHA tries to eliminate it with the argument that if there is no upper limit, the distance that engineers allow will eventually creep up to the point where trench collapse will occur. The following citation is seen by OSHA as being a requirement that the engineer must incorporate into the design:

> 1926.652(e)(2) Additional requirements for support systems for trench excavations.

> 1926.652(e)(2)(i) Excavation of material to a level no greater than 2 feet (.61 m) below the bottom of the members of a support system shall be permitted, but only if the system is designed to resist the forces calculated for the full depth of the trench, and there are no indications while the trench is open of a possible loss of soil from behind or below the bottom of the support system.

Since manufactured trench jack rails never extend more than 2 ft below the cylinder, the above requirement forces the jack cylinder to be no more than 4 ft above the bottom of the trench. There does not seem to be any rule preventing the engineer from designing a stronger rail or sheeting that extends more than 2 ft from the cylinder as long as all the elements are structurally adequate.

In addition to increased spacing, design by an engineer is required whenever trench jacks are used outside the tabulated data such as when the surcharge load is greater than 72 psf or when they are used in conjunction with other types of shoring systems, shield, braced sheet pile, or other braced stiff sheeting.

Example 9.1 Trench jack design problem

A 3-ft-wide × 2-ft-thick × 3000-ft-long concrete duct bank is to be constructed in a street at a depth of 14 ft. The contractor plans to use trench jack shoring and will place the concrete from the truck shoot using 12-yd^3 (or 12-cy) concrete trucks. Right-of-way and traffic constrictions limit the concrete truck access so that the wheels must get within 2 ft of the trench edge, and a Boussinesq analysis indicates that the surcharge loading will be 450 psf in the first 5 ft of trench depth. The soil and trench configurations are shown in Fig. 9.22.

The OSHA Appendix A soil classification for the soil is type C-60. From previous experience the contractor has determined that the soil will stand long enough to get the trench jacks installed. Due to the high surcharge loads the contractor needs a design by an engineer, and the contractor would also like to get jack spacing greater than tabulated data would allow. Determine the trench jack spacing.

Step 1. Determine soil loading.

Use the Peck Hanson, and Thornburn, apparent earth pressure diagram for cuts in noncohesive soils (Fig. 6.9)

$$p_a = 0.65K_a\gamma H = 0.65 \times 0.26 \times 125 \times 14 = 300 \text{ psf}$$

where $K_a = \tan^2(45 - \phi/2) = 0.26$. From Table 5.3 for 16 blows per foot $f = 36.5°$. Use $g = 125$ pcf.

$$H = 14 \text{ ft}$$

Step 2. Determine jack spacing.

Limit trench jack cylinder loading to 18,000 lb. At the first 5 ft from the surface the loading is the maximum due to the surcharge load. The load is

$$300 \text{ psf for soil} + 450 \text{ psf for concrete truck} = 750 \text{ psf}$$

Total load on one cylinder is loading × 4 ft × jack spacing ≤ 18,000 lb, or

$$\text{Spacing} = \frac{18,000}{750 \times 4} = 6 \text{ ft on center}$$

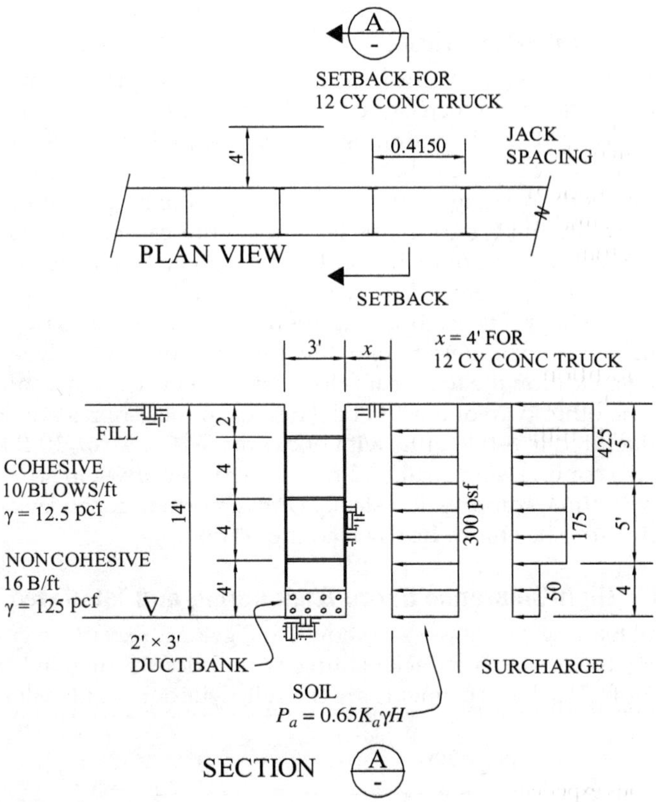

FIGURE 9.22 Trench jack design example 9.1.

Conclusions

If this plan were developed using tabulated data from a manufacturer (and ignoring the actual surcharge load from the concrete truck), using type B soil the trench jack spacing would be 7 ft on center and type C soil 5 ft on center. Under OSHA Appendix A the soil would most likely be classified as type C because it is noncohesive. Due to the surcharge load tabulated data cannot be used, and an engineered design is required.

Note that the surcharge load from the concrete truck is highest at the axle and reduces as one looks at the load at any distance away from the wheel along the trench. Averaging the load over 8 ft would probably result in a continuous surcharge load of approximately 300 psf. Calculations would yield a safe jack spacing of 8 ft on center. For a 3000-ft trench an engineered plan results in 375 trench jack sets and removals plus 3 ft extra of open space between jacks while a tabulated data plan would have 600 jack sets and less working space.

If the concrete truck could be set back enough to give a normal surcharge load, 100 psf, the resulting engineered design load would be 400 psf and jack spacing would be 8 ft on center whereas the load from type C soil would be 840 psf and the spacing 5 ft on center.

9.4 High-Clearance Shores

The need for greater vertical clearance between the bottom of the trench and the first trench jack cylinder drove the development, starting in the 1990s and continuing today, of a high-clearance vertical shore. The basic principle behind this shore is that the stiffness of the leg extends the arching point to within 4 ft of the bottom of the excavation. Today the typical high-clearance shore can be set where the cylinder is a maximum of 7 ft off the bottom. Jack spacing still does not exceed 8 ft on center. The jack leg is a made from a heavy-duty aluminum waler (Fig. 9.23), and the bottom cylinder varies from three 2-in aluminum cylinders to one 3-in aluminum cylinder. Due to the loading condition the top strut is in tension and is usually chained or has some other type of tension tie. High-clearance shores are usually manufactured in 8-ft lengths with one compression strut, 10-ft length with two compression struts, 12-ft with two compression struts, and 16 ft with three compression struts. Unlike normal trench jacks, each manufacturer has his or her own configuration.

9.4.1 High-Clearance Shore Engineering and Tabulated Data

Typical loading conditions are shown in Fig. 9.24. Except for cylinder spacing, the basic trench jack loading assumptions shown in Fig. 9.18 also apply. The basic elements are the rail, cylinder, and tension tie.

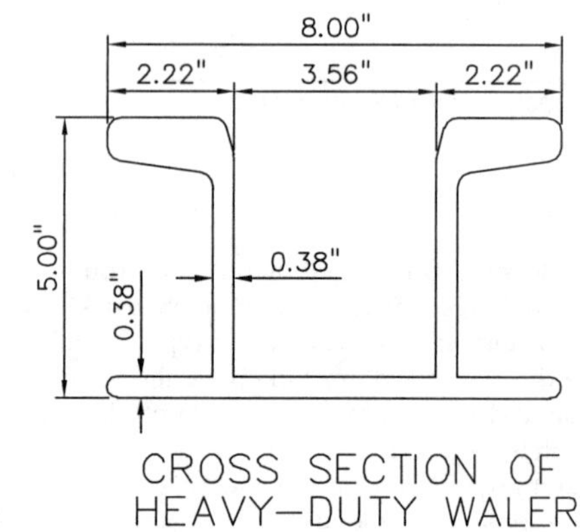

MATERIAL	ALUMINUM
ALLOY	6061–T6
AREA	9.76 in^2
WEIGHT	11.72 plf
SECTION–MODULUS – TOP (LEG SIDE)	14.50 in^3
SECTION–MODULUS – BOTTOM (BLADE SIDE)	14.40 in^3

FIGURE 9.23 Heavy-duty waler rail.

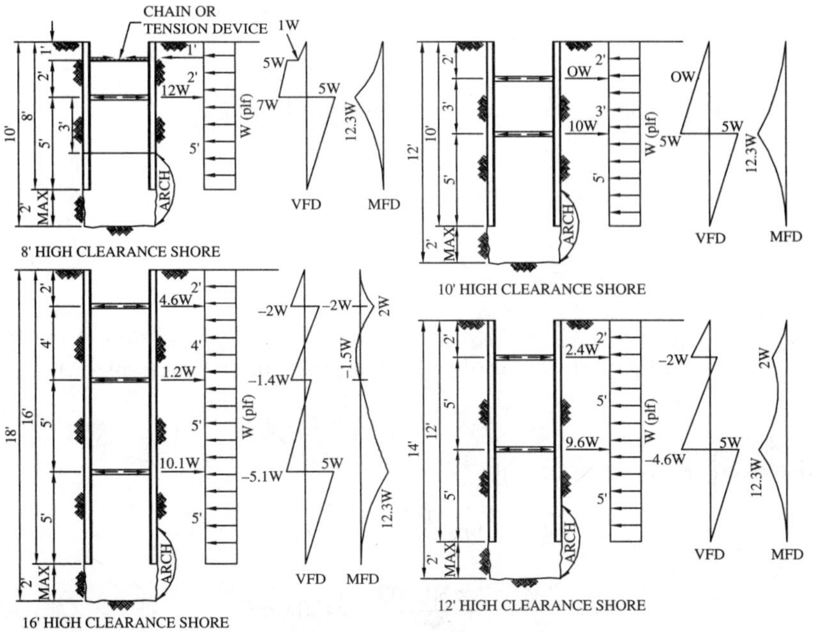

FIGURE 9.24 High-clearance shore, 8, 10, 12, and 16 ft.

Rail Design

In order for the shore to maintain basic trench jack arching assumptions, maximum 4-ft vertical and 8-ft horizontal spacing, the leg of the shore must have enough bending strength to provide an arching point within 4 ft of the bottom. Therefore the design of these shores is controlled by the cantilever leg bending moment $M = WL^2/2$, for a 5-ft leg length $M = 12.5W$ (Fig. 9.24). The hydraulic cylinders then need to have enough strength to carry the load delivered to them. In keeping with the concept that rail failure would not necessarily instigate a progressive failure whereas cylinder failure would, manufacturers have used the 1.33 stress increase for shoring applications on the rail design but not on the cylinder design.

Determine the allowable loading for heavy-duty waler rail with 5-ft leg extension.

From Fig. 9.7, compression in components of beams, formula 17

$$s = \frac{b}{t} = \frac{5}{0.375} = 13.3$$

where s = slenderness (Fig. 9.7 aluminum design formulas)
$b = 5.0$ in heavy-duty waler rail
$t = 0.375$ in heavy-duty waler rail

$$\therefore \quad 8.9 < s < 19 \qquad \text{and} \qquad F_{bc} = 40.5 - 1.41\frac{b}{t} = 21.7 \text{ ksi}$$

From Fig. 9.7, tension in components of beams, formula 2

$$F_{bt} = 19 \text{ ksi}$$

Allow one-third stress increase for shoring application:

$$F'_{bt} = 19 \times 1.33 = 25 \text{ ksi}$$

Since $M = SF_{b,}$

$$M_{std} = 25 \text{ ksi} \times 14.4 \text{ in}^3 = 360 \text{ k·in}$$
$$= 30 \text{ k·ft}$$

where $S = 14.4$ in³ section modulus of waler rail, blade side (tension)

$$F_b = 25 \text{ ksi}$$

Also from cantilever moment

$$M = \frac{WL^2}{2}$$

$$W = \frac{2M}{L^2} = \frac{2 \times 30{,}000}{5^2} = 2400 \text{ plf}$$

where M = cantilever moment in bottom leg, ft·lb
W = load on shore, plf
L = 5 ft cantilever length, ft

Manufacturers' tabulated data for high-clearance shores are presented by stating allowable spacing in OSHA soil types for given depths. Solve for allowable spacing at depth, using

$$W = w \times \text{shore spacing}$$

and $$w = \text{OSTF} \times H + \text{surcharge}$$

$$\text{Shore spacing} = \frac{W}{\text{OSTF} \times H + \text{surcharge}}$$

where W = 2400 plf load on shore leg
OSTF = OSHA soil type factor
OSTF = 25 lb/ft of depth for type A soil
= 45 lb/ft of depth for type B soil
= 60 lb/ft of depth for type C soil
H = full depth of trench, ft
H = 5 to 8, 10, 12, 14, 16, and 20 ft for 8-ft rails
H = 9, 10, 12, 14, 16, and 20 ft for 10-ft rails
H = 11, 12, 14, 16, and 20 ft for 12-ft rails
H = 14, 16, and 20 ft for 16-ft rails
Surcharge = 72 psf represents 2 ft of soil or lateral effect of 20,000 lb equipment

Depth (ft)	Type A-25 Shore Length				Type B-45 Shore Length				Type C-60 Shore Length			
	8	10	12	16	8	10	12	16	8	10	12	16
5	8				8				6			
8	8				6				4			
9		8				5				4		
10		7				5				4		
11			7				4				3	
12			8				4				3	
14				6				3				3
16				5				3				2
18				5				3				2
20				4				2				2

Notes:
1. This chart is based on allowable rail loading $W = 2400$ plf.
2. Maximum trench width is 12 ft.
3. Use plywood in trenches over 8 ft deep and as needed to prevent sloughing or raveling.

TABLE 9.12 Allowable Spacing for High-Clearance Shore

For example, for any rail in OSHA type B-45 soil at 14 ft deep

$$\text{Allowable shore spacing} = \frac{2400}{45 \times 14 + 72} = 5.7 \text{ ft}$$

Table 9.12 presents typical spacing and depth resulting from these calculations.

Cylinder Design

In cases where the shores use 2-in aluminum cylinders, the cylinders are stacked one above the other. Although the stiffness of the rail does not necessarily support this theory, a simplifying assumption is that the bottom cylinder takes the soil load that travels up the leg, while the top cylinder takes the load that travels down the leg from above. A slightly more conservative approach would be to apportion at least two-thirds of the load at the reaction to the bottom cylinder. In the case of one single 3-in cylinder, the entire load goes to that cylinder.

From Sec. 9.3.2, trench jack cylinder design, we know that by allowable stress calculations 2-in aluminum hydraulic cylinders have a safe working load of 18,000 lb, and from test results they have a safe working load of 22,000 lb with a 1.5 factor of safety.

The other common cylinders in use are 3½-in -OD × ¼-in wall 6061T-6 aluminum and 3½-in-OD × ¼-in wall steel A500 grade C, $F_y =$ 46 ksi and $F_u = 58$ ksi. For aluminum cylinders

$$p = \frac{2St}{D}$$

where p = bursting or bulging pressure, psi
$S = F_{tu} = 42{,}000$ psi for bursting
$S = F_{ty} = 35{,}000$ psi for bulging
$t = 0.25$ in
$D = 3.5$ in

Calculate the ultimate bursting pressure and force for a 6061-T6 aluminum hydraulic cylinder:

$$p_{bursting} = \frac{2 \times 42{,}000 \times 0.25}{3.5} = 6000 \text{ psi}$$

Calculate the ultimate force for bursting:

$$\text{Force } P = p_{bursting} \times \text{area} = 42{,}000 \text{ lb}$$

where

$$\text{Area } A = \frac{3^2 \pi}{4} = 7 \text{ in}^2$$

$$p_{bursting} = 6000 \text{ psi}$$

Using a 1.5 factor of safety on bursting, the safe working load of the aluminum cylinder is

$$\text{SWL} = \frac{42{,}000}{1.5} = 28{,}000 \text{ lb} \qquad \text{aluminum}$$

For a 3½-in × ¼-in wall steel cylinder

$$p_{bursting} = \frac{2 \times 58{,}000 \times 0.25}{3.5} = 8286 \text{ psi}$$

$$P = 8286 \times 7 = 58{,}000 \text{ lb}$$

$$\text{SWL} = \frac{58{,}000}{1.5} = 38{,}000 \text{ lb} \qquad \text{steel}$$

Check cylinder load on high-clearance shores.
From Fig. 9.24 the maximum reaction is

$$P = 12W = 28{,}800 \text{ lb} \qquad \text{for 8-ft shore}$$

$$P = 10W = 24{,}000 \text{ lb} \qquad \text{for 10-ft shore}$$

$$P = 9.6W = 23{,}000 \text{ lb} \qquad \text{for 12-ft shore}$$

$$P = 10.1W = 24{,}100 \text{ lb} \qquad \text{for 16-ft shore}$$

where $W = 2400$ plf

$$\therefore \quad \text{Two 2-in cylinders or one 3-in cylinder} \qquad \text{OK}$$

If two 2-in cylinders are used on the 8-ft shore, the bottom cylinder will see

$$2/3 \times 28{,}800 = 19{,}200 \qquad \text{OK}$$

The maximum reaction for the upper cylinders is

$$P = 4.6\text{W} = 12{,}240 \text{ lb} \qquad \text{for 16-ft shore}$$

and

$$P = -4W = -9600 \text{ lb} \quad \text{for 8-ft shore (needs chain or cable with } SWL = 9600 \text{ lb)}$$

Also from Sec. 9.3.2 the allowable jack extension for 2-in cylinders is 9 ft 4 in. If a 3.5 × 3.5 × 3/16-in steel oversleeve is used, the extension can be taken to 15 ft maximum. Most manufacturer's tabulated data for high-clearance shores allow a maximum 12-ft extension.

Checking the 3.5-in × ¼-in wall cylinder for buckling when extended 12 ft without oversleeve gives

$$\text{Slenderness s} = \frac{L}{r} = 62$$

where $L = 72$ in effective length
$r = 1.1524$ in radius of gyration
$\therefore$ s is between s_1 and s_2

$$\text{Allowable stress } F_{cy} = 20.2 - 0.126\frac{L}{r} = 12.38 \text{ ksi}$$

$$f_c = \frac{P}{A} = 11.3 \text{ ksi} \qquad \text{OK}$$

where $P = 28{,}000$ lb and $A = 2.5525$ in^2.

Manufacturers also develop their tabulated data around the maximum cylinder loading, using the assumption that the soil at the bottom will arch to the point where the leg can carry the load and the cylinder strength is the controlling factor. Table 9.13 gives high-clearance shore data based on this formulation.

The rules for plywood are to use plywood to stop sloughing or raveling at any depth and always use it in depths over 8 ft in any soil type. Essentially the only case where plywood would not be used is when an 8-ft high-clearance shore is used in a trench 8 ft or less deep.

	Type A-25				Type B-45				Type C-60			
Depth (ft)	Shore Length				Shore Length				Shore Length			
	8	10	12	16	8	10	12	16	8	10	12	16
5	8				8				8			
8	8				6				5			
9		8				6				5		
10		7				5				4		
11			7				5				4	
12			6				5				4	
14				6				4				3
16				5				4				3
18				4				3				2
20				4				3				2

Notes:
1. This chart is based on allowable cylinder loading = 28,000 plf.
2. Maximum trench width is 12 ft.
3. Use plywood in trenches over 8 ft deep and as needed to prevent sloughing or raveling.

TABLE 9.13 High-Clearance Shore Allowable Spacing Based on Cylinder Strength

Also note that prevention of sloughing and raveling overrides allowable jack spacing. The sheeting must be spaced closely enough to prevent sloughing or raveling even if that means continuous jacks to be set a maximum of 2 ft on center.

9.4.2 High-Clearance Shore Use and Safety Issues

Definitely the major reason for deciding to use high-clearance shores is to get up to 7 ft clear by 12 ft wide clear working space at the bottom of the trench. Trenches that use high-clearance shores are typically for pipe larger than 36-in diameter, utility vaults, and buried concrete structures. The same required soil conditions for use of trench jacks apply to high-clearance shores.

The major safety issue with high-clearance shores is that they are much heavier than trench jacks and do more damage when they fall on workers. The lightest shore with plywood weighs approximately 500 lb and would be life-threatening or damaging to any workers or production work that it fell on. The following safety precautions should be observed:

- If high-clearance shores lose pressure or are not properly seated to the trench wall due to an uneven wall or sloped trench wall, they will drop vertically or fall forward on their face. Installers should check the jacks after installation and at the start of each shift by tapping them with a bar or hammer to see if they sound solid—they will sound different if they are loose.

Make sure that hydraulic fittings are protected and that workers are aware that if they are broken, the jack will lose pressure and fall.

- When high-clearance shores are stacked in the trench one above the other, the leg of the top shore should be connected to the top leg of the bottom shore, and the hydraulics should be connected so that both shores work as one. The connection should be strong enough to hold the weight of the bottom jack plus a 1.5 factor of safety.
- Durability is an issue when moving and lifting high-clearance shores. Make sure lift connections are secure, and raise them from a flat to vertical position symmetrically. Uneven pressure will bend the legs and stress the connections to the point where the pins and blocks can become damaged and fall apart when the jack is being lifted into position.

9.4.3 High-Clearance Shoring Design by a Registered Engineer

Shoring design by an engineer will most often produce greater shore spacing and a safer installation because it is designed around the actual soil conditions based on soils information collected from the site and on engineering properties of the soil. High-clearance shores are no exception. With an engineered design the depth of the excavation, the surcharge loads and the soil engineering parameters, cohesion, and internal angle of friction are given, and the allowable shore spacing needs to be determined.

Based on the shore waler rail strength and a rectangular loading diagram consisting of soil + surcharge, the high-clearance shore spacing SS is determined from

$$\mathrm{SS} = \frac{\mathrm{srs}}{w + \mathrm{sc}}$$

where SS = shore spacing, ft
w = soil load, psf
sc = surcharge load, psf
srs = 2400 plf shoring rail strength for waler rail

Other materials can be used, and rail strength can be calculated and used here.

Based on the shore waler cylinder strength and a rectangular loading diagram consisting of soil + surcharge, the high-clearance shore spacing is determined from

$$\mathrm{SS} = \frac{P}{12 \times (w + \mathrm{sc})} \qquad \text{for 8-ft shore}$$

and

$$\mathrm{SS} = \frac{P}{10 \times (w + \mathrm{sc})} \qquad \text{for 10- to 16-ft shore}$$

where SS = shore spacing, ft
w = soil load, psf
sc = surcharge load, psf
P = 28,000 lb SWL of 3-in aluminum jack cylinder
SWL = 36,000 lb for two 2-in aluminum cylinders
SWL = 38,000 lb for steel cylinder

Designs using cylinder strength greater than 28,000 lb will result in the soil arching from the bottom to more than 4 ft up. If the first cylinder is set 7 ft from the bottom, the arching distance will not be less than 7 ft.

In soils that work best for trench jacks, most often the soil loading can be determined from NAVFAC or Terzaghi and Peck formulas [Eq. (6.26)] for apparent soil pressure

$$p_a = 0.65 k_a \gamma H \qquad \text{for noncohesive soil}$$

$$\sigma_{h_1} = 0.2\gamma H \quad \text{to} \quad \sigma_{h_2} = 0.4\gamma H \qquad \text{for medium to stiff cohesive (Fig. 6.9)}$$

With trenches that use high-clearance shores, the surcharge loads are usually higher than normal due to larger equipment handling larger pipe and materials. Surcharge loads of 300 psf for the first 10 ft and 150 psf from 10 to 20 ft deep should be a minimum.

For example, determine the spacing for high-clearance shores set 12 ft deep in medium stiff clay based on rail strength. Use

$$w = 0.3\gamma H = 0.3 \times 120 \text{ pcf} \times 12 \text{ ft} = 432 \text{ psf}$$

$$\text{Surcharge sc} = 150 \text{ psf}$$

$$\text{waler rail strength} = 2400 \text{ plf}$$

$$\mathrm{SS} = \frac{\mathrm{srs}}{w + \mathrm{sc}} = \frac{2400}{432 + 150} = 4 \text{ ft on center}$$

If a steel cylinder is used on a 10-ft shore, the spacing will be

$$\mathrm{SS} = \frac{P}{10 \times (w + \mathrm{sc})} = \frac{38{,}000}{10 \times (432 + 150)} = 6.5 \text{ ft on center}$$

Soil strength and surcharge assumptions should always be backed up by data and calculations.

9.5 Waler Rails

Waler rail systems are made up by placing vertical sheeting with horizontal wales and hydraulic struts. OSHA and manufacturers have developed systems and tabulated data around use of trench jack rails, heavy-duty waler rails, 2- and 3-in hydraulic cylinders, and timber or plywood sheeting. This is essentially the same as any two-wale strutted shoring system such as sheet pile or pile and plate except for the following differences:

- The manufactured systems rely on soil arching vertically between the wales.
- Both OSHA and manufacturers allow no more than 4 ft vertically between the wales.
- The sheeting does not extend below the bottom of the excavation.
- With a sheet pile and wale or pile plate and wale system, the installation procedure is to drive the sheeting completely and then place the wales as the excavation proceeds. With waler systems the sheeting is set and driven against the wales as the excavation proceeds downward; or the excavation is completed first, and then the sheeting and wales are set in and pressurized from the surface.

The major reasons to use waler rails are as follows:

1. They get more horizontal spacing between the cylinders, as much as 12 ft with OSHA tabulated data.
2. Waler rail and sheeting combinations can be used to advance excavations in soils that will not stand up long enough to install trench jacks or shoring shields.
3. Sheeting can be fastened to pairs of waler rail sets to construct a shoring box. The box can be open- or closed-ended and constructed at the trench side or off-site.
4. Waler rail systems can straddle utilities in bad soils better than trench jacks. See Fig. 9.25. The trench jacks with plywood cannot be spaced 6 ft on center next to the existing 48-in line. The waler rail allows for the angled crossing of the pipe and for there to be lagging under the pipe.

The OSHA standard tabulates for three different aluminum wale sections, S = 3, 7, and 14 in^3. Most manufacturers have available a standard duty wale with a least section modulus = 3.62 in^3, and a

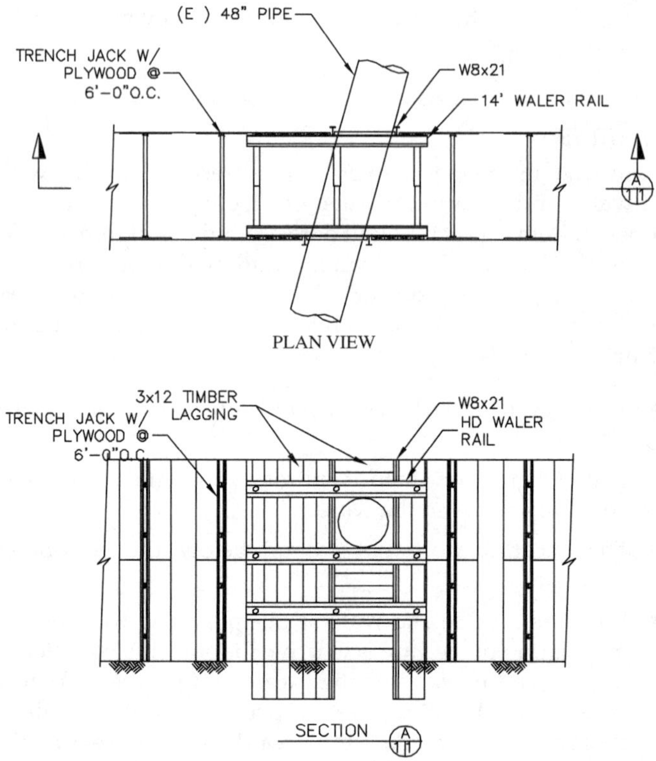

FIGURE 9.25 Waler rail application.

heavy-duty rail with $S_{min} = 14.4$ in^3 (Fig. 9.26). OSHA tabulates for use with 2- and 3-in hydraulic cylinders, and manufacturers generally only supply 2-in cylinders.

Each manufacturer of aluminum shoring equipment has her or his own die at the aluminum casting plant for the waler rails, and each is slightly different from the competitors'. However, their properties are similar to those shown in Fig. 9.26. There is nothing to prevent manufacturers from developing larger and stronger wales; however, the ones available are balanced to 2- and 3-in aluminum cylinders having a safe working load of 18,000 and 28,000 lb, respectively. Larger wales would need a wider gap between the webs to allow larger hydraulic cylinders.

9.5.1 Waler Rail Engineering and Tabulated Data

With a horizontal wale system the sheeting carries the load to the wales, and then the wales carry the load to the struts. In type A , B, and C-60 soils if the sheeting were not there, arching would still carry the load to the wales although sloughing or raveling would still be possible and require sheeting to prevent it. In these soils arching

WALER SPECIFICATION SHEET

SECTION PROPERTIES	STANDARD RAIL	HEAVY-DUTY RAIL
MATERIAL	ALUMINUM	ALUMINUM
ALLOY	6061–T6	6061–T6
AREA	4.87 in^2	9.76 in^2
WEIGHT	5.84 plf	11.72 plf
SECTION–MODULUS – TOP (LEG SIDE)	3.62 in^3	14.50 in^3
SECTION–MODULUS – BOTTOM (BLADE SIDE)	4.52 in^3	14.40 in^3
EQUIVALENT TIMBER SIZE * (#2 DOULAS FIR)	8x10 (ON EDGE)	12x16 (ON EDGE)

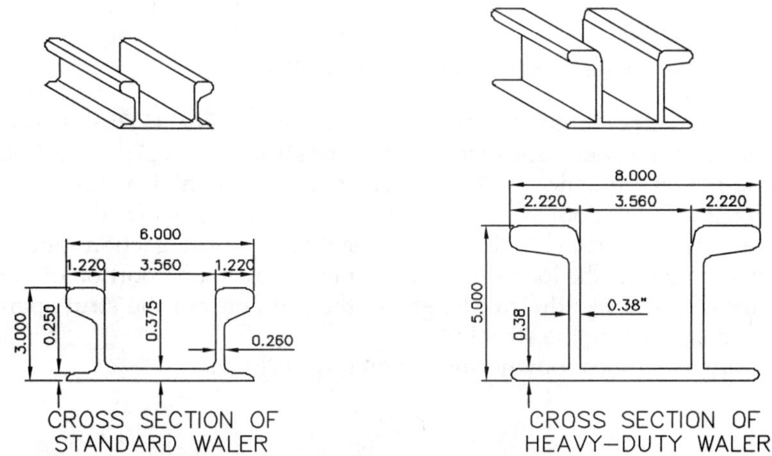

FIGURE 9.26 Standard and heavy-duty waler rail.

would still carry as much as 12 ft to the struts, but the arch void would be much larger and present a serious hazard if it fell out. Also in this case some of the soil load would arch directly to the struts and bypass the wale.

In the case of C-80 soils the arch shears before it gets to the arch support, and therefore the sheeting has to be strong enough to carry the soil load to the wales and the wales have to be strong enough to carry the entire wale load to the struts. In this case if the waler rail is designed for maximum bending between the struts, there can be significant deflections, 2 to 4 in with aluminum rails, in spans 8 to 12 ft.

Apparent earth pressure diagrams should be used to design these systems, and it is reasonable to apply a 1.33 shoring use factor to wales and sheeting, because they do not fail progressively; not struts however, this assumption could lead to an unacceptable level of risk in C-80 soils. Usually in C-80 soils the sheeting is driven in stages while being guided by the rails. This method of installation can deliver additional lateral forces beyond soil and surcharge to the rail due to misalignment and deflection of the sheeting. For this reason the author does not believe that the shoring use factor should be used in C-80 soils.

Calculating Strut Spacing and Tabulated Data

In developing tabulated data for waler rail systems, the soil + surcharge load, the wale strength, and the vertical spacing are given, and the maximum allowable strut spacing is calculated. The aluminum cylinders are then checked to see if their capacity is exceeded.

The following note is critical to the development of OSHA tabulated data:

> Aluminum Hydraulic Shoring for Trenches —1926 Subpart P Appendix D— (g) Footnotes, and general notes, for Tables D-1.1, D-1.2, D-1.3, and D-1.4.
>
> (9) Wales are calculated for simple span conditions.

This is always the case for a wale with two struts [Fig. 9.27(a)]. This is not true for a continuous wale with three struts [Fig. 9.27(b) and (c)]. The major difference is that the strut reaction is higher, $^{10}/_{8} \times WL$ [Fig. 9.28(b)]. Another minor difference in all cases is that the struts are never set directly at the ends of the rail. The strut reaction should always include the load that comes from the overhang portion of the wale. Always use the full length of the rail and not the strut span when calculating the strut load.

By using the moment calculation (Fig. 9.28),

$$M = \frac{W\ell^2}{8}$$

where M = maximum moment, ft·lb
W = wale load, plf
l = horizontal strut spacing, ft

The horizontal spacing can be determined. The cylinder load can be determined from

$$P = \frac{W\ell}{2} \qquad \text{Fig. 9.28(a)}$$

or

$$P = \frac{10}{8}W\ell \quad \text{or} \quad P = \frac{3}{8}W\ell \qquad \text{Fig. 9.28(b)}$$

where P = strut load, lb
W = wale load, plf
l = horizontal strut spacing, ft

Calculate the maximum wale moment $M = SF_b$. The results are given in Table 9.14.

From Fig. 9.7, tension in components of beams, formula 2

$$F_{bt} = 19 \text{ ksi}$$

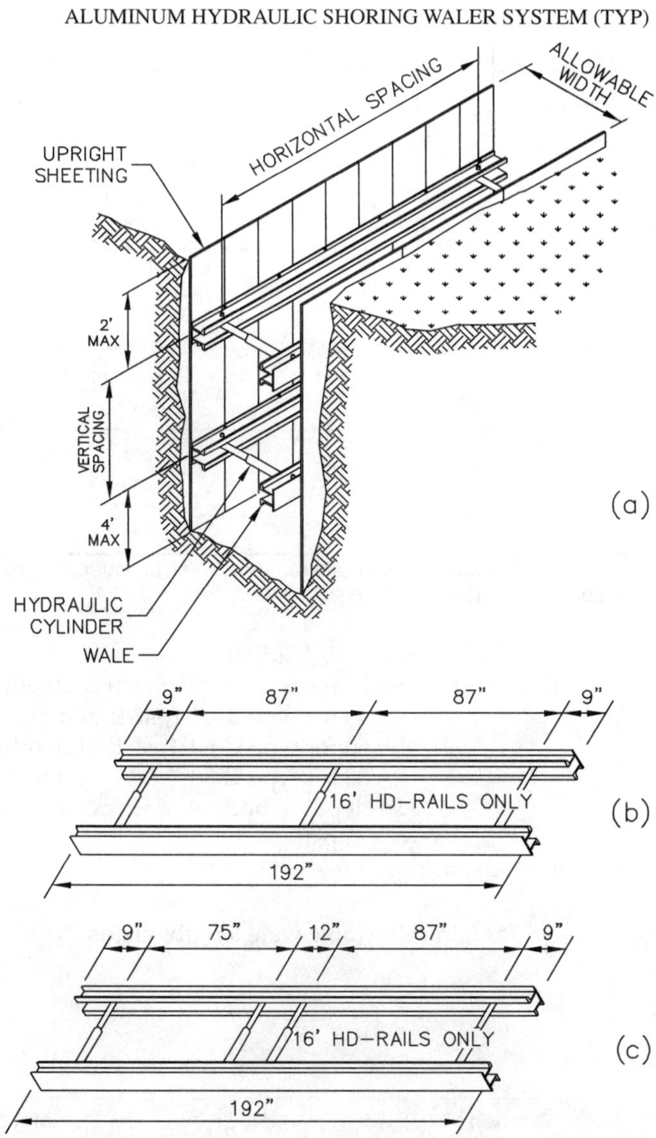

FIGURE 9.27 (a) Single-span waler; (b) and (c) double-span waler system.

Allow one-third stress increase for shoring application

$$F'_{bt} = 19 \times 1.33 = 25 \text{ ksi}$$

Calculating wale load from OSHA soil types gives

W = (OSHA soil type × H + surcharge) × vertical wale spacing

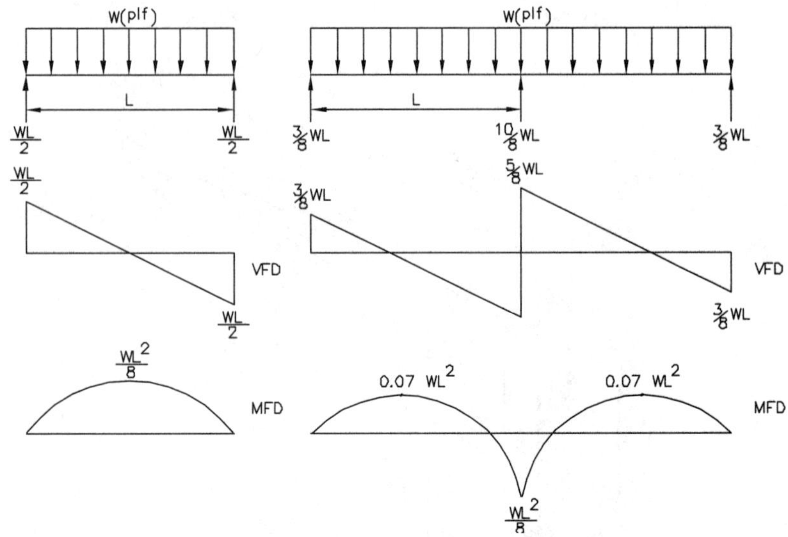

FIGURE 9.28 (a) Shear and moment diagram for two-strut waler rail; (b) shear and moment diagram for three-strut wale.

where W = wale load, plf
OSHA soil type = 25 for type A-25 (psf/ft of depth)
= 45 for type B-45 (psf/ft of depth)
= 60 for type C-60 (psf/ft of depth)
= 80 for type C-80 (psf/ft of depth)
H = depth to bottom of excavation, ft
Surcharge = 72 psf
Vertical wale spacing = 4 ft

Solving $M = \frac{W\ell^2}{8}$ for l, the allowable wale length, yields

$$\ell = \sqrt{\frac{8M}{W}}$$

Source	Section Modulus (in³)	F_b (ksi)	$M = SF_b$ (k•in)	M (k•ft)	$M \times 1.33$ (k•ft)
OSHA	3.5	19	66.5	5.5	7
OSHA	7	19	133	11	15
OSHA	14	19	266	22	29
Mfg.	3.62	19	69	5.8	8
Mfg.	14.4	19	273	23	30

TABLE 9.14 Allowable Aluminum Waler Rail Moment

Sample Calculation In C-80 soil at 15 ft deep what is the allowable strut spacing using a 16-ft-long heavy-duty waler rail with a section modulus of 14.4 in^3?

Because it is C-80 soil, the moment without the 1.33 shoring use factor is being used.

$$\ell = \sqrt{\frac{8 \times 23{,}000}{(15 \times 80 + 72) \times 4}} = 6 \text{ ft}$$

Space at 6 ft on center, leaving 2 ft overhang at the ends.

The cylinder load at each end is

$$P = \frac{3}{8} \times [(15 \times 80 + 72) \times 4] \times (6 + 2) = 15{,}260 \text{ lb at each end}$$

and

$$P = \frac{10}{8} \times [(15 \times 80 + 72) \times 4] \times (6) = 38{,}160 \text{ lb in the middle}$$

Therefore, use 2-in cylinders at the ends and two 2-in cylinders, SWL = 36,000 lb, at the center or use a 3-in steel cylinder with SWL = 38,000 lb.

By separating depths into 5-ft increments Table 9.15 was developed for the heavy-duty waler rail.

9.5.2 Sheeting Requirements for Waler Rail System

Because arching will take place in type A, B, and C-60 soils, plywood sheeting will work. In C-80 soils the sheeting has to carry the load over the 4-ft vertical span to the wales. In timber design always use simple span formulas.

Find wood sheeting requirement for 20-ft maximum C-80 depth.

We equate $M = \frac{w\ell^2}{8}$ ft·lb × 12 in/ft and $M = SF_b$ in·lb

and

$$S = \frac{w\ell^2 \times 12}{8F_b} = \frac{(20 \times 80 + 72) \times (4)^2 \times 12}{8 \times 1500} = 26.75 \text{ in}^3$$

where S = section modulus, in^3

w = soil + surcharge load at depth, psf

ℓ = 4-ft span

F_b = 1500 psi allowable bending strength for construction grade Douglas fir.

This allows 1.33 shoring use factor. A 3 × 12 timber has

$$S = \frac{bd^3}{6} = \frac{12 \times 3^3}{6} = 54 \text{ in}^3$$

Depth *H* (ft)	Soil Type A-25 (lb/ft of *H*)	Surcharge (psf)	Wale Load *W* (plf)	Wale Section *S* (in³)	Allowable Spacing (ft)	Cylinder Load (lb)	Middle Cylinder Three-Strut System (lb)
5	25	72	788	30	16	6,304	15,760
10	25	72	1,288	30	14	8,791	21,977
15	25	72	1,788	30	12	10,358	25,894
20	25	72	2,288	30	10	11,717	29,292

Depth *H* (ft)	Soil Type B-45 (plf of *H*)	Surcharge (psf)	Wale Load *W* (plf)	Wale Section *S* (in³)	Allowable Spacing (ft)	Cylinder Load (lb)	Middle Cylinder Three-Strut System (lb)
5	45	72	1,188	30	14	8,443	21,107
10	45	72	2,088	30	11	11,193	27,982
15	45	72	2,988	30	9	13,390	33,474
20	45	72	3,888	30	8	15,274	38,184

Depth *H* (ft)	Soil Type C-60 (plf of *H*)	Surcharge (psf)	Wale Load *W* (plf)	Wale Section *S* (in³)	Allowable Spacing (ft)	Cylinder Load (lb)	Middle Cylinder Three-Strut System (lb)
5	60	72	1,488	30	13	9,449	23,622
10	60	72	2,688	30	9	12,700	31,749
15	60	72	3,888	30	8	15,274	38,184
20	60	72	5,088	30	7	17,472	43,681

Depth H (ft)	Soil Type C-80 (plf of H)	Surcharge (psf)	Wale Load W (plf)	Wale Section S (in^3)	Allowable Spacing (ft)	Cylinder Load (lb)	Middle Cylinder Three-Strut System (lb)
5	80	72	1,888	22	10	9,114	22,786
10	80	72	3,488	22	7	12,388	30,971
15	80	72	5,088	22	6	14,962	37,406
20	80	72	6,688	22	5	17,154	42,886

Notes:
1. Vertical spacing is 4 ft maximum.
2. Strut spacing allows 1.33 factor for shoring use in calculations for type A, B, and C soil.
3. The 2-in cylinders shall have a 3.5-in × 3.5-in × 3/16-in wall oversleeve or equivalent for trench widths from 9 to 15 ft. The 3-in cylinders can only be used to 12 ft wide unless an engineered oversleeve is provided.
4. In A and B soils use plywood to prevent sloughing and raveling. In C-60 soils use plywood beyond 10 ft deep. In C-80 soils use 3-in timber or equivalent sheeting.
5. This table is for preliminary design purposes only. Waler systems being constructed in the field shall be in accordance with OSHA Appendix D, manufacturers' tabulated data, or design by an engineer.

TABLE 9.15 Design Tabulated Data For Heavy-Duty Waler Rail

Depth *H* (ft)	Soil Type (plf of *H*)	Surcharge (psf)	Min *S* (in³)	Timber	Surfaced Lumber
5	80	72	7.6	2 × 12	3 × 12
10	80	72	14.0	2 × 12	3 × 12
15	80	72	20.4	3 × 12	3 × 12
20	80	72	26.8	3 × 12	4 × 12

Notes:
1. Table is based on wood min $F_b = 1500$ psi, $F_v = 140$ psi.
2. The sheeting section requirement is per foot so that the section modulus required for 24-in-wide sheeting is 2*S* and for 6-in-wide sheeting it is ½*S*.

Table 9.16 Sheeting Requirements for Waler Rail Spaced at 4 ft Vertical in C-80 Soil

and 3 × 12 surfaced lumber has

$$S = \frac{11.5 \times 2.5^3}{6} = 29.9 \text{ in}^3$$

Check the shear.

$$f_v = \frac{3V}{2bd} = \frac{3 \times 3344}{2 \times 12 \times 3} = 139 \text{ psi} < 140 \text{ psi} \qquad \text{OK}$$

This is not OK for surfaced lumber. If sheeting is spaced, the soil is arching, and it still receives the full load from the spaced area. For instance, if the sheeting is spaced at 24 in on center, it carries 2 times the load it would carry if spaced at 12 in on center. The sheeting section requirement is per foot so that the section modulus required for 24 in on center is 2 times.

Table 9.16 shows the results of the above calculations for tight sheeting in 5-ft incremental depths in C-80 soil.

9.5.3 Hydraulic Shoring Boxes Constructed from Waler Rails

Open-ended shoring boxes can be constructed using tables calculated for type A, B, or C-60 soil. Careful attention should be paid to how sheeting is attached, and lifting points should be symmetric because the assembly is intended to be lifted into the excavation and removed by lifting.

If boxes have end panels, the waler rails have axial end loading added to them that needs to be taken into consideration; combined stress proportions have to total less than 1, by Eq. (9.2).

$$\frac{f_b}{F_b} + \frac{f_a}{F_a} \le 1$$

where f_b = computed bending stress
F_b = allowable bending stress
f_a = computed axial stress
F_a = allowable axial stress

As a general rule, the added axial stress will down rate the bending strength by approximately 10 percent;

9.5.4 Waler Rail Installation and Safety Issues

Waler rail systems can be installed completely from outside the excavation. At the start of the installation operation the total amount of sets for the entire excavation depth should be stacked in the starter excavation so that they do not have to be inserted between already set wales. If the excavation is dug in 4-ft increments as the wales are being set, where sheeting is required, it should be driven a couple of feet below the bottom, especially in C soils.

In C-80 soils the cantilever bending moment is $w\ell^2/2$ and at 20-ft depth exceeds the allowable bending strength of 3-in-thick timber at a 2-ft cantilever. If the sheeting is not driven 2 ft below the bottom, the wale should be set within 2 ft of the bottom.

The safety issues associated with waler rails in addition to those listed for trench jacks are as follows:

- Driving sheeting against waler rails as an excavation proceeds to depth can develop higher loading than the soil and surcharge predict. Large deflections, over 2 in with aluminum rails, are an indication of overstress and should be avoided. If excessive deflection occurs, shorten strut spans or redrive sheeting.
- In C-80 soil, drive sheeting a minimum of 2 ft below the bottom or as required to prevent sheeting deflection prior to letting workers inside the shoring.
- If the hydraulics are broken or fail, the entire waler rail section will fall vertically. To guard against this, the wale should be prevented from falling by nailing wood blocks to the sheeting below the wale location.

9.5.5 Waler Rail Design by a Civil Engineer

The main benefit of engineered waler rail design is to take advantage of soil strength parameters that fall between OSHA soil type levels. There is no alternative to engineered design when the surcharge loads are greater than 72 psf.

Waler rail systems can be designed with vertical spacing greater than 4 ft; however, beyond 4 ft vertical the soil arching assumption should be looked at closely or ignored so that the sheeting is designed with enough strength to carry the load to the wales. Light aluminum and steel sheet pile sections work best. Spacing beyond what is shown in tabulated data will require stronger wale sections and hydraulics.

For larger systems the terminology changes from *waler rail* to *sheet and brace systems*. Larger wale and hydraulic strut systems have been in use in Europe and to a small extent on the east coast of the United States since the 1970s. These systems are heavy and expensive; however, they are now becoming economically feasible because large shoring rental companies are making the investment in promoting them and stocking them in their inventories.

9.6 Shoring Shields

Shoring shields and boxes are a staple of the shoring industry. The primary advantage of shields is that they provide a clear open space as opposed to trench jacks where struts are continually obstructing the workers and the production work. Shields provide continuous sheeted protection from the trench wall that prevents raveling or caving soils from injuring workers and disrupting the production work area. Trench boxes are moved and set into place by the same equipment that is being used to excavate the hole. Shields are most often used for pipeline and other linear utility installations owing to their ability to be dragged along with the production operation. Other major uses are shoring for manhole and other rectangular belowground structures, and within the last 20 years they have become popular as the least expensive option for entry structures to trenchless utility work (bore pits) and work pits for pipe bursting and relining work. Where shields can be used, they are several times less expensive than the next alternatives which are slide rail and then sheet and brace systems. They often win the cost and functionality competition between open cut, trench jack, and shoring shield. In deeper excavations over about 16 ft workers tend to feel far more comfortable working inside shoring shields than trench jacks.

Becaue of their protracted use in pipeline and pit construction, low purchase cost relative to long-term rental, and practically no required maintenance, contractors own more shoring shields than any other piece of shoring equipment.

9.6.1 Shoring Shield Conditions for Use

Shoring shields can be used in any soil type, but the best condition is stiff cohesive soils with blow counts above 8 and medium dense sands and gravels with some silts and clays or subangular particles and blow counts above 10. Shoring shield use is generally limited to depths above 30 ft due to weight and handling requirements of the stacked boxes, availability of soil that will stand at those depths, and the strength of the box required.

There are two conditions that can eliminate shields as a shoring option:

1. Shoring boxes are considered a passive shoring system. They protect workers by waiting for the soil to fall and then shielding the workers from the collapse and do not protect the surrounding ground and existing structures from movement. The preferred ground condition for shield use is where the ground stands up long enough to set the box in at depth and stays that way until the production work is completed. If the owner will not allow the existing facilities to be at risk, the shielded option is pretty much a nonstarter.

   To some extent shields can be made to work as an active system through the installation process. Large excavators help make this possible because they can dig fast and set the box before ground collapse. By placing excavated soil or crushed rock between the box wall and the excavation wall, the soil is prevented from moving. There is a risk period, which is the time from when the excavation is started to when the shield is set and the filler material is placed. To eliminate the risk period or in soils that will not stand, boxes can also be set by digging inside the box and below the bottom of the box so they can be pushed or pounded down. Knife-edge shields are available for this purpose (Fig. 9.29); however, the knife

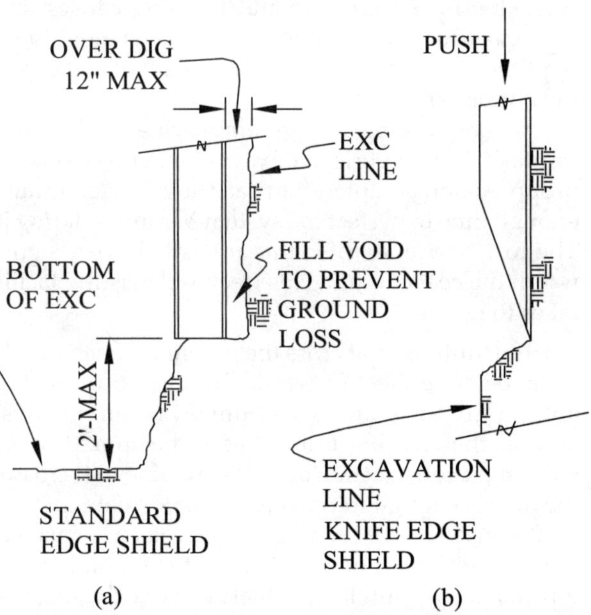

FIGURE 9.29 (a) Standard shield bottom; (b) knife edge.

edge does not make it possible to advance the box more than a few inches below the bottom of the excavation. This method still leaves unprotected soil where the excavation is below the bottom of the box; also driving the box into the ground can put excessive stress and deflection into the shield walls. Since there is soil arching from the bottom edge of the box to the bottom corner of the excavation, it is not unreasonable to consider that this type of installation meets the requirements of an active shoring system.

Problems can also result when the box is being removed or pulled forward. Backfill or else subsequent shoring such as steel plate braced with trench jacks has to be in place prior to moving the box. If backfill is in place, there is a void resulting from the shield wall thickness that is left behind when the box is moved.

As a general rule in cohesive soil with blow counts above 6 and cohesion greater than 750 psf, the digging in method works; however, if a soft layer of soil is within 8 ft of the surface, this method usually fails because the heavy equipment and pounding break down the surface layer, resulting in settlement and movement of surrounding facilities. But note that pile driving in these conditions can have the same effects.

In noncohesive soils with blow counts above 6 when the excavation is kept roughly not more than 2 ft below the bottom of the box, digging in can work. This 2-ft limitation exists because pounding and pushing causes the soil arch to break down, and this limitation is not due to OSHA requirements to have no more than 2 ft of soil exposed below the bottom of the box.

Pipeline shoring systems employing dig-in shield systems can take time to perfect, days and even weeks; so if they are proposed and accepted for trial, the contractor must be given enough time to perfect the system before declaring it a failure. The cost incentives to do this are usually very significant and worth the cost of damaged street and existing facilities that it takes to perfect it.

2. Existing utilities that cross the pipeline trench prevent shields from being pulled forward. Pulling and resetting shields around crossing utilities is time-consuming plus the area around the crossing utility has to be shored in some other way. In pit construction any existing line that crosses through the hole is enough to prevent use of a shield.

   Because most water, gas, and electric installations are within the top 8 ft of the surface, it is possible to remove the top shield and pull lower shields under the utilities. Shoring or sloping still needs to be provided above the boxes.

9.6.2 Shoring Shield Size and Nomenclature

The basic elements and nomenclature of a shoring shield are shown in Fig. 9.30. Shoring shields and boxes are described by height × length × wall thickness. The wall thickness is nominal. Most shields are constructed with 3/16-in plate "skin" over tube steel ribs so the actual wall thickness is nominal size +3/8 in. Standard heights are 2, 4, 6, and 10 ft. Standard lengths are in 2-ft increments from 10 to 32 ft. Wall thickness is critical to wall strength and deflection. In terms of strength efficiency, the ratio of the weight to the psf rating, of the box there is a practical length to each wall thickness. In other words at certain lengths a box that weighs less can be built with a higher strength rating and less deflection by increasing the wall thickness. Thicknesses and practical lengths are 3 in and 12 ft, 4 in and 16 ft, 6 in and 24 ft, 8 in and 27 ft, and 10 in and 30 ft. Every additional 2 in of wall thickness translates into an additional 4 in of excavation and backfill quantity, and so thinner walls are desirable. Required length, next strength, and then wall thickness usually control the selection of the box. If boxes are going to be dug in or used to protect existing facilities, thicker walls should be used to control deflection. Custom boxes can be ordered from manufacturers in any size and strength. The sky is the limit, but the practical limit is hit when the weight of the box exceeds the handling strength of the excavator being used to move the box.

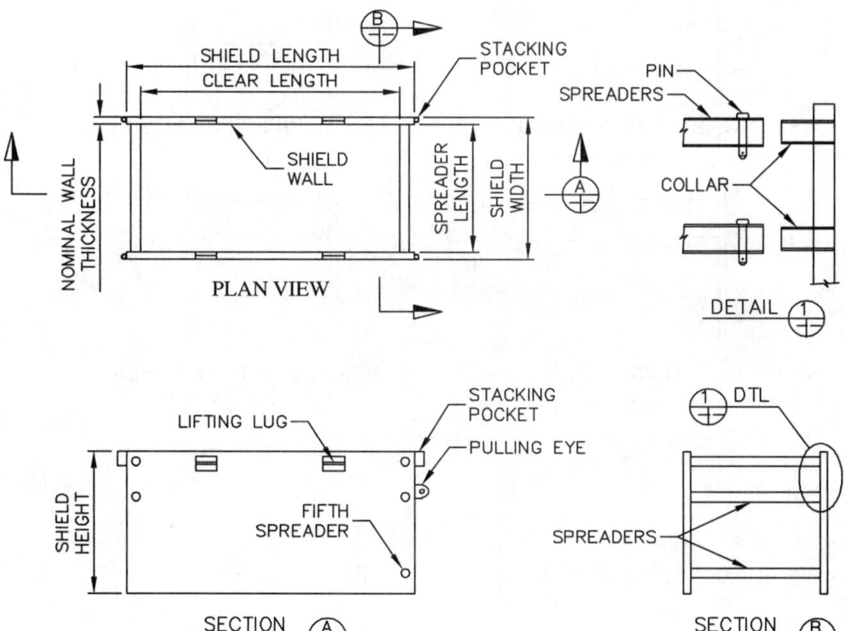

Figure 9.30 Shoring shield nomenclature.

Other features that affect the use of shields are as follows:

- *Single- or double-wall construction.* Single-wall construction is light, has low psf ratings, and is generally designed to be handled with backhoes and small excavators to a practical depth of approximately 12 ft.
- *Material (steel or aluminum).* Aluminum boxes are lighter and compete with the single-wall boxes. Aluminum box walls can be bent and punctured by excavator teeth more easily than steel walls. Another downside to aluminum boxes is that due to their light weight and the value of aluminum they are a prime target for theft.
- *Knife edge.* A knife edge [Fig. 9.29(b)] is used to help dig and push a box into the ground. Also when pipe bedding and initial backfill are placed inside the box before pulling or lifting, the pipe can get locked in and dragged along with the box. To prevent this, the knife edge helps to release soil at the bottom of the box as it is being lifted or pulled.
- *Spreader configuration.* The most common spreader configuration is two schedule 80 steel pipes as shown in Fig. 9.31(a).

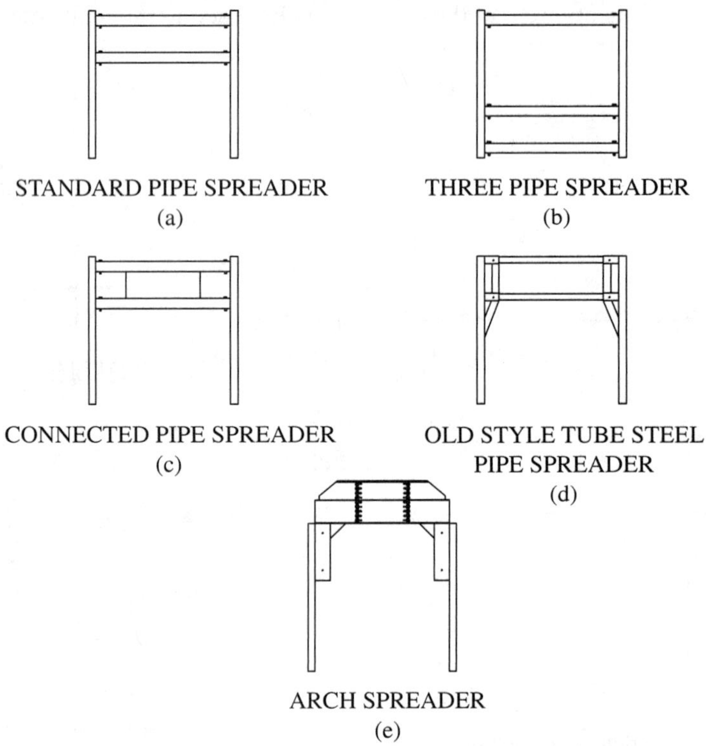

Figure 9.31 Spreader configurations.

These spreader pipes slide over tube steel size pipe collars that leave room for wobble. For instance, an 8-in Schedule 80, 7.625-in inside diameter, fits over a 7-in-OD collar. Two advantages to having wobble room are that it is easier to install the spreaders on the collar and that one wall can be moved a couple of inches relative to the other wall. This allowable movement can cut the pulling force needed to overcome friction or suction from shield wall to soil in half by dislodging one wall at a time.

The three-pipe spreader setup [Fig. 9.31(b)] can be used at the leading edge of the shield to keep it from toeing in when it is being pulled forward. This configuration is not used on the trailing end because the production work, usually pipe, takes up the space.

The third pipe spreader can also be used as a beam to support steel plates or undriven sheet piles set at the ends of the shield to close off the ends.

Spreaders tend to get bumped vertically by excavation equipment digging inside the shield. By connecting the pipe spreaders [Fig. 9.31(c)] there is greater strength in the *X-X* axis, and the buckling length in that axis is shortened. The connection adds no strength or shortened buckling length in the *Y-Y* (horizontal) direction.

The old style spreader [Fig. 9.31 (d)] was developed for use with tube steel construction. Since shields are mostly constructed from tube steel, it most likely made sense to take the off fall 6 × 6 × 3/8 tube steel and build spreader frames from it. Tube steel spreaders 5 × 5 × any thickness conveniently fit inside the 6 × 6 × 3/8 spreader pockets. The main advantage to this configuration is that the end leg cantilever span is cut by a significant distance, resulting in less toe-in deflection and overstress of the end beam. The downside of this system is that there is very little wobble, making the shield hard to assemble and hard to break loose when seized up in soil. The old style spreader was popular in shield manufacturing during the early 1990s but is rarely used on new shields today.

The arch spreader [Fig. 9.31(e)] was developed commercially in the late 1990s for the purpose of getting greater clearance between the spreaders and the bottom of the ditch. This spreader is essentially a large beam proportioned for large bending moments combined with axial loading. The practical height of these spreaders is limited by weight and stability. The arch is usually only used on the trailing end of the shield, and at some point the weight causes that end to drag into the bottom of the trench. Also buckling and torsions at the top become a factor as the arch is extended higher into the air. It is possible to stack another shield on top and connect the spreader to it to gain stability.

9.6.3 Shoring Shield Manufacturing and Engineering

Structurally a shoring shield is a beam that carries a distributed load to girders that are cantilevered. The cantilever is supported by a column and a tension member. The basic design methodology is to determine the wall strength and then size the end beam and spreaders to accommodate the wall loading. Manufacturers' design is in accordance with American Institute of Steel Construction (AISC) allowable stress design requirements except for the following differences:

1. A shoring use factor of 1.33 is applied to all members.
2. Orthotropic plate theory is used to determine effective flange widths for stiffened compression members instead of AISC A-B5-8, effective flange formula.
3. Shoring shields are designed for strength and ignore deflection.

Shoring manufacturers have not agreed on strict design parameters or apparent loading diagrams used for determining soil load ratings and tabulated data. This is evident by comparing the ratings from different manufacturers for shields constructed from essentially the same materials and configuration. The bottom line is that shields are well constructed and to anyone's knowledge have never failed when used within the tabulated ratings; in fact they rarely fail catastrophically. The author has seen one case where spreader collars failed in tension on the line of the pinholes. The incident happened overnight and would have resulted in life-threatening injury if workers had been inside the box at the time. The loading built up overnight in soft marine clays and was estimated to be over 150 percent of the shield rating. When the shield wall fails from the soil, there is usually visual evidence of excessive deflection prior to failure. Generally shield walls fail by permanently deflecting; however, they do not completely fold in two because after an initial soil failure the load decreases, since the soil is no longer restrained by the box. At that point the box overstress resulting in plastic deformation is relieved, and the box wall goes back into an elastic deformation range. Overload from surcharge such as a crane set up opposite the middle of the box could cause a rapid unannounced wall failure that continues to progress into the trench.

9.6.4 Orthotropic Plate Design

A shoring shield wall panel is similar to a steel plate bridge deck with floor ribs that carry deck loads to floor beams (Fig. 9.32). The steel deck is considered to be the top flange of the rib beams. The bottom of the beam is either the bottom flange of a steel shape such as an I, T,

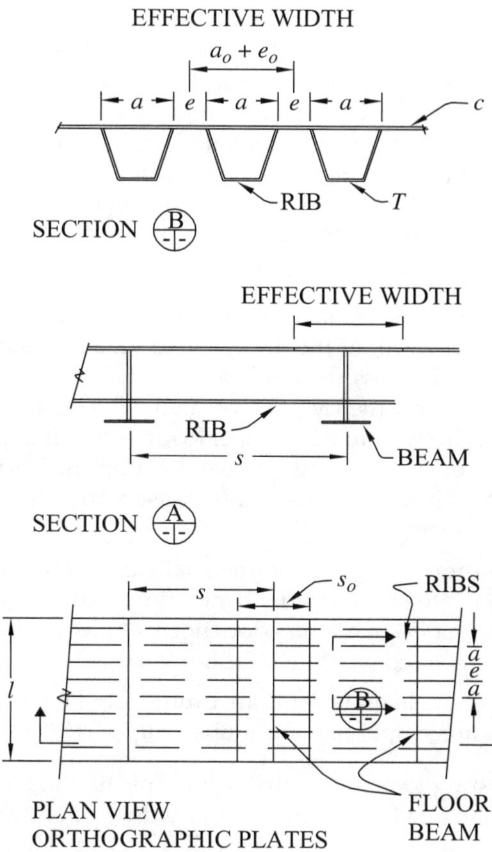

FIGURE 9.32 Orthotropic bridge deck system.

angle, or channel in an open system, or the bottom wall of a steel tube or trapezoidal shape in a closed system. The closed system provides greater rigidity in the longitudinal and transverse directions than the open system where the tension flange is not prevented from moving laterally.

These deck systems work similar to a thick steel plate except that they have different elastic properties in the transverse and longitudinal directions. Because of this property they are called *orthotropic plate*. The American Institute of Steel Construction has developed a design method for orthotropic plate, and AASHO has adopted it for highway bridge deck design. Testing using a simulated wheel load on the ribs has determined that the system is several times stronger than what is predicted by flexural theory; as a result, the design procedure uses modified methods for computing moments and deflection. Stresses are then computed based on section properties. The section properties depend on the effective width of the deck plate which in turn is affected by the span of the ribs.

For the purpose of shoring panel design, the only part of the orthotropic design used is the procedure for determining the effective flange width; everything else is designed using conventional theory. The reason for this is due to some basic differences between orthotropic bridge decks and shoring shields:

1. Orthotropic bridge decks distribute wheel loads, point/area, across the width of the deck. Shield loads are almost always distributed across the entire width and length of the panel. Tests that the theory was verified on and design method derived from simulated wheel loading to failure. The design method does address distributed loading and has no objection to it. It is just easier and more conservative to use ordinary flexural theory.
2. Orthotropic bridge decks are usually continuous over several spans while shoring shields are almost always single-span. Again the design method does not exclude this, but it is much easier to use conventional design.
3. Shoring shield end beams are cantilevered while bridge floor beams are simply or continuously supported.

It is interesting to note that the orthotropic method for determining effective flange width is empirical and does not rely at all on the thickness of the steel deck plate. In bridge decks the thickness is determined by deflection criteria, and shield wall thickness is generally determined by durability and optimum section modulus. In actuality most manufacturer shield wall designs assume that the effective flange is one-half the distance to the next web without producing a calculation. The orthotropic calculations confirm this assumption except in shields with length-to-depth ratios approximately less than 1, combined with rib-to-depth spacing greater than 3:1.

The objective in shoring shield design is to obtain the strongest shield wall with the least amount of steel, similar to bridge design where the longest, lightest span is the goal. A lighter shield is less expensive to manufacture, but the real advantage carried through the long life of the shield is less weight for transporting and handling in the ditch. Use of the orthotropic effective flange can boost the strength calculation for the shield wall as much as 60 percent. It is beyond the scope of this book to go more deeply into orthotropic design theory; however, review of *Design Manual for Orthotropic Steel Plate Deck Bridges* should convince anyone reviewing the structural adequacy of a shield that shoring panel design relying on it is a safe, conservative design.

Shield Wall				1 Wall Panel			
Wall	Description	Material	$^{*}F_y$ (ksi)	Qty. (ea)	Length (ft)	Weight/ft (plf)	Weight (lb)
Skin	Plate 3/16" × 8 ft	A572 GR 60	60	2	20	61.2	2449
Pound plate	PL 1/2 × 6"	A36	36	1	20	7.0	140
Top tubes	TS 4 × 3 × 3/16	A500 GR B	55	2	20	8.2	326
Bottom tube	TS6 × 4 × 3/16	A500 GR B	55	1	20	12.0	239
Middle tubes	TS6 × 4 × 3/16	A500 GR B	55	2	17.5	12.0	419
End beams	TS6 × 6 × 3/8	A500 GR B	55	4	7.34	27.5	807
Collars	Tube 7" OD × 1/2"	A500 GR C	46	2	1.25	35.0	88
Weld metal	Wire feed	E-70xx		1			50
						Total	4517
* Mill certification required for all F_y greater than minimum for material grade.							

FIGURE 9.33 Shield wall materials and weight for 8-ft × 20-ft × 6-in wall.

Example 9.2 The following engineering goes through the design steps for an 8-ft × 20-ft × 6-in wall shoring shield. The materials used and design procedure are typical of what most manufacturers use, See Figs. 9.33 to 9.35.

Materials

Wall Design

Step 1. Determine allowable stress.

Due to beam web $F_y = 55$ ksi and beam flange $F_y = 60$ ksi. The stress at the flange has to be apportioned to not exceed the stress in the web (Fig. 9.35).

$$\text{Check } \frac{b}{t} = \frac{26.7}{3/16} = 142 > \frac{95}{\sqrt{55}} = 12.8 \therefore \quad \text{noncompact}$$

From AISC F1-4 or F1-5

$$F_b = 0.6\ F_y = 0.6 \times 58.5 = 35.1 \text{ ksi}$$

Allow a 1.33 shoring use factor.

$$F_b = 35.1 \times 1.33 = 46.7 \text{ ksi}$$

Step 2. Determine the effective flange width from Fig. 9.36 and Table 9.17.

Step 3. Calculate the CG, moment of inertia I, and section modulus S (Fig. 9.37).

From Fig. 9.37, the moment of inertia and least section modulus for the entire shield wall is as follows:

	Moment of inertia $I_{x\text{-}x}$	**Section modulus $S_{x\text{-}x}$**
Top beam	79.7 in^4	25.0 in^3
Middle beam	127.3 in^4	39.9 in^3
Bottom beam	79.7 in^4	25.0 in^3
Total	414 in^4	129.8 in^3

The total section area is 50 in^2.

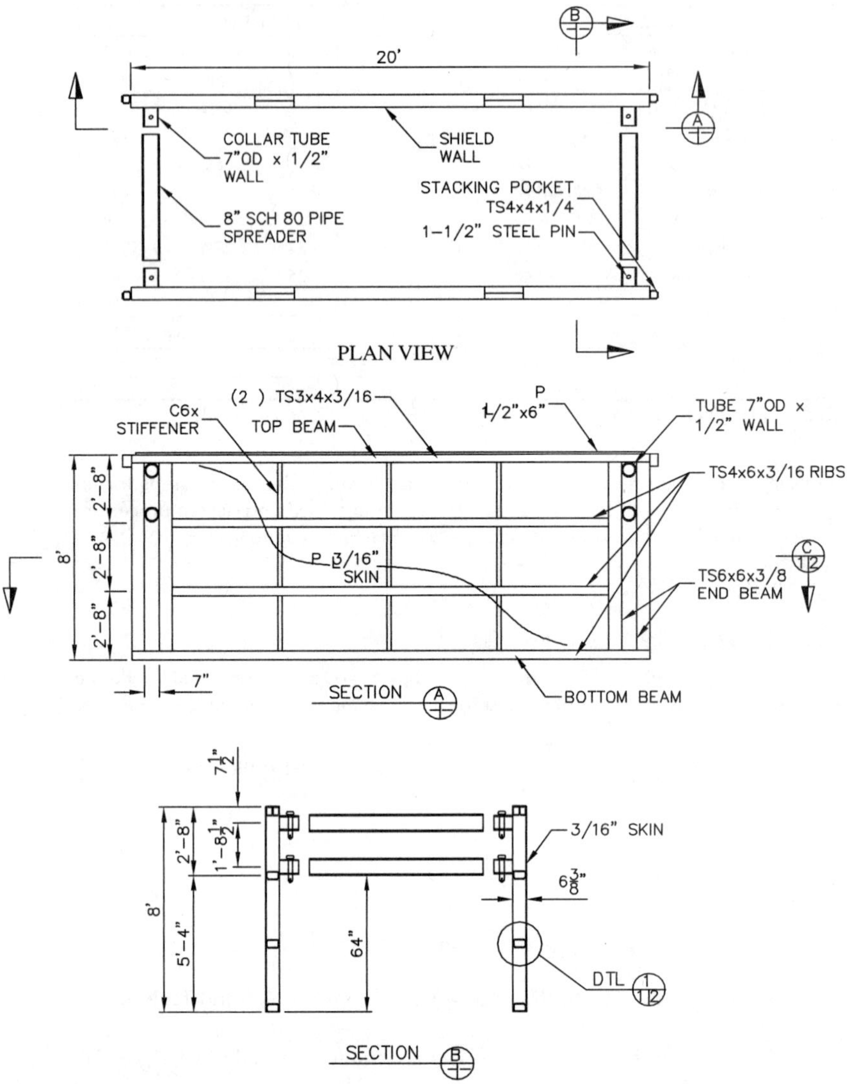

FIGURE 9.34 Drawing for an 8-ft × 20-ft × 6-in wall shield.

The radius of gyration $r = \sqrt{\frac{I}{A}} = \sqrt{\frac{414}{50}} = 2.87$ in.

Step 4. Determine maximum allowable moment from

$$M = SF_{bc} = 6062 \text{ k·in}$$

$$M = 505.1 \text{ k·ft}$$

where $S = 129.8$ in^3 and $F_b = 46.7$ ksi.

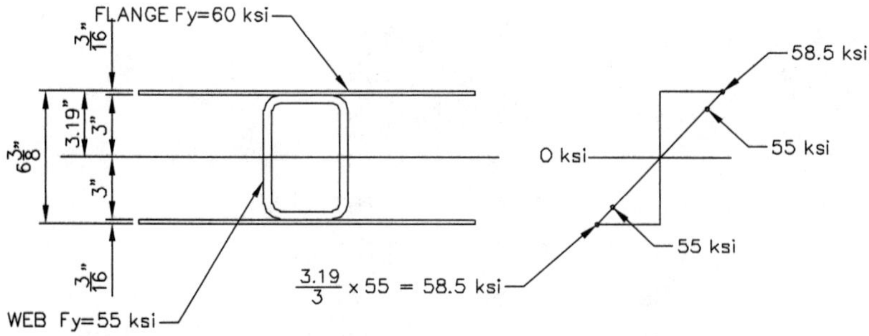

FIGURE 9.35 Allowable stress at extreme fiber.

Step 5. Determine the psf rating from

$$M = \frac{w\ell^2}{8}$$

$$w = \frac{8M}{\ell^2} = \frac{8 \times 505.1}{18.4^2} = 11{,}935 \text{ plf}$$

where w = distributed load on beam, plf
M = 505.1 k·ft allowable moment in beam
ℓ = 18.4-ft beam length

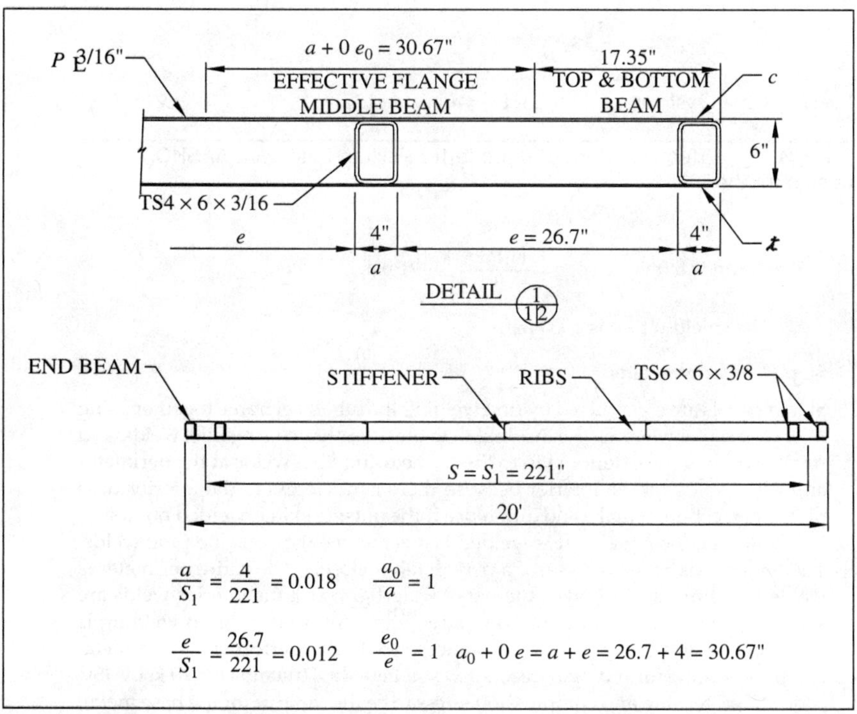

FIGURE 9.36 Effective flange width calculation.

a_e/s_e e_e/s_e s_f/l_e	a_0/a_e e_0/e_e s_0/s_e	a_e/s_e e_e/s_e s_f/l_e	a_0/a_e e_0/e_e s_0/s_e
0.000	1.100	0.400	0.809
0.050	1.090	0.450	0.757
0.100	1.080	0.500	0.722
0.150	1.050	0.550	0.671
0.200	1.010	0.600	0.622
0.250	0.961	0.650	0.590
0.300	0.921	0.700	0.540
0.350	0.870	0.750	0.512

Notes:
For closed rib effective flange

$$a_0' + e_0' = \frac{a_0}{a_e} a_e + \frac{e_0}{e_e} e_e$$

For closed-end beam effective flange

$$s_0 = \frac{s_o}{s_e} s_e$$

For single-span systems a_e, e_e, s_e, and $\ell_e = a, e, s,$ and ℓ.

Table 9.17 Effective Width of Plate (after Pelikan-Esslinger, AASHO, design method)

The beam is 8 ft wide so $w = \dfrac{11{,}935}{8} = 1491$ psf.

∴ The shield rating is 1500 psf.

Step 6. Design welds.

Shield panels are constructed by first welding the tube steel frame together using fillet welds, then welding the inside skin to the tubes again using fillet welds, and finally welding the exterior skin to the frame using fillet welds at the perimeter and plug welds on the interior because there is no access to the interior tube plate intersection. It makes no difference if the outside skin is welded on first or last; however, the panel that is welded last is the one that gets the plug welds. The plug welds are easy to see if a person looks closely. The entire perimeter is welded continuously to make the panel watertight, and the interior welds are skip welds because continuous weld strength is not needed. Shop welding is all wire feed and today in many instances robotically performed. Weld metal is typically some form of wire feed 70-xxx, where the 70 stands for 70 ksi. AISC code allows tension and compression stress to be the same as for the base metal; for steel tube it would be $0.6F_y = 0.6 \times 55$ ksi $= 33$ ksi. For 1/16 in of 70 ksi weld

Top and Bottom Beam		Thick (in)	Wide (in)	Area (in^2)	Distance From CG d (in)	Ad^2 (in^3)	I (in^4)	$I+Ad^2$ (in^4)
Compression flange	3/16" plate	0.1875	17.35	3.3	3.1	31.14	0.0	31.1
Web	TS 6 × 4 × 3/16	6	4	3.52	0.0	0.00	17.4	17.4
Tension flange	3/16" plate	0.1875	17.35	3.3	3.1	31.14	0.0	31.1
		6.375		10.0				79.7

Beam depth d =	6.375	in	Moment of inertia $I_{x\text{-}x}$	79.69	in^4
CG =	3.2	in from bottom	Section modulus compression, $S_{x\text{-}x\ c}$	25.0	in^3
	3.2	in from top	Section modulus tension, $S_{x\text{-}x\ T}$	25.0	in^3
			Statical moment of compression flange, Q	10.1	in^3
			Statical moment of tension flange, Q_T	10.1	in^3

Middle Beam		Thick (in)	Wide (in)	Area (in^2)	Distance From CG d (in)	Ad^2 (in^3)	I (in^4)	$I+Ad^2$ (in^4)
Compression flange	3/16" plate	0.1875	30.6	5.7	3.1	54.92	0.0	54.9
Web	TS 6 × 4 × 3/16	6	4	3.52	0.0	0.00	17.4	17.4
Tension flange	3/16" plate	0.1875	30.6	5.7	3.1	54.92	0.0	54.9
		6.375		15.0				127.3

Beam depth d =	6.375	in	Moment of inertia $I_{x\text{-}x}$	127.26	in^4
CG =	3.2	in from bottom	Section modulus compression $S_{x\text{-}x\ c}$	39.9	in^3
	3.2	in from top	Section modulus tension $S_{x\text{-}x\ T}$	39.9	in^3
			Statical moment of compression flange	17.8	in^3
			Statical moment of tension flange	17.8	in^3

FIGURE 9.37 Calculation for moment of inertia, section modulus, and static moment of flanges.

metal, the allowable strength is 1/16 × 0.707 × 33,000 = 1.47 k/in and for 3/16 welds it is 3 times that or 4.4 k/in. The 3/16 skin thickness limits the size of the fillet weld to 3/16 in. The maximum allowable weld on thicknesses over ¼ in is thickness less 1/6 in.

The basic equation for weld design is

$$P = \frac{VQ}{I}$$

where P = weld strength required, k/in
V = beam shear, k
Q = static moment of flange, in^3
I = moment of inertia, in^4

For the middle beams the distributed load is 1500 psf × 30.6 in/12 = 3.8 k/ft. The maximum shear is

$$V = \frac{W\ell}{2} = 3.8\text{ k/ft} \times 18.4\text{ ft}/2 = 35\text{ k}$$

The weld required is

$$P = \frac{VQ}{I} = \frac{25 \times 17.8}{127.3} = 3.7\text{ k/in}$$

where P = weld strength required, k/in
V = 25 k
Q = 17.8 in^3
I = 127.3 in^4

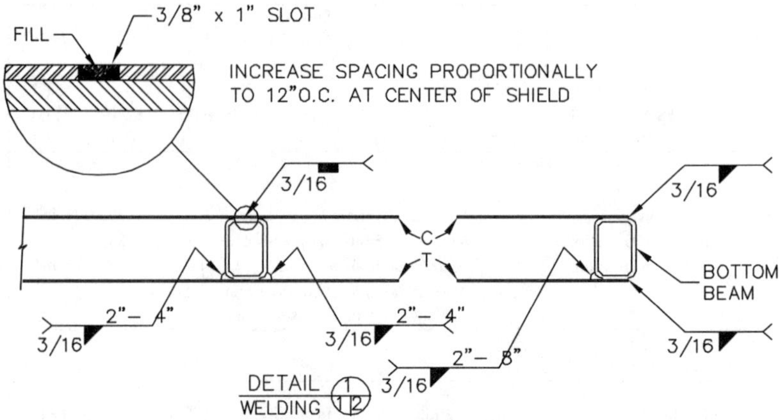

FIGURE 9.38 Shield panel beam weld details.

On the inside tension skin, use skip fillet welds on both sides of the tube (Fig. 9.38).

$$\text{Weld required per side} = (3.7\ \text{k/in})/2 = 1.85\ \text{k/in}$$

Using 3/16 fillet that provides 4.4 k/in, weld 2 in at 4 in on center providing

$$(2\ \text{in} \times 4.4\ \text{k/in})/4\ \text{in} = 2.2\ \text{k/in} > 1.85\ \text{k/in} \qquad \text{OK}$$

On the outside compression skin use a 1-in × ½-in-wide slot 3/16 in welded every 4 in on center. This provides an equivalent 2-in fillet weld every 4 in on center. At the end beam provide continuous 3/16-in fillet at the outside edge of tension skin, 3/16-in fillet on 8-in center on the inside, and 3/16-in continuous fillet at the outside edge of compression skin (Fig. 9.38).

$$4.4\ \text{k/in} + (4.4\ \text{k/in})/8\ \text{in} = 4.95\ \text{k/in} > 3.7\ \text{k/in} \qquad \text{tension side}$$

$$= 4.40\ \text{k/in} > 3.7\ \text{k/in} \qquad \text{compression side}$$

The shear decreases linearly to zero in the middle of the panel so the weld spacing can be increased proportionately.

Step 7. Check the deflection.

The shield deflection is not reported on tabulated data sheets; however, it can be used in the field as a gauge of the load the shield is supporting. If computed deflection is exceeded, the shield is overloaded. For a simple beam the deflection is

$$\Delta_{max} = \frac{5w\ell^4}{384EI} = \frac{5 \times \frac{1500}{12} \times (18.4 \times 12)^4}{384 \times 29{,}000{,}000 \times 51.75} = 2.6\ \text{in} \qquad \text{in}$$

where $w = 1500$ psf

$\ell = 18.4$ ft

$E = 29{,}000$ ksi   modulus of elasticity for steel

$I = \dfrac{414\ \text{in}^4}{8\ \text{ft wide beam}} = 51.75\ \text{in}^4$ per ft   shield moment of inertia per ft of beam width

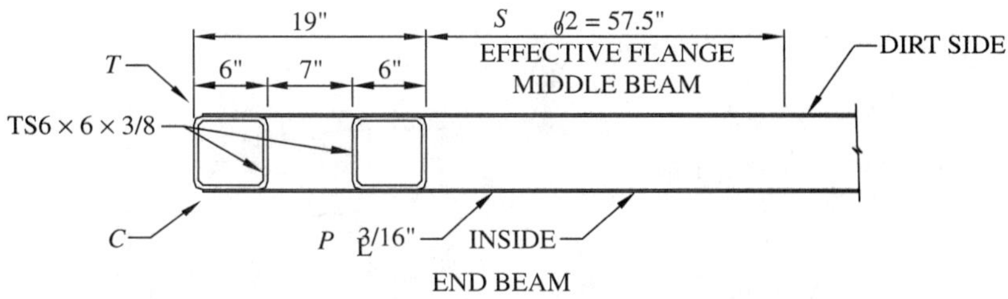

FIGURE 9.39 Shield end beam.

The deflection is independent of the material strengths used and only dependent on the geometric properties and modulus of elasticity for steel.

Step 8. Check end beam strength.

The end beam receives the soil loading from the ribs and carries it up to the struts. The end beam load, in kips per lineal foot (klf), is

$$W = \frac{w\ell}{2} = \frac{1500 \times 20}{2} = 15 \text{ klf}$$

Where W = end beam load, klf
w = 1500 psf soil loading
ℓ = 20 ft shield length

Determine the effective flange. The end beam is cantilevered 60 in, and the rib span is 225 in. For a cantilever beam

$$\ell_e = 2\ell_{cantilever} \qquad \text{and} \qquad \frac{s_f}{\ell_e} = \frac{225}{120} > 0.75$$

From Fig. 9.39 and Table 9.17,

$$\frac{s_0}{s_f} = 0.512 \qquad \text{and} \qquad s_0 = 0.512 \times 225 = 115 \text{ in}$$

A ½ flange width is 57.5 in, and the total flange is 77.1 in. Using the same method as in Fig. 9.37, the moment of inertia is 318 in^4, the section modulus is 99.8 in^3, and Q = 44.7 in^3.

The end beam moment from Fig. 9.40 is 236 k·ft, and the end beam bending stress is

$$\frac{M}{s} = f_b = \frac{236 \times 12}{99.8} = 28.4 \text{ ksi} < 46.7 \text{ ksi} \qquad \text{OK}$$

The maximum shear from Fig. 9.40 is 146 k, and shear stress is

$$f_v = \frac{V}{dt_w} = \frac{146}{6 \times 4 \times .375} = 16.2 \text{ ksi} < 22 \qquad \text{OK}$$

where V = maximum shear, k
d = depth of beam, in
t_w = thickness of beam web, in

Per AISC (F3-3) allow $F_v = 0.4\ F_y = 0.4 \times 55$ ksi = 22 ksi. The 1.33 shoring use factor could also be used here if needed. Welding design is similar to wall rib design.

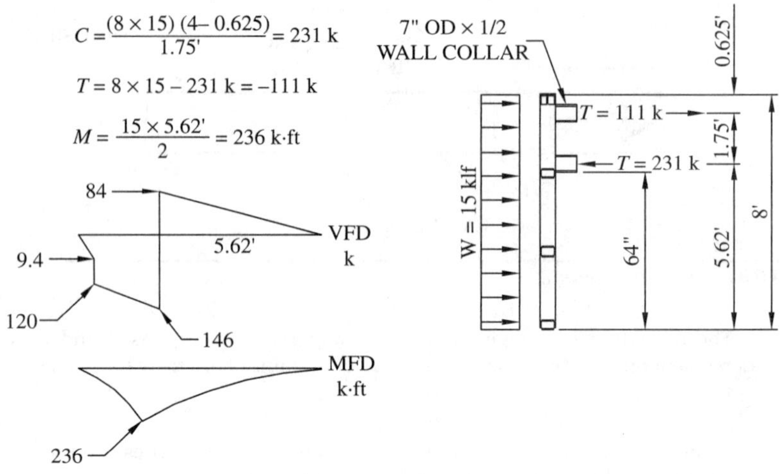

FIGURE 9.40 Shield end beam loading with shear and moment diagrams.

Step 9. Strut design

Struts tend to see a lot of abuse because they are in the way of the excavator arm and bucket during excavation inside the shield, and they are convenient to use for lifting and pulling by the excavator bucket. The strut is usually grabbed near the shield wall to avoid bending the strut, and therefore the collar has to be able to resist the shear. Manufacturers provide built in pulling eyes and lifting pockets to prevent abuse of the spreaders; however, their use requires another step to set cables and connect them to the excavator and therefore are not always used.

The strut receiver, sometimes referred to as a collar, is made in accordance with tubing specification A500 grade B or C. The reason for using tubing is threefold: to get an outside diameter that will fit well inside a pipe size spreader; to get a higher-strength material that is more compatible in strength with the shield wall tubing to which it is being welded; and to be more compatible with the strength of the pins used to connect to the spreader. The pin and collar strength is only necessary at the top spreader where they resist a tension force. The bottom spreader is in compression against the end beam and does not bear on the pin or collar. The pin material varies by manufacturer and is usually higher-strength steel, 60 to 100 ksi, in order to keep the pin diameter and weight down. Pins are designed for double shear and the pinholes designed for bearing with bearing strength of the collars controlling the design. In short shields with very high wall strengths it is possible to cause the receiver to fail in tension along a line that intersects the pinholes.

Spreaders are almost universally among U.S. manufacturers schedule 80 ASTM A53, $F_y = 35$ ksi and $F_u = 60$ ksi. The end condition of the spreader can be considered fixed, even though it can wobble slightly on the 7-in-OD collar. It is possible for there to be translation in the vertical and longitudinal directions; however, unlike normal multiple column loading conditions there is a tension strut that is fixed and also resisting translation so that there is 2 times as much translation resistance as normal. AISC Table C-C2.1, Fig. 9.8, condition (c) with a recommended K value of 1.2 most closely fits this fixity condition, except that it anticipates less transition resistance. For that reason, a K value of 1.0 seems reasonable. It is also possible to reason that when the shield is completely loaded, it is gripped by the soil and translation is prevented so that Table C-C2.1 condition (a), fixed with no translation, exists and the K value should be 0.65. Failure of a strut would cause some load to move to the strut at the other end of the shield and could cause that strut to fail . The apparent pressure diagrams were developed to cover the possibility of this type of progressive failure, and for that reason it is the author's opinion that the 1.33 shoring use factor should not be used on strut design. An argument for the shoring use factor in strut design would be that since a shoring shield supports lateral loads from the soil, immediately after failure the soil load is rapidly decreased, and therefore the strut material would move quickly from the plastic state back into the elastic state, thereby preventing total failure of the system. In vertical load resisting systems the gravity load never decreases and stays the same even after the structure is pancaked. Manufacturers generally work with a K factor of 0.5 and the 1.33 shoring use factor.

From Fig. 9.40, the maximum strut load is 231 k. The area of an 8-in Schedule 80 pipe is 12.8 in^2, and the radius of gyration r is 2.88 in. The axial stress is

$$f_a = \frac{P}{A} = \frac{231}{12.8} = 18.01 \text{ ksi}$$

where P = axial load,k, and A = area of spreader pipe. in^2. From AISC (E2-1)

$$C_c = \sqrt{\frac{2\pi^2 E}{F_y}} = 127$$

where C_c = column slenderness ratio (coefficient)
E = 29,000 ksi, modulus of elasticity for steel
F_y = 35 ksi yield strength for A53 pipe

$$\text{When } \frac{Kl}{r} \leq C_c, \quad F_a = \frac{[1 - \{(K\ell/r)^2/2C_c\}]F_y}{(5/3) + [3(K\ell/r)]/8C_c - [(K\ell/r)^3]/8C_c^3} \qquad \text{AISC (E2-1)}$$

By setting $f_a = F_a$ and solving, AISC (E2-1) can be solved for ℓ. The easy way to solve this is to go to AISC, 9th ed., Table C-36 next to F_a = 18 ksi and find $\frac{K\ell}{r}$ = 53, and by using Eq. (9.3) solve for

$$\ell = \frac{53r}{K} = \frac{53 \times 2.88}{1 \times 12} \cong 13 \text{ ft} \qquad \text{if } K = 1 \text{ or } 26 \text{ ft} \qquad \text{if } K = 0.5$$

where ℓ = allowable strut length, ft
r = 2.88 in, radius of gyration for 8-in Schedule 80 pipe
K = theoretical length modification factor for strut end condition

The tabulated data sheet for the shield should state the allowable strut size and length at maximum load. At lighter loadings the strut length can be longer and can be determined by using the calculations shown in Fig. 9.40 and AISC Table C-36 with formula (9.3).

Step 10. Develop tabulated data for 8 × 20 shield.

Given the shield rating 1500 psf, depth ratings are developed from

Shield rating = soil type factor × depth + 72 psf surcharge

where Soil type factor = 25 psf/ft of depth for type A soil
= 45 psf/ft of depth for type B soil
= 60 psf/ft of depth for type C-60 soil
= 80 psf/ft of depth for type C- soil

Allowable loading = 1500 psf

Maximum allowable surcharge = 72 psf

Spreader material 8-in Schedule 80 steel pipe ASTM A53 grade B

Maximum spreader width = 13 ft

Soil Type	Allowable Depth
A	57 ft
B	32 ft
C-60	24 ft
C	19 ft

The competent person should be capable of calculating the expected loading on the shield. For example, if the box is going to be placed in a 25-ft-deep excavation in C-60 soil with a 200 psf surcharge, the shield load is

$$25 \text{ ft} \times 60 \text{ psf/ft of depth} + 200 \text{ psf surcharge} = 1700 \text{ psf}$$

A box with a rating of 1700 psf or greater should be selected.

OSHA allows the shield loading to be calculated based on the depth that the box is going to be set at. This is reasonable because the box is considered as protection from the soil falling into the excavation and not as supporting the existing soil. Because in cohesive soils the loading tends to be more trapezoidal with the highest loading at mid-depth, in cases where the shield is intended to support the soil and surrounding facilities, the stacked boxes should be strong enough to carry the full-depth load. Also the competent person should recognize that surcharge loads are higher near the surface, and that boxes that are pounded into or pulled through weak soils can develop considerably higher loads than calculated by soil mechanics or OSHA soil types.

The wall strength of shoring shields with the same material configuration and different lengths can be calculated by proportioning the square of their lengths

$$\frac{\ell_1^2}{\ell_2^2} \times \text{wall strength}$$

where ℓ_1 = length between spreaders of rated shield, ft, and ℓ_2 = length between spreaders of shield to be calculated, ft.

For example, from the 8 × 20 × 6-in wall, a 1500 psf shield calculated above a 24-ft shield constructed from the same materials would be rated at $(18.4^2/22.4^2) \times 1500 =$ 1012 psf, and a 16-ft-long shield from the same materials would be $(18.4^2/14.4^2) \times$ 1500 = 2450 psf. The end beams and receiver connections should then be sized for the wall strength.

9.6.5 Shoring Shield Safety Issues

Shoring shields definitely protect workers from being buried in a trench wall collapse when they are working inside them; however, there are serious safety hazards associated with assembly, installing them in the excavation, and worker access and egress that workers should be aware of. The following are major safety issues:

- Shield panels can weigh several thousand pounds and can cause major injuries if they are dropped or swung into workers as the panels dangle from a cable while workers are trying to assemble them. Assembly requires heavy lifting equipment such as backhoe, boom truck, excavator, or crane depending on the size of the box being assembled. Careful consideration should be given to the lifting capacity and size of rigging needed to assemble the box. Larger equipment may be required to assemble and lift the box in place than is needed to pull the box when it is in the trench. Trying to make do with light equipment increases the risk of harm to workers and should not be attempted. At some point during the assembly and installation process, the full weight of the assembled box will have to be lifted, and therefore the equipment should have the capacity to safely lift it at the radius required to place it in the trench. Inspect the equipment and rigging prior to starting the assembly operation. There should be a minimum of two workers on the ground and one on the lifting rig to assemble the box. Make a plan and have necessary equipment such as ladders and pin driving hammers available before starting the work. When lifting and setting the box, workers cannot be inside or under the box. Use tag lines. Do not exceed overhead power line safety clearances.
- When stacked boxes are set, there must be a method of disconnecting the lift cables from the box or lift connection without workers entering an unshored excavation. This can be accomplished with a separate set of lift cables for each box so that the bridle can be disconnected from above at the hook. Stacking pins should be set from inside the boxes after they have all been set.
- The end space of a shoring box, approximately 2 ft at the front and back of the box, is dangerous. At either end of the box the unsupported trench wall can collapse, with the soil falling vertically and spilling into the end of the box and inundating workers. When trench walls are excavatable rock or shale, they can fall out on the angle of their bedding planes, giving them momentum in the horizontal as well as vertical direction. Safety precautions include instructing workers to not stand within 2 ft of the ends of the box for any length of time. In deep stacked box excavations, 1-in-thick × 8-ft-high end plates should be set on the ends of the box. Plates can be hung by chain or welded to the ends.

- Access and egress from the shoring box should be planned out. There must always be a ladder for workers to get to the top of the box. In deep, narrow excavations the ladder can be too steep and not easily accessible because the bottom of the ladder is tight against the opposite wall. In these cases a vertical climbing ladder, in accordance with OSHA ladder requirements, should be attached to one wall. At the top there has to be safe access from the ground to the box wall. Shoring rental companies have developed an assortment of bridges and catwalks that can be attached to the top edge of the box and travel along with it. The fall distances and consequences of a fall are too great to tolerate make-do approaches to solving the problem. Plan ahead for access and egress, and if the plan changes, stop the operation until a safe, permanent solution is in place.
- Shoring boxes can move transverse to the trench line if there is space between the box wall and the trench. If the space is more than the distance between the inside wall and the pipe being laid, a worker can be crushed between the wall and the pipe. As a general rule, if there is to be no fill between the box wall and trench wall, there should be no more than 6 in between the outside of the box and the trench wall and never enough room for the box to shift enough to crush a worker between the production work and the interior box wall. The solution to this is to place excavated soil between the box wall and the trench wall at the front and back of the box.

  In situations where the trench wall collapses before the box is set but the slope is still not safe, boxes are set into the excavation to protect workers from further wall collapse and rocks rolling off the slopes. In these cases the bottom shield should still have soil at the bottom half of the shield to prevent movement and tipping. If shoring boxes are not in some way braced to prevent it, they can be tipped sideways by collapsing trench walls. Stacked boxes should always have stacking pins in place, and also in these cases access and egress are more of a problem.
- Pulling boxes forward or lifting them vertically in soft clays can be a problem. More than one box has been abandoned in place because it could not be moved. Under normal conditions workers can remain inside a box when it is being moved forward; however, the banging and pulling that take place around trying to move seized boxes present too much of a risk for workers to be inside them. When pulling connections or cables break, the rebound can easily kill anyone who is struck by it. Unfortunately this is the time when everyone likes to stand closeby and watch the battle between the excavator and the earth.

9.6.6 Shoring Shield Plan by a Registered Engineer

In the following situations an engineer is most often called in to design an excavation using a shield:

1. OSHA soil loading formulas are too conservative for the available shield rating, or surcharge loads are greater than 72 psf.

   In many cases OSHA soil loading is far more excessive than soil mechanics and apparent pressure diagrams would predict. For example, in a noncohesive medium dense, say 20 blows per foot, sandy soil, OSHA would call it soil type B and possibly C. At 25 ft deep the soil loading is 25 × 45 = 1125 psf. Using apparent soil loading from Peck, Hanson, and Thornburn, $f = 34$ and $g = 120$ psf, $K_a = 0.28$ and $P_a = 0.65k_a gH = 0.65 \times 0.28 \times 120 \times 25 = 550$ psf. This is one-half the soil loading predicted by OSHA formulas. The lower soil loading allows the selection of longer, thinner-walled shields or longer shields than would be available under the OSHA rules. The use of soil mechanics and apparent pressure diagrams is always more efficient and accurate than using OSHA soil type formulas.

2. Four-sided excavations where the ends of the shield are bracing the shoring at the end wall.

   The shoring pit shown in Fig. 9.41 has beams that load the end of the shoring shield axially. In this case the combined stress equation, Eq. (9.2), must be satisfied.

$$\frac{f_b}{F_b} + \frac{f_a}{F_a} \leq 1$$

where f_b = computed bending stress
F_b = allowable bending stress
f_a = computed axial stress
F_a = allowable axial stress

Using geometric properties calculated in Fig. 9.37, the bending moment on the shield is

$$M = \frac{wl^2}{8} = \frac{(960 + 200) \times 8 \times 18.4^2}{8} = 394 \text{ k·ft}$$

and bending stress is

$$f_b = \frac{M}{S} = \frac{394 \times 12}{129.8} = 36.4 \text{ ksi}$$

The axial load is

$$P = \frac{(960 + 200) \times 8 \times 20}{2} = 92.8 \text{ k}$$

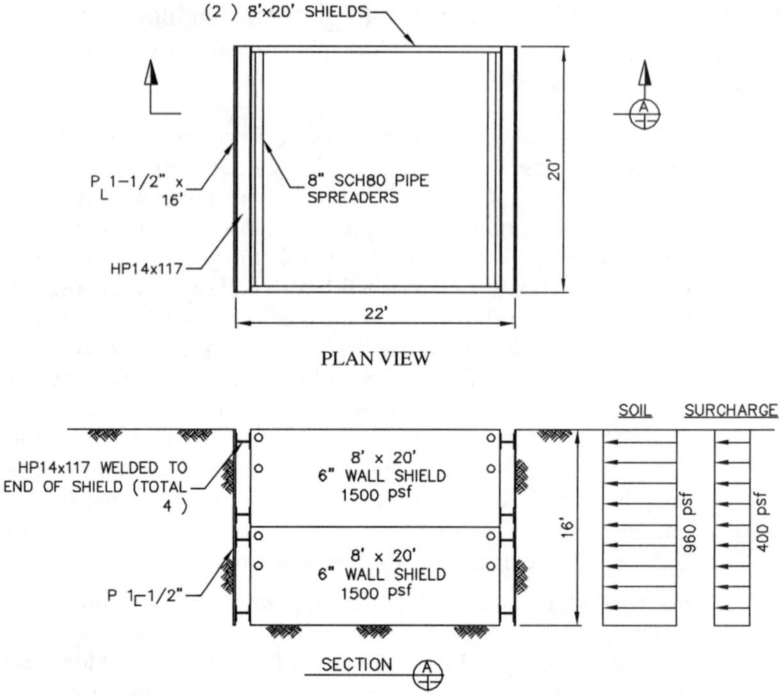

FIGURE 9.41 End loaded shield.

And the axial stress is

$$f_a = \frac{P}{A} = \frac{92.8}{50} = 1.85 \text{ ksi}$$

From Example 9.2

$$F_b = 46.7 \text{ ksi}$$

To determine F_a,

$$\frac{K\ell}{r} = \frac{1 \times 20 \times 12}{2.88} = 83$$

where $K = 1$ per AISC Table C-C2.1(d) pinned each end
$\ell = 20$ ft shield length
$r = 2.88$ in radius of gyration generated from Fig. 9.37

From AISC Table C-50, $F_y = 50$ ksi, under $Kl/r = 83$,

$$F_a = 18.41 \text{ ksi}$$

$$\therefore \quad \frac{f_b}{F_b} + \frac{f_a}{F_a} = \frac{36.4}{46.7} + \frac{1.85}{18.41} = 0.88 \leq 1 \qquad \text{OK}$$

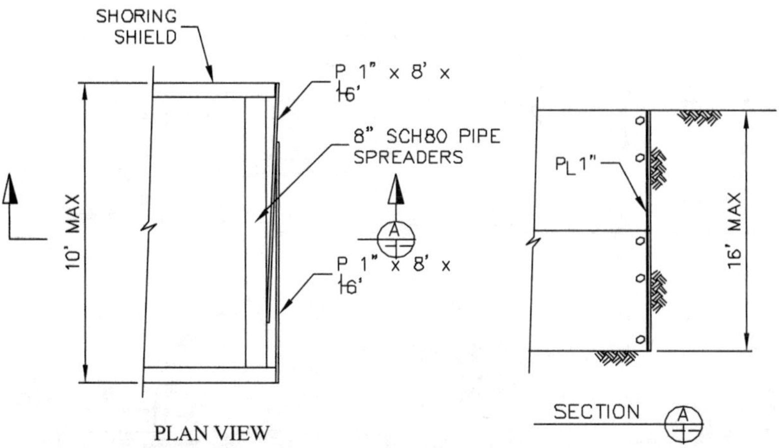

FIGURE 9.42 End plates on 10-ft-wide shoring shields.

In fact the compression component of this ratio is 0.1 so that the maximum bending stress could be $f_b = (1 - 0.1) \times 46.7 = 42$ ksi. This can be back-calculated to find an allowable shield wall loading of 1340 psf, 89 percent of the shield rating without axial loading. The soil loading in Fig. 9.41 is fairly high, OSHA C-60 soil × depth, and the surcharge load, 200 psf, is higher than normal. As a general rule under normal soil loading and using industry standard shoring shields, axial loads placed on the ends of shoring shields down rate the shield by approximately 10 percent.

Bore and jack operations, not microtunnel, usually require an approximately 10-ft-wide bore pit. Due to this and the fact that an 8- to 10-ft-wide excavation is also fairly common for structural and mechanical excavations it has become common practice to place steel plates or sheet piles at the end of shields (Fig. 9.42).

Usually 2- to 8-ft-wide × 1-in steel road plates are set against the end of the box and resting on the spreader pipes close to the collars. Calculations in OSHA type A and B soils show that this arrangement is within AISC stress limitations. Some manufacturers have allowed this arrangement in their tabulated data, and in the field inspectors and engineers often look the other way when it is used without site-specific engineering. At these shoring widths and depths the axial loading on the shield is inconsequential, and the added bending load on the compression struts also turns out to be fairly inconsequential. As a general rule if the shield is loaded to less than 80 percent of capacity and the shield is not more than 10 ft wide, then end loading from sheeting and beams will have little effect on the shield and struts.

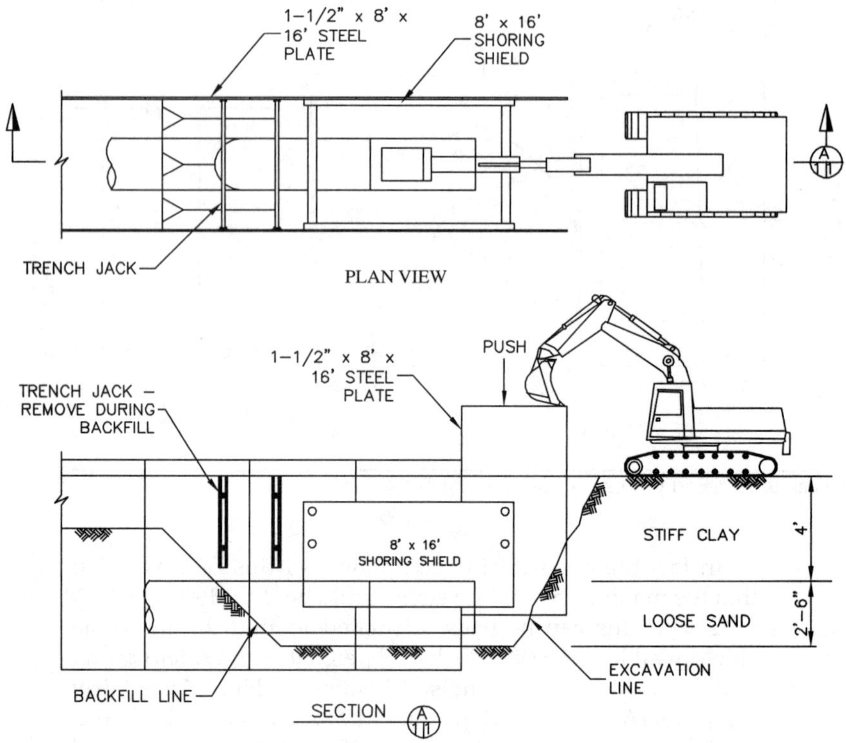

FIGURE 9.43 Shield and plate pipeline shoring system.

3. Shoring shield and steel plates used as a poor person's slide rail system.

   The shield plate system shown in Fig. 9.43 has been used successfully many times over in loose sands and soft clays by pipeline contractors. The shield acts as a wale, and the plates are incrementally advanced below the bottom of each excavation lift until the bottom is reached. Backfill or trench jacks hold the plates at the trailing end, and steel plates are leapfrogged and driven as the shield is pulled forward. The system works best where there is solid ground within 10 ft of the surface and soft clay or loose sand at the bottom; otherwise there is considerable settlement at the surface, and surrounding facilities are easily damaged. Using these systems also tears up the streets and utilities when the soft layer starts within 6 ft of the surface.

   The incentive for a contractor to use this system is considerable. The cost of a couple of shoring shields and some plates runs about $8000 per month, and many times the contractor owns them and it costs him or her nothing. A comparable slide rail system can run $20,000 to $30,000 per month, and driving and pulling sheet piles can run $40,000 to $60,000 per month.

Short shields with 6- to 8-in-thick walls combined with 1½-in steel plate will have enough strength to work this system to depths of 16 ft. The whole excavation process—excavation, plate driving, pipelaying, and backfill—has to be well equipped and coordinated to get maximum production and minimum destruction at the surface. Production rates of 60 to 100 ft/day are possible.

9.7 Manhole Boxes

Steel manhole boxes (Fig. 9.44) are designed and rated and manufactured using the same materials as shoring shields. The calculations are slightly different because the boxes are four-sided and have corner connections that can be considered fixed, and have axial as well as bending forces. Due to shorter spans, usually 8 to 10 ft, and fixed ends they are usually 2- to 4-in wall thickness. Larger 12- and 16-ft hexagonal boxes were developed in the 1990s and are now starting to become popular. Manhole boxes are almost always set in to excavations that stand long enough to get them in and are rarely dug inside and pushed down. They are not designed for it, and there is not enough room inside the box to excavate very deep. If the box needs to be dug in, it is better to use a four-sided slide rail or shield system.

There are also round manhole shields manufactured from corrugated metal pipe. These shields are lighter and require less excavation.

FIGURE 9.44 Steel manhole box 8 × 8 × 10 × 4 -in wall.

The potential downside to these shields is that they could buckle or take on an egg shape when struck by a trench wall collapse on one side only. Fill around the base will help to prevent this.

Manhole boxes are tabulated for wall strength in pounds per square foot of soil load and allowable depth in OSHA soil types with a 72 psf surcharge load. For the competent person, or engineer, calculating the manhole box strength required is no different than for shoring shields.

9.7.1 Manhole Box Safety Issues

In addition to those listed for shoring shields, safety issues are as follows:

- In pipeline construction the manhole is usually set in after the pipe laying operation has passed through. The trench width does not always fit the box and usually needs to be widened at the manhole. Widening the trench is awkward for the excavator, and as a result, there can be a lot of space between the box wall and the soil. There can also be no trench wall at all in the direction of the pipeline. The box needs to be anchored with fill at the bottom of the excavation so that it cannot tip over if it is struck by a caving trench wall.
- Access from the trench bank and ladders must be provided. Manhole construction is a separate operation from pipe laying and needs to be treated that way as regards planning and safety equipment.

9.8 Aluminum Shields and Build-a-Box

These shoring systems are lightweight and modular (Fig. 9.45). Aluminum weighs approximately 14.1 lb per inch thick square foot while steel weighs 40.8 lb per inch thick square ft, making aluminum shoring boxes almost 3 times lighter than steel. Their major application is in shallow work to approximately 12 ft deep where rubber-tired backhoes and small excavators are used. The build-a-box system can be carried in the back of a truck and rapidly constructed by hand. Both L- and T-shaped configurations and walls with openings are possible (Fig. 9.45). The wall materials are from extruded shapes, and the engineering uses the same principles as steel shield design. Because the modulus of elasticity of aluminum $E = 10{,}100$ ksi is almost 3 times less than that of steel, at 29,000 ksi, the wall deflection of aluminum boxes is 3 times that of steel boxes. The strutting is a screw jack system manufactured from steel specifically for the system. This system is by far the best and safest solution for working around T and corner trenches that have outside corners that can collapse (Fig. 9.10).

FIGURE 9.45 Aluminum build-a-box.

9.8.1 Aluminum Shields and Build-a-Box Safety Issues

Safety issues specific to use of aluminum build-a-box are relatively few:

- Due to their light weight they can shift more easily when trench walls fall on them. Anchor them with soil at the bottom or wood-block them to trench walls.
- With long boxes, 10 to 16 ft, the deflection after trench wall collapse can be 3 to 6 in, enough to crush legs and arms between the box wall and production work being installed.
- Even though these boxes are lighter than their steel counterparts, the competent person and the operator of hoisting equipment should be sure that they are working within the safe range of the equipment.

9.9 Arch Spreaders

Arch spreaders function to increase the vertical clearance from the bottom of the trench to the underside of the first spreader. Typical arch spreaders available today raise the clearance to at least the top of the box. The spreader is essentially a beam that is proportioned for bending and axial forces. Higher clearances generate larger bending moments and therefore require deeper, stronger beams. The practical limit to height from the bottom of the trench is directly related to beam strength and weight limitations. Clearances of 1.5 ft above the top of a 10-ft box and 2.5 ft above the top of an 8-ft box are within the commercial range (manufacturing in quantities, shipping, yarding, and handling) of manufacturers. Any size arch spreader is possible; however, the weight and size can easily exceed the weight of the boxes they are used on and the handling capacity of the equipment being used to set them. Another aspect of commercially manufactured arch spreaders is that they need to be adaptable to different trench widths and to many different shields. This requires a bolted connection at the point of maximum bending in the high-clearance strut beam which can be bulky on the tension side and time-consuming to bolt up in the field.

9.9.1 Arch Spreader Design and Tabulated Data

Arch spreader strength is completely independent of the shoring shield that it is attached to. This means that any shield size can be attached to it, and neither the strength of the shield nor the strength of the arch spreader will be changed. The psf rating of the shield may be higher or lower than the psf rating of the arch spreader. Whichever piece of equipment is rated lower controls the allowable depth. The arch spreader shown in Fig. 9.46 was developed by the author and has never been built. This design is unique in that it uses a 1¾-in, 150 ksi threaded rod for the tension connection and a 12-in round × ¾-in wall pipe for the compression member. The tension rod and compression pipe are low-cost and readily available so that they can be easily field- or shop-fabricated and changed out for different trench widths. The allowable tension strength of the rod controls the design, and all other parts are designed to meet that strength. Ultimately a psf rating can be assigned to the spreader, and allowable depth can be tabulated for different shield sizes.

Example 9.3 **Arch Spreader Design**

The following design example calculates the strength and develops the tabulated data for the arch spreader shown in Fig. 9.46.

Materials

Step 1. Determine tension capacity and compression requirement.

From Fig. 9.46

$$\sum M_T$$

$$C = \frac{8W(4.5 + 8/2)}{3} = 22.7W$$

where W = load on shield end beam, klf

W = shield load rating, psf × shield length/2

and

$$\sum \text{Forces} = 0$$

$$T = 8W - 22.6W = -14.6W$$

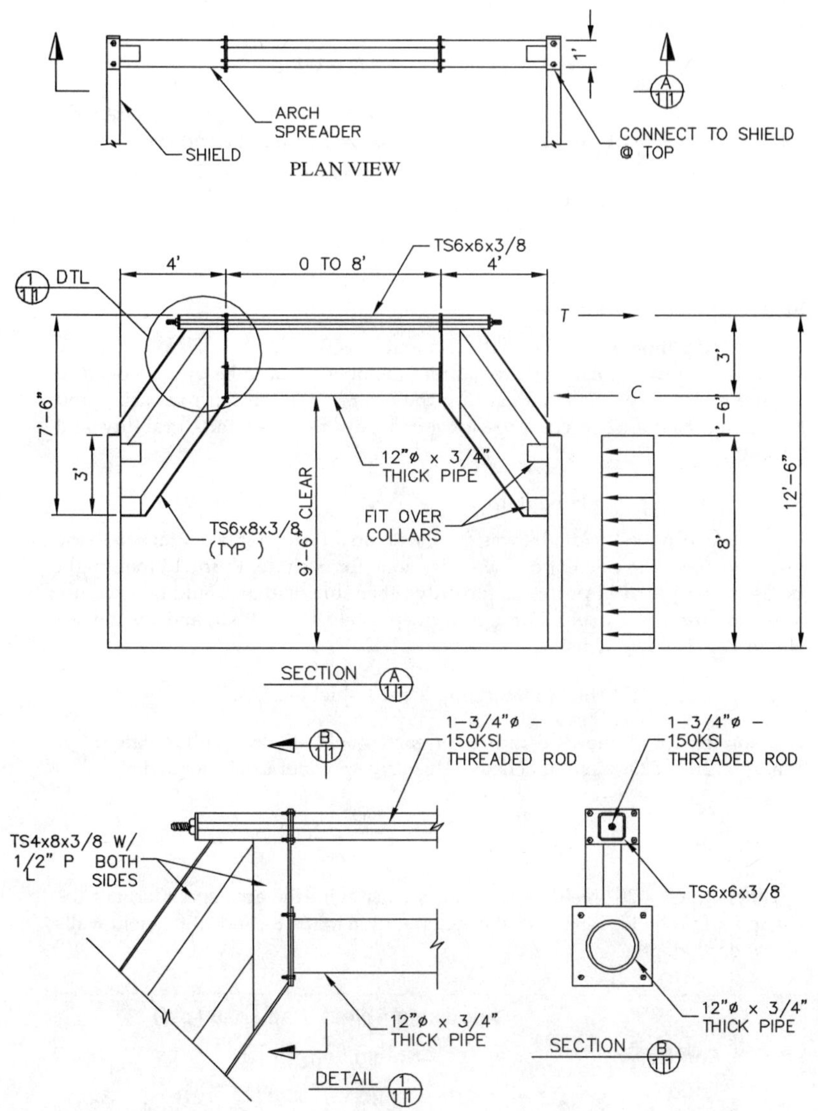

FIGURE 9.46 Arch spreader.

From Williams *Ground Anchor System Manual No. 103* published data for 1¾-in × 150 ksi thread rod, and AISC D1-for tension, allow $0.5F_u$, and $0.6F_y$ (this is a progressive failure type member, do not allow 1.33 shoring use factor)

$$P_u = 400 \text{ k} \times 0.5 = 200 \text{ k}$$

$$T = P_y = 320 \text{ k} \times 0.6 = 192 \text{ k}$$

$$W = \frac{192 \text{ k}}{14.6} = 13.42 \text{ klf}$$

and

$$C = 22.7W = 22.7 \times 13.42 = 304 \text{ k}$$

A 12-in round × ¾-in wall pipe has the following geometric properties: Area = 28.27 in^2, moment of inertia $I = 510.9$ in^4, and $r = 4.25$ in. The ends are fixed so $K = 1$ and $\frac{K\ell}{r} = \frac{8 \times 12}{4.25} = 22.6$ and from AISC Table C-36, $F_a = 17.14$ ksi.

$$\therefore \quad f_a = \frac{304}{28.27} = 10.75 \text{ ksi} < 17.4 \qquad \text{OK}$$

Step 2. Design vertical members and connections.

The bending moment in the vertical member is 3 × 192 k = 576 k·ft, and the maximum shear is 192 k. The remaining members should be proportioned to handle these forces. Bolts connecting the 6 × 6 × 3/8 tension rod casing and the flange at the 12-in compression need only be sized for durability and handling.

Step 3. Develop tabulated data.

The above result $W = 13.42$ klf was done with an 8-ft-deep shield attached to the arch spreader. The calculation has to be done for each depth shield that might be used. For practical purposes the only other shields that would be used are a 6- and 10-ft-deep shield. For a 10-ft-deep shield $W = 8.9$ klf, and for the 6-ft shield $W = 21.3$ klf. By using

$$W = \text{shield load rating } w \text{ psf} \times \text{shield length}/2$$

and solving for w, the load rating for each shield size can be tabulated. For example, an 8 × 20 shield attached to this arch spreader can be loaded to

$$w = \frac{2W}{\text{shield length}} = \frac{2 \times 13.42}{20} = 1342 \text{ psf}$$

This is the 8 × 20 shield load rating for use with the arch spreader, not the rating for the shield. Table 9.18 shows tabulation for 6-, 8-, and 10-ft shield walls at standard lengths.

		Allowable Shield Wall Load (psf)					
		Shield Length (ft)					
Shield Depth (ft)	End Beam Load (klf)	12	16	20	24	28	32
6	21.3	3550	2663	2130	1775	1521	1331
8	13.4	2233	1675	1340	1117	957	838
10	8.9	1483	1113	890	742	636	556

Notes:

1. The lowest strength rating (psf), the shield tabulated data, or arch spreader rating is to be used to determine allowable shield loading when the arch spreader is used.
2. Arch spreader weight is approximately 5000 lb.

TABLE 9.18 Arch Spreader Tabulated Data for Spreader Shown in Figure 9.46

For 6-in wall shields these arch spreader ratings are slightly less than shield wall ratings. To more closely match or exceed the shield wall ratings of thicker-walled shields, either a stronger arch has to be designed or the clearance height has to be lowered. It is possible to lower the clearance height by attaching the arch outside the line of the shield spreaders (weld it to the box). Building a stronger arch spreader is also possible. There is a 2½-in × 150 ksi thread bar with P_y = 622 k, allowing a 373-k tension. The strength of an arch spreader designed around this bar would be 373/192 = 195 percent higher. Normally the types of projects that require arch spreaders have large sums of money involved, and therefore the cost of design and construction of the arch is insignificant compared to the benefit.

9.9.2 Arch Spreader Safety Issues

Due to the weight and consequences of failure of an arch spreader, it is important to pay attention to details. Safety issues include these:

- When the arch spreader selection and use are done with tabulated data, the competent person should make sure that the proper loading is being used. Usually heavy lifting of large pipe or precast structures is taking place around the shield/arch installation. Surcharge loads for these types of operations usually result in lateral loading of 200 to 500 psf. Placement of lifting equipment can have a large effect on this outcome. Use the proper loading, plan the setbacks for equipment, and make sure they are adhered to.
- Arch spreaders are heavy, sometimes as heavy as the box. Because they are usually used only on one end of the box and above the top of it, short boxes can be easily tipped on one end. Lifting cable assemblies should take eccentric loading into consideration.
- Make sure that cabled loads or excavator arms and buckets do not strike the fully loaded compression side (bottom flange) of the arch strut; the result can be buckling and rapid failure of the arch.

Arch Spreader 16 ft wide				Approximate Weight			
Parts	Description	Material	*F_y (ksi)	Qty. (ea)	Size	Weight unit	Weight (lbs)
Tension rod	1-3/4", 150 ksi rod	A-722-98	150	1	10	3.1	31
Rod nuts	1-3/4" R-73 hex nut	A-722-98		2	3.46" × 3.5"	5.0	10
Compression pipe	12" rnd × 3/4" wall	A53 Gr B	35	1	8	96.0	768
Tension rod casing	TS6 × 6 × 3/8	A500 Gr B	55	1	8	27.5	220
Middle tubes	TS8 × 4 × 3/8	A500 Gr B	55	1	50	27.5	1375
Plate	PL 1/2	A572	55	4	30	20.4	2448
Weld metal	Wire feed	E-70xx		1			200
						Total	5052

FIGURE 9.47 Arch spreader materials.

- Arch spreaders are adaptable to different trench widths and shield sizes. Changing spreader widths involves a bolt-up tension connection. Make sure that the specified bolt strength is being used and that they are properly torqued if that is also a requirement.
- The connection at the shield wall may be different for different types of shields. Manufactured stock arch spreaders are designed to be as universal as possible; however, shield spreader locations and sizes do vary. Alterations such as cutting and welding at the connection to the shield do not change the overall strength of the arch. The installer should not hesitate to make changes to ensure the shield is securely attached. Due to the off-center mass of the spreader, the forces involved in moving and setting the shield can be greater than those in the loaded condition. Have altered connections certified by a registered engineer.

9.10 Slide Rail

The slide rail shoring system represents a major innovation in shoring technology. The system is most efficient in "bad ground" where dig and set shoring systems such as trench jack and shoring shield will not work because the ground collapses before the shoring can be installed. The alternative in this type of ground prior to slide rail was to advance piling, H-pile or sheet pile, first and then dig and brace. Both the dig and set and the set and dig systems can be considered somewhat static in that the parts never move once they are set in place. One unique thing among many with the slide rail system is that the system parts: posts, panels, struts, boogie car, and roller are all moved during the excavation process [Fig. 9.48]. Prior to slide rail the closest thing to this was timber and wale sheeting systems where the timbers are advanced periodically as the bottom is being dug, and digging in shields. The slide rail system features a vertical rail post and panel that are manufactured and designed to slide relative to each other specifically for the purpose of digging the system into the ground. Originally the struts were fixed to one location but pined so that they could rotate as the system walked itself into the ground. Further evolution of the system brought a "boogie car strut system" that allowed the struts to be moved up and down the rail post during installation. The slide rail system in most respects represents an equivalent alternative to interlocked sheet pile and H-pile and lagging shoring.

Slide rail shoring was first developed in Europe about 35 years ago as an alternative to sheet piling. Pile driving in European cities was destructive to buildings and infrastructure that was older and more fragile than found in the United States. The streets are not as

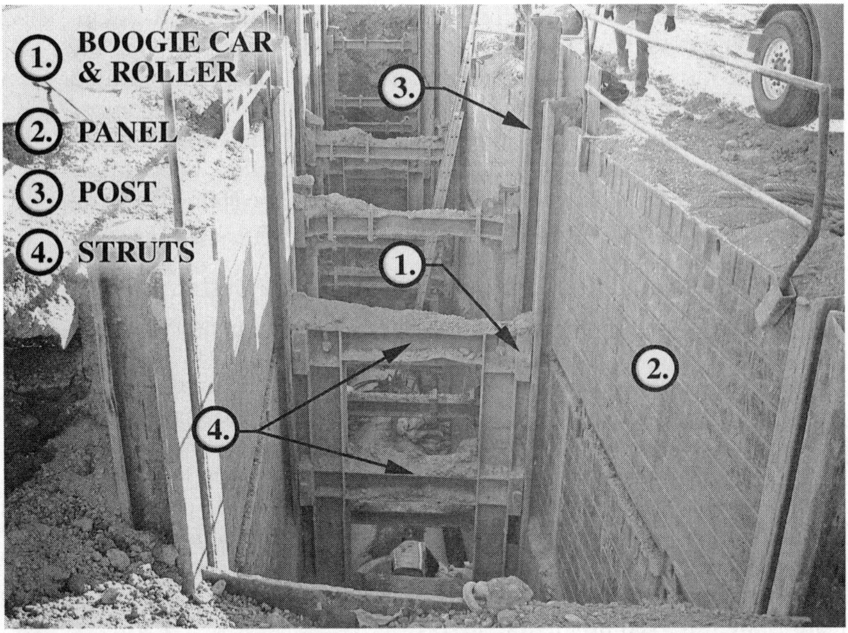

FIGURE 9.48 (a) Slide rail pipeline shoring application to 26 ft deep; (b) linear application 35 ft deep.

wide, and working room is limited. First developed as a linear system, the slide rail system was more productive and far less damaging to surrounding facilities.

Introducing the slide rail system to the U.S. market took some visionary and dedicated individuals. The system was expensive to purchase, and the installation process had to be properly equipped and learned by contractors who were good at and invested in other tried and true shoring systems. Utility owners and their engineers were nervous about accepting new technology in place of totally predictable sheet pile, H-pile, and tight sheeted timber and wale systems that they were familiar with. The system was first brought to the East coast and Gulf states in the early 1980s. Both of these regions have a large share of soft marine clays that are well suited to slide rail use. After a large financial investment and promotional effort by a very few shoring rental companies, the system became accepted as an equal alternative to the other systems. In the early 1990s, Jim Dalton of D&E Steel Plate Shoring, Inc., in California invested heavily in the system and dedicated his efforts to making it known and accepted on the West coast. Dalton saw the potential and spared no expense including abandoning millions of dollars' worth of original equipment that he had invested in when a better, stronger system came along. In the United States, large pipe, large excavating equipment, and deeper excavations called for bigger and more durable slide rail systems. Major U.S.

manufacturers of shoring equipment took up the challenge, developing their own version of the European slide rail system. By the late 1990s slide rail was becoming known and used in most of the United States. A trend at that time toward larger nationwide shoring companies and their commitment to innovate and bring the latest technology to the industry has elevated slide rail to the position that it now holds as the best and least expensive alternative shoring system when dig and then set shoring systems are no longer an option.

9.10.1 Slide Rail Use and Comparison to Pile Systems

Slide rail is a unique shoring system, and its use should be evaluated alongside other systems as an equal and not neccssarily as an alternative to them. They all have something unique to offer. Where slide rail works well, it works better and in some cases can do things that no other system can do. In Sacramento, California, a several mile long 25- to 35-ft-deep pipeline project in cobbles could only be performed with slide rail. The cobbles prevented effective sheet pile diving and H-pile drilling or driving. Open cut was precluded because of right-of-way and maintaining traffic. Microtunneling was tried and failed. All contractors working on separate reaches of the contract ended up using slide rail to complete the project.

The following is a comparison of the attributes of slide rail to those of sheet pile and H-pile shoring systems:

- All three systems work best in "bad ground," loose sands and gravels and soft cohesive soils, and become harder to drive in dense sands and stiff clays where dig and set systems work best and are less expensive.
- Slide rail and H-pile shoring will not cut off water flow at the bottom of the excavation. The only thing that will cut off water flow is an impervious wall below the bottom; interlocked sheet piles accomplish this.
- Slide rail provides no protection from bottom heave. Control of bottom heave requires a cutoff mechanism below the bottom of the excavation. To some extent arching around H-piles does this, however, not very effectively because the soft bottom condition that creates bottom heave also allows shearing of the soil arch between the piles. Sheet piles driven at a proper depth below the bottom or some sort of base soil stabilization is the only thing that can prevent bottom heave with certainty. If bottom heave or water cutoff is a serious issue eliminate slide rail as a shoring option.
- Slide rail installation is a parallel operation simultaneous with the excavation work and uses the same equipment. This allows for a short smooth and efficient linear excavation and pipelaying operation. Pile operations require special driving

or drilling equipment and therefore necessitate a tandem operation that can add more time to the operation. Even if pile driving is performed ahead of excavation, the wale and strutting operation stops the excavation work for a time.

- Standard slide rail equipment is effective to approximately 30 ft deep. Triple slide rail (there is limited supply and knowledgeable experience available) can go to 40 ft deep. H-pile and sheet pile excavations can work theoretically to any depth. The normal and practical limit is 60 ft.
- Slide rail can deal with large rock or dense cells of cobbles where the other systems will fail. This is due to the ability to dig out below the bottom of the slide rail as the excavation is being advanced. To some extent this can be done with sheet pile; however, it requires separate pile driving equipment at the site simultaneously with the excavation operation.
- Often predriven interlocked sheet piles are specified because they prevent the possibility of soil loss behind the sheeting during excavation and soil squeezing out between the sheeting after the excavation is opened up. For these two purposes H-pile and steel plate (not timber lagging) and slide rail should be considered equivalent to tight interlocked predriven sheeting because it can be advanced and kept always below the bottom of the excavation; it does not allow wet or squeezing soil to pass through the panels or around the edges. In difficult soils the pressure seals the panel to the rail post similar to the way an interlock works on a sheet pile.
- Slide rail is the only shoring system that is designed to leave no void in the ground after it is removed. The rail posts and panels can be raised up nearly simultaneously with the backfill operation so that the compaction lifts are confined by in situ soil and not sheeting that will be removed later. Sheet pile and H-pile cannot be coordinated in this way and will always leave a void behind unless they are cut off and left behind.
- There is a learning curve and installation equipment mix associated with all three systems. Pile driving equipment is separate, specialized, and used only for that purpose. The labor force needs to be trained and experienced, and they even have their own trade union categories, pile bucks, crane operators, and oilers. The welded waling and strutting operations also require trained and specially equipped welders.

  Slide rail installation uses excavators, loaders, and laborers the same as with the excavation and backfill operation. A minimum size for the excavator is 225, and the loader should be minimum 5 yd^3. Techniques and small tools for moving, setting, driving, and pulling slide rail parts have to be developed.

Slide rail equipment suppliers usually provide an experienced installer on the first project. The time it takes to peak the learning curve is probably the same for each type of system.

- Of the three systems slide rail is the only one that is manufactured and put together as a system for the specific purpose of providing temporary shoring. It is the only one that is tabulated and can be used off the shelf without a site-specific engineered plan. A variety of parts are manufactured so that the system can adapt to different configurations without destroying parts of them.

  Sheet and H-piles can be used and reused in many configurations; however, the waling and strutting are cut and fit and then reused each time in a lesser capacity until they are scrapped. By OSHA requirements every sheet pile and H-pile installation is site-specific and requires an engineered plan. Although generalized plans using these materials can be developed, they are generally developed by contractors for their specific repetitive purposes and have not been developed by manufacturers and rental yards.
- As a general rule, slide rail is faster and less expensive to install but quite expensive to own and maintain. This makes it the first choice for short-term excavations such as underground tank replacement or repetitive linear applications, e.g., pipeline. Rental companies can make the major investment and then by cycling it between contractors keep it in use most of the time. In the case of bore pits and pump station construction where the shoring can be in the ground for several months, the rental cost usually reaches the purchase cost after 3 to 6 months, and the added installation time becomes insignificant.

  As an installation time comparison, with an experienced contractor a 15-ft × 25-ft × 25-ft-deep pit can be shored and excavated using slide rail in a 10-h day. The same pit in H-pile or sheet pile will take approximately 3-10-h days with an experienced crew and can take up to 10 days. In a linear application the ratio can be even better.

9.10.2 Slide Rail Components and Installation

The slide rail system has three basic components: sheeting, slide posts, and a strutting system (Fig. 9.49). Each part can be moved relative to the other by crane or excavator at any time during the life of the shored excavation. The strut frame [Fig. 9.49(Section A)] is pushed toward the bottom of the slide posts to prevent toe-in during the dig-in process. After the bottom is reached, the struts are moved toward the top to allow pipe installation or working room for foundations and wall formwork. The strut frame is pushed or pulled by the excavator and

APPENDIX 1

OSHA Subpart P, Excavations and Commentary

AP1.1 Book Chapters Relating to Subpart P

The entire subject matter of this book is developed around the material contained in OSHA Subpart P, Excavations. Chapters specific to OSHA sections are as follows:

1926.650—Scope Application, and Definitions Applicable to This Subpart	Chapter 1
1926.651—Specific Excavation Requirements	Chapter 4
1926.652—Requirements for protective systems	Chapter 9
Appendix A Soil Classification	Chapters 5, 6, 7, Appendix 2
Appendix B Sloping and Benching	Chapter 8
Appendix C Timber Shoring for Trenches	Chapter 9
Appendix D Aluminum Hydraulic Shoring for Trenches	Chapter 9
Appendix E Alternatives to Timber Shoring	Chapter 9
Appendix F Selection of Protective Systems	Chapter 9

AP1.2 OSHA Subpart P, Excavations, with Commentary

The following is the text from 29CFR 1926 Safety and Health Regulations for Construction, Subpart P, Excavations. The commentary is strictly the author's interpretation and not to be interpreted as OSHA policy or consensus of opinion from any group or organization. The comments are intended to provide a clearer understanding of the OSHA text and to highlight topics that should be emphasized in competent person training seminars. It is the author's conviction that this regulation is an excellent tried and true approach to regulating excavation safety, and comments are never intended as criticism.

The requirements of this Subpart P, or a similar state-administered program, must be met whenever employees are required to work inside an excavation. If there is a citation given for a violation of Subpart P, the section numbers violated will be listed on the ticket; for instance, violation of 1926.651(b)(1)... will be listed . The requirements are clearly stated in the document. There is no policy or added rules elsewhere that needs to be conformed to; in other words one does not need to look outside this document for worker safety requirements specific to excavation work. General safety rules apply anyplace the work is being conducted including inside excavations.

Federal OSHA provides an excellent website www.osha.gov that contains the entire text presented here. This is an excellent resource for gaining insight to the development and interpretation of this standard. By clicking on blue section headings a listing of inquiries and interpretations comes up. Further clicks bring up the text of the inquiries and interpretations.

Disputes or questions about wording of this document are first interpreted by OSHA. The contractor who does not agree with the OSHA conclusion is free to go to the courts for an interpretation and ruling. This process can take months and years. OSHA will cede to the most recent court decision on the issue and enforce accordingly. Rule additions and changes involve a hearing process with industrywide input that can take years. It is important to remember that OSHA officers are public employees and have absolutely no authority to make or change rules; this process is the only way that it can be changed. No amount of badgering or intimidation can change that. OSHA personnel have only one mission—to prevent accidents in the workplace—and they are inclined to do this using every and all approaches that work to this end.

State-administered OSHA safety programs meet or exceed the requirements of the federal program. The state programs were adopted from the federal standard and in most cases read exactly the same. Note that a threshold requirement for federal acceptance of additional or reworded state rules is that they do not add undue burden, restrictions, and costs without sufficient return. Due to the similarity and abundance of state programs, the federal rules are presented here. Commentary is presented after statement of the rule.

1926.650—Scope application, and definitions applicable to this subpart.

1926.650(a)—Scope and application. This subpart applies to all open excavations made in the earth's surface. Excavations are defined to include trenches.

Commentary

Subpart P is in addition to all safety requirements in 29 CFR 1926 Safety and Health Regulations for Construction. The rules in "sub P" are specific to problems that arise with excavation work, and only apply to worker safety when workers are in the excavation.

1926.650(b)—Definitions applicable to this subpart.

"**Accepted engineering practices**" means those requirements which are compatible with standards of practice required by a registered professional engineer.

Commentary

There is some confusion around standard practice in excavation safety work because it is temporary in nature, varies regionally, and until 1989, when the standard was adopted, this type of engineering work did not necessarily require an engineer's stamp (unlicenced engineers working for the contractor could do the work). Standard practice requirements for permanent work are defined by building codes and associations and require the engineer to follow them and stamp the work.

At a minimum for construction engineering, the engineer should have previous experience and be able to back up engineering assumptions with accepted or proven theories. Drawing seat-of-the-pants conclusions with nothing to back them up is unacceptable.

"Aluminum Hydraulic Shoring" means a pre-engineered shoring system comprised of aluminum hydraulic cylinders (crossbraces) used in conjunction with vertical rails (uprights) or horizontal rails (wales). Such system is designed specifically to support the sidewalls of an excavation and prevent cave-ins.

"Bell-bottom pier hole" means a type of shaft or footing excavation, the bottom of which is made larger than the cross section above to form a belled shape.

Commentary

Bell-bottom pier hole is defined because workers can be required to be lowered down into a drilled shaft and clean out the bottom of the excavation. This is an example of a specific practice within a specific excavation specialty industry where the normal rules are waived because it has been shown to be safe and shoring would interfere with the work. Another exception is swimming pool excavation work.

"Benching (Benching system)" means a method of protecting employees from cave-ins by excavating the sides of an excavation to form one or a series of horizontal levels or steps, usually with vertical or near-vertical surfaces between levels.

"Cave-in" means the separation of a mass of soil or rock material from the side of an excavation, or the loss of soil from under a trench shield or support system, and its sudden movement into the excavation, either by falling or sliding, in sufficient quantity so that it could entrap, bury, or otherwise injure and immobilize a person.

"Competent person" means one who is capable of identifying existing and predictable hazards in the surroundings, or working conditions which are unsanitary, hazardous, or dangerous to employees, and who has authorization to take prompt corrective measures to eliminate them.

Commentary

The term *competent person* is used in other parts of the construction safety rules with a different definition. For example, there is a "competent person" definition for crane work and scaffold erection work.

OSHA inspection uses this definition to determine the presence of a competent person at the site. If there is no person at the site with these qualities when a function requiring a competent person must be performed, there is no competent person at the site. It does not matter that a person has had training and may even have a record of that training, if he or she is not capable of identifying, predicting, and taking action, including stopping the work; it is considered prima facie (self-evident from the facts) evidence that he or she is not a competent person, and a citation will be issued.

The contractor is required to provide training for her or his workers, including competent person training for at least one person on an excavation project. Proof that the training was provided and that the person was tested for comprehension is critical to the contractor's case that he or she took steps to position a competent person at the site. After all, the contractor would argue, the contractor cannot be responsible for the competent person's retention of knowledge after the course was provided. A quality course with good record keeping is essential to the success of this argument.

There is no OSHA requirement that a competent person have a certificate of completion of a training course, nor is there any indication of how often the person must be refreshed in the training. OSHA does not define the elements of an acceptable course or certify trainers. The results of the training will only speak to the fact that the contractor has done due diligence in seeking to place a competent person at the site. The OSHA requirement is that the competent person possess the knowledge and authority described in the definition.

The competent person and any person performing critical work in the field should be aware that they may be personally held responsible for negligence in doing their work while the contractor might not. The reasoning goes that the contractor who has done everything possible to provide training and assistance to his or her workers cannot be held accountable for the negligent acts of these workers in the field. *Negligent* means with knowledge of what is required to be done by the law and then willfully not doing it. It is far better to have done it wrong than to have to say, "I knew I had to do it, but I did not do it."

At a minimum the competent person should

- Have a concept of hazards associated with the excavation activity that is being engaged in at the time, whether it be risk of existing facilities impacting safety, hazardous atmospheres, cave-in, etc.
- Understand the concept of the worker protection system. When asked to specify the worker protection system, the answer should be either "My protection system is a relief system in the form of sloping and benching," or "My protection system is a support system in the form of shielding, trench jacks, etc."
- Be familiar with Appendix A soils identification, specifically type A, B, and C soils and that she or he must first identify the soil before utilizing tabulated data for worker protection.
- Understand that even with excavations under 5 ft deep, a rational (based on explainable reasoning) decision must be made that the excavation is safe for workers to enter.
- Have clear authority; saying that he or she has it does not make it so. One suggestion to make this demonstrable is to assign a person outside the production staff, preferably a safety officer or officer of the company, whom the competent person can call if his or her decisions are being minimized at the job site. The regulation is looking for someone to be responsible and held accountable. This should be understood by all from the start.

The competent person must have knowledge in the particular activity being performed. The competent person required when dewatering systems are operating need only have knowledge of the dewatering system and pump operation and maintenance. The competent person performing daily inspections of an engineered shoring system must have enough knowledge of the particular system to anticipate and identify problems that are associated with the protective system. Instruction from the design engineer is recommended.

The term *qualified person* is used in the standard and should not be confused with competent person. 29 CFR 1926.32(l) states:

> "Qualified" means one who, by possession of a recognized degree, certificate, or professional standing, or who by extensive knowledge, training and experience, has successfully demonstrated his ability to solve or resolve problems relating to the subject matter, the work, or the project.

If this person does not possess a license, degree, or certificate of training, a contractor should have some sort of written rationale for why that person is considered qualified. The contractor should be able to convince a judge that he or she did the due diligence to make sure there was a qualified person fulfilling the requirements.

"Cross braces" mean the horizontal members of a shoring system installed perpendicular to the sides of the excavation, the ends of which bear against either uprights or wales.

"Excavation" means any man-made cut, cavity, trench, or depression in an earth surface, formed by earth removal.

"Faces" or "sides" means the vertical or inclined earth surfaces formed as a result of excavation work.

"Failure" means the breakage, displacement, or permanent deformation of a structural member or connection so as to reduce its structural integrity and its supportive capabilities.

"Hazardous atmosphere" means an atmosphere which by reason of being explosive, flammable, poisonous, corrosive, oxidizing, irritating, oxygen deficient, toxic, or otherwise harmful, may cause death, illness, or injury.

"Kickout" means the accidental release or failure of a cross brace.

"Protective system" means a method of protecting employees from cave-ins, from material that could fall or roll from an excavation face or into an excavation, or from the collapse of adjacent structures. Protective systems include support systems, sloping and benching systems, shield systems, and other systems that provide the necessary protection.

"Ramp" means an inclined walking or working surface that is used to gain access to one point from another, and is constructed from earth or from structural materials such as steel or wood.

"Registered Professional Engineer" means a person who is registered as a professional engineer in the state where the work is to be performed. However, a professional engineer, registered in any state is deemed to be a "registered professional engineer" within the meaning of this standard when approving designs for "manufactured protective systems" or "tabulated data" to be used in interstate commerce.

Commentary

The reason for engineering registration in the state where work is being performed is so that lawsuits are confined to the state and registration requirements within that state. The problem associated with manufactured equipment is that it is often manufactured in a state other than where it is being used. Some states and many times specifying engineers require that the equipment be certified by an engineer in the state where it is being used to keep lawsuits within the state. This places a huge burden on manufacturers because in theory they need to have available and pay for a certifying engineer in all 50 states or an engineer with licenses in all 50 states, thereby multiplying the engineering cost by 50. To further complicate matters, for manufactured equipment the certifying engineer's license expires usually every 2 years so that the tabulated data need to be recertified every 2 years. Add to that the requirement that every separate serial number within a model line needs a separate data sheet. In the old days one data sheet was issued for each model by the manufacturer with the stamp of the state in which the design engineer practiced and had no expiration date on the stamp.

"Sheeting" means the members of a shoring system that retain the earth in position and in turn are supported by other members of the shoring system.

"Shield (Shield system)" means a structure that is able to withstand the forces imposed on it by a cave-in and thereby protect employees within the structure.

Shields can be permanent structures or can be designed to be portable and moved along as work progresses. Additionally, shields can be either premanufactured or job-built in accordance with 1926.652(c)(3) or (c)(4). Shields used in trenches are usually referred to as "trench boxes" or "trench shields."

Commentary

A shield system is considered a passive shoring system, one that does not support the soil, as it protects the worker in the event of a cave-in.

1926.652(c)(3) or (c)(4) refers to tabulated data developed by an engineer for an entire project, defined region, or specific site within a project.

"Shoring (Shoring system)" means a structure such as a metal hydraulic, mechanical or timber shoring system that supports the sides of an excavation and which is designed to prevent cave-ins.

"Sides." See "Faces."

"Sloping (Sloping system)" means a method of protecting employees from cave-ins by excavating to form sides of an excavation that are inclined away from the excavation so as to prevent cave-ins. The angle of incline required to prevent a cave-in varies with differences in such factors as the soil type, environmental conditions of exposure, and application of surcharge loads.

"Stable rock" means natural solid mineral material that can be excavated with vertical sides and will remain intact while exposed. Unstable rock is considered to be stable when the rock material on the side or sides of the excavation is secured against caving-in or movement by rock bolts or by another protective system that has been designed by a registered professional engineer.

Commentary

It seems reasonable that rock that has been drilled and blasted could be included in this definition.

Virtually all rock strata have joints and bedding planes; the spacing could be so far apart it is not apparent. The key to stability in rock is the strength of the rock, the degree of the dip in the bedding planes, and the friction along those planes. If the bedding planes are level, the rock may still fail because it is weak and cannot hold up the rock and soil above it. The author has observed rock that had to be excavated using a rock trencher, and yet the trench walls would explode and collapse at a depth of 16 ft because it could not hold the rock above. Aside from very hard rock with practically flat bedding planes, determining stable rock for vertical slopes or stable rock sloping is very complicated and needs to be done by experienced engineers. By OSHA standards if rock is not stable, the slopes automatically must be 1:1 because it is noncohesive. There is no possibility of steeper slopes unless an engineer is involved. This is for good reason because rock slope failure, even if it is one small rock falling, can easily injure workers struck by it, unlike workers getting hit by ravel from sands and gravels.

"Structural ramp" means a ramp built of steel or wood, usually used for vehicle access. Ramps made of soil or rock are not considered structural ramps.

"Support system" means a structure such as underpinning, bracing, or shoring, which provides support to an adjacent structure, underground installation, or the sides of an excavation.

"Tabulated data" means tables and charts approved by a registered professional engineer and used to design and construct a protective system.

"Trench (Trench excavation)" means a narrow excavation (in relation to its length) made below the surface of the ground. In general, the depth is greater than the width, but the width of a trench (measured at the bottom) is not greater than 15 feet (4.6 m). If forms or other structures are installed or constructed in an excavation so as to reduce the dimension measured from the forms or structure to the side of the excavation to 15 feet (4.6 m) or less (measured at the bottom of the excavation), the excavation is also considered to be a trench.

Commentary

One of the reasons that OSHA has defined a trench in this way is so that a distinction about access and egress of the excavation can be made. In a trench there must be access to a ladder within 25 ft of travel. In an excavation there needs to be a way to get into and out of it; otherwise an excavation the size of a city block would need a lot of ladders.

OSHA purposes aside, there is a linear aspect to a trench, and it is also associated with pipe and utility work. The word is most often used in that context.

"Trench box." See "Shield."

"Trench shield." See "Shield."

"Uprights" means the vertical members of a trench shoring system placed in contact with the earth and usually positioned so that individual members do not contact each other. Uprights placed so that individual

in accordance with the conditions and requirements set forth in Appendices A and B to this subpart.

Commentary

Appendix A is soils identification, and Appendix B is OSHA developed tabulated data for open cut systems less than 20 ft deep. Both appendices were developed and adopted by federal OSHA .

Option 2 can be used in any soil condition found on the planet. This leads to the conclusion that it is generalized and conservative in most situations. Otherwise no one would sign off on it. The system has been proved to be safe and effective since it was formally adopted in 1989. It is the standard of the industry in the United States.

See Article 8.4.1 for a complete discussion of Appendix B.

1926.652(b)(3)—Option (3)—Designs using other tabulated data.

1926.652(b)(3)(i)—Designs of sloping or benching systems shall be selected from and in accordance with tabulated data, such as tables and charts.

1926.652(b)(3)(ii)—The tabulated data shall be in written form and shall include all of the following:

1926.652(b)(3)(ii)(A)—Identification of the parameters that affect the selection of a sloping or benching system drawn from such data;

1926.652(b)(3)(ii)(B)—Identification of the limits of use of the data, to include the magnitude and configuration of slopes determined to be safe;

1926.652(b)(3)(ii)(C)—Explanatory information as may be necessary to aid the user in making a correct selection of a protective system from the data.

1926.652(b)(3)(iii)—At least one copy of the tabulated data which identifies the registered professional engineer, who approved the data, shall be maintained at the jobsite during construction of the protective system. After that time the data may be stored off the jobsite, but a copy of the data shall be made available to the Secretary upon request.

Commentary

Option 3 is best suited to regional soils or industry-specific situations. For instance, a large industrial site such as an oil refinery may have specific soil conditions along with surcharge and access problems that are unique to that site. Under this option a sloping and benching system that addresses these issues which may be in some places more conservative and in some places more efficient than option 2 can be developed for every contractor working at the site. Important elements of this system are the following:

- Definition of the region where it applies
- Basis for selection of options, usually a soil classification system; however, it could also be more specific about surcharge

and buried infrastructure limitations and distances, seasonal variations, length of time the excavation will be open, etc.

- Clear depiction of slopes
- Step-through selection procedure
- Complete tabulation and current stamp of engineer who developed it

The data serve three specific functions:

1. They enable the user to select and utilize the sloping and benching systems.
2. They enable reviewing and inspection personnel to confirm that the system is properly constructed.
3. They enable the courts to determine liability in the event of an accident.

1926.652(b)(4)—Option (4)—Design by a registered professional engineer.

1926.652(b)(4)(i)—Sloping and benching systems not utilizing Option (1) or Option (2) or Option (3) under paragraph (b) of this section shall be approved by a registered professional engineer.

1926.652(b)(4)(ii)—Designs shall be in written form and shall include at least the following:

1926.652(b)(4)(ii)(A)—The magnitude of the slopes that were determined to be safe for the particular project;

1926.652(b)(4)(ii)(B)—The configurations that were determined to be safe for the particular project;

1926.652(b)(4)(ii)(C)—The identity of the registered professional engineer approving the design.

1926.652(b)(4)(iii)—At least one copy of the design shall be maintained at the jobsite while the slope is being constructed. After that time the design need not be at the jobsite, but a copy shall be made available to the Secretary upon request.

Commentary

This is essentially the same as option 3 except for two important differences:

1. The plan is site-specific. The person at the site does not select the slopes, a registered engineer does.
2. The slopes are determined by the engineer who develops the plan, usually based on soil parameters developed from a soils investigation and accepted engineering principles used in soil mechanics. The design calculations should always reflect these two elements.

Important elements of the plan include

- Definition of the location and extent where it applies
- Depiction of existing site conditions
- Clear depiction of slopes
- Step-through excavation and backfill procedure
- Complete tabulation and current stamp of engineer who developed the plan

The plan serves three specific functions:

1. It is a plan that can be used for sloping and benching at a defined location.
2. It enables reviewing and inspection personnel to confirm that the system is properly designed and constructed.
3. It enables the courts to determine liability in the event of an accident.

This option is often seen as a way to get around the more stringent option 2 requirements. Many would like to read in the term *whichever is more stringent* when reviewing slopes proposed under this option. Federal OSHA put this option in the standard specifically for the purpose of promoting innovation and efficiency while achieving safe conditions in excavation work. Engineered design is safer because it is based on site-specific soils investigations and engineering principles used by an experienced engineer. In the office, review of these plans should focus on confirming that the engineer is experienced, standard engineering principles were used , and the plan is clear. In the field, inspection should be focused on establishing the fact that it was constructed in accordance with the plan.

1926.652(c)—Design of support systems, shield systems, and other protective systems. Designs of support systems, shield systems, and other protective systems shall be selected and constructed by the employer or his designee and shall be in accordance with the requirements of paragraph (c)(1); or, in the alternative, paragraph (c)(2); or, in the alternative, paragraph (c)(3); or, i the alternative, paragraph (c)(4) as follows:

Commentary

Four options are available:

- Option 1: Identify soil in accordance with Appendix A and utilize Appendix C, Timber, or Appendix D, Aluminum Hydraulic Shoring.
- Option 2: Identify soil in accordance with Appendix A and utilize manufactured shoring equipment.

- Option 3: Engineered shoring design for a region.
- Option 4: Engineered shoring design for specific site.

1926.652(c)(1)—Option (1)—Designs using appendices A, C and D. Designs for timber shoring in trenches shall be determined in accordance with the conditions and requirements set forth in appendices A and C to this subpart. Designs for aluminum hydraulic shoring shall be in accordance with paragraph (c)(2) of this section, but if manufacturer's tabulated data cannot be utilized, designs shall be in accordance with appendix D.

Commentary

See Article 9.2, Timber Shoring, and Article 9.3, Aluminum Hydraulic Shoring, for an in-depth discussion of Appendices C and D.

The last sentence of this rule essentially says that if there are no tabulated data available for the trench jacks being used, then Appendix D can be used. It is not an either/or situation. Situations where tabulated data for trench jacks cannot be utilized occur when

- There is use of old trench jacks where the manufacturer is no longer in business.
- There are trench jacks that cannot be identified to a specific manufacturer or have replacement parts from supplier other than the manufacturer.

It is important to establish that the manufacturer's data are not available before utilizing Appendix D. The manufacturer's tabulated data are going to be more favorable. Most manufacturers' tabulated data allow slightly greater jack spacing and depths up to 24 ft. Also by working within the manufacturer's data the user has the manufacturer to back him or her up in the event of mechanical problems or an accident.

1926.652(c)(2)—Option (2)—Designs using manufacturer's tabulated data.

1926.652(c)(2)(i)—Design of support systems, shield systems, or other protective systems that are drawn from manufacturer's tabulated data shall be in accordance with all specifications, recommendations, and limitations issued or made by the manufacturer.

1926.652(c)(2)(ii)—Deviation from the specifications, recommendations, and limitations issued or made by the manufacturer shall only be allowed after the manufacturer issues specific written approval.

1926.652(c)(2)(iii)—Manufacturer's specifications, recommendations, and limitations, and manufacturer's approval to deviate from the specifications, recommendations, and limitations shall be in written form at the jobsite during construction of the protective system. After that time this data may be stored off the jobsite, but a copy shall be made available to the Secretary upon request.

Commentary

See Chap. 9 for a complete discussion of manufactured shoring equipment including engineering design, tabulated data, engineered designs, and safety issues.

1926.652(c)(3)—Option (3)—Designs using other tabulated data.

1926.652(c)(3)(i)—Designs of support systems, shield systems, or other protective systems shall be selected from and be in accordance with tabulated data, such as tables and charts.

1926.652(c)(3)(ii)—The tabulated data shall be in written form and include all of the following:

1926.652(c)(3)(ii)(A)—Identification of the parameters that affect the selection of a protective system drawn from such data;

1926.652(c)(3)(ii)(B)—Identification of the limits of use of the data;

1926.652(c)(3)(ii)(C)—Explanatory information as may be necessary to aid the user in making a correct selection of a protective system from the data.

1926.652(c)(3)(iii)—At least one copy of the tabulated data, which identifies the registered professional engineer who approved the data, shall be maintained at the jobsite during construction of the protective system. After that time the data may be stored off the jobsite, but a copy of the data shall be made available to the Secretary upon request.

Commentary

This option would be used when a standardized shoring system utilizing a combination of shoring elements is developed. The elements may be a combination of manufactured equipment and standards structural elements such as H-pile, sheet pile, plate, and timber. The plan is usually sponsored by a contractor in order to use his or her own standard shoring system in many different locations; however, shoring suppliers or manufacturers of materials that go into the system, such as H-piles, might sponsor the plan to stimulate sales. The engineer would be responsible for the accuracy of the design, and the contractor or supplier would be responsible for the quality and condition of the materials.

Important elements of this system are

- Definition of the region where it can be used
- Basis for selection of options, usually a soil classification system; however, it could also be more specific about surcharge and buried infrastructure limitations and distances, seasonal variations, length of time the excavation will be open, etc.
- Clear depiction of the elements in the data
- Step-through installation and removal procedure

- Complete tabulation and current stamp of engineer who developed the data

The data serve three specific functions:

1. They enable the user to select and utilize the shoring system.
2. They enable reviewing and inspection personnel to confirm that the system is properly designed and constructed.
3. They enable the courts to determine liability in the event of an accident.

1926.652(c)(4)—Option (4)—Design by a registered professional engineer.

1926.652(c)(4)(i)—Support systems, shield systems, and other protective systems not utilizing Option 1, Option 2 or Option 3, above, shall be approved by a registered professional engineer..

1926.652(c)(4)(ii)—Designs shall be in written form and shall include the following:

1926.652(c)(4)(ii)(A)—A plan indicating the sizes, types, and configurations of the materials to be used in the protective system; and

1926.652(c)(4)(ii)(B)—The identity of the registered professional engineer approving the design.

1926.652(c)(4)(iii)—At least one copy of the design shall be maintained at the jobsite during construction of the protective system. After that time, the design may be stored off the jobsite, but a copy of the design shall be made available to the Secretary upon request.

Commentary

Every shoring system that does not have tabulated data falls into the category of design by a registered engineer, specifically sheet pile, pile and plate, timber shoring that is not installed under tabulation in Appendix C, and manufactured equipment that is not used in accordance with tabulated data.

This is essentially the same as option 3 except for two important differences:

1. The plan is site-specific.
2. The user does not select a plan from an array of options.

Important elements of the plan are

- Definition of the location and extent where it applies
- Depiction of existing site conditions
- Clear depiction of the shoring plan
- Step-through installation and removal procedure

- Complete drawings and current stamp of engineer who developed the plan

The plan serves three specific functions:

1. It is a plan that can be used for shoring at a defined location.
2. It enables reviewing and inspection personnel to confirm that the system is properly designed and constructed.
3. It enables the courts to determine liability in the event of an accident.

1926.652(d)—Materials and equipment.

1926.652(d)(1)—Materials and equipment used for protective systems shall be free from damage or defects that might impair their proper function.

Commentary

New materials and equipment are rarely used and most commonly used in shoring work. During installation of this material it should be checked for quality and damage before installing it. If there is any doubt about adequacy, it should be verified by an engineer.

Damage to in-place shoring systems should be checked by the competent persons during daily inspections.

1926.652(d)(2)—Manufactured materials and equipment used for protective systems shall be used and maintained in a manner that is consistent with the recommendations of the manufacturer, and in a manner that will prevent employee exposure to hazards.

Commentary

The contractor will be found liable for accidents resulting from mishandling, abuse, and poor rigging of manufactured shoring equipment. The manufacturer will be liable for defects and failure to supply adequate safe handling and use information.

1926.652(d)(3)—When material or equipment that is used for protective systems is damaged, a competent person shall examine the material or equipment and evaluate its suitability for continued use. If the competent person cannot assure the material or equipment is able to support the intended loads or is otherwise suitable for safe use, then such material or equipment shall be removed from service, and shall be evaluated and approved by a registered professional engineer before being returned to service.

Commentary

Every incidence of damage of in-place shoring systems should be cleared.

The competent person must be familiar with the system and able to identify the damage. A person with experience in evaluating structural capacity, a *qualified person*, should determine if it is suitable for continued use.

> 1926.652(e)—Installation and removal of support
>
> 1926.652(e)(1)—General.
>
> 1926.652(e)(1)(i)—Members of support systems shall be securely connected together to prevent sliding, falling, kickouts, or other predictable failure.
>
> 1926.652(e)(1)(ii)—Support systems shall be installed and removed in a manner that protects employees from cave-ins, structural collapses, or from being struck by members of the support system.

Commentary

Workers are never allowed to work in unprotected areas. Safe installation includes using proper rigging and lifting equipment.

> 1926.652(e)(1)(iii)—Individual members of support systems shall not be subjected to loads exceeding those which those members were designed to withstand.
>
> 1926.652(e)(1)(iv)—Before temporary removal of individual members begins, additional precautions shall be taken to ensure the safety of employees, such as installing other structural members to carry the loads imposed on the support system.
>
> 1926.652(e)(1)(v)—Removal shall begin at, and progress from, the bottom of the excavation. Members shall be released slowly so as to note any indication of possible failure of the remaining members of the structure or possible cave-in of the sides of the excavation.

Commentary

Shoring systems develop loading over time. Soil arching also develops over time. At the time of removal the loads are the greatest; given this coupled with the fact that shoring element removal can be similar to removing columns that support arches, there is a high potential for cave-in during the removal process. An installation and removal process should be thought through prior to installing manufactured shoring, and the installation and removal process should be included in engineered designs.

> 1926.652(e)(1)(vi)—Backfilling shall progress together with the removal of support systems from excavations.

Commentary

Removal of lower struts and wales shifts part of the load to the struts and wales above. Also a collapse of soil in the unshored lower portion of an excavation can cause failure of the entire system above it. Backfill shortly after removal mitigates these effects.

1926.652(e)(2)—Additional requirements for support systems for trench excavations.

1926.652(e)(2)(i)—Excavation of material to a level no greater than 2 feet (.61 m) below the bottom of the members of a support system shall be permitted, but only if the system is designed to resist the forces calculated for the full depth of the trench, and there are no indications while the trench is open of a possible loss of soil from behind or below the bottom of the support system.

Commentary

These requirements also apply to engineered designs under shoring options 3 and 4, design by a registered engineer. The 2-ft rule is hard and fast with OSHA and shows up in several places, timber shoring, trench jack data, and below as it applies to shields. It is also common to specify on plans for H-pile, wale and lagging, and sheet pile and wale plans that the wales must be installed before excavating more than 2 ft below their location. The concern is that shoring that is undermined from below can create a failure in the shoring above. Also an exposure to a bank of soil especially deep can inundate a worker if it collapses.

1926.652(e)(2)(ii)—Installation of a support system shall be closely coordinated with the excavation of trenches.

Commentary

The more time that passes while soil is unsupported, the higher the possibility of collapse and the higher the load will be on shoring. Trench collapse near a line of shoring can extend into the shored area, or the collapse can spill into the shored area.

1926.652(f)—Sloping and benching systems. Employees shall not be permitted to work on the faces of sloped or benched excavations at levels above other employees except when employees at the lower levels are adequately protected from the hazard of falling, rolling, or sliding material or equipment.

1926.652(g)—Shield systems

1926.652(g)(1)—General.

1926.652(g)(1)(i)—Shield systems shall not be subjected to loads exceeding those which the system was designed to withstand.

Commentary

Shoring shield tabulated data have a load rating, usually in pounds per square foot, for this purpose in addition to allowable depth. Allowable depth is usually based on the psf rating by subtracting

72 psf and then dividing by the soil type loading (A-25, B-45, C-80). If surcharge loads are greater than 72 psf, the depth must be reduced to make the difference.

1926.652(g)(1)(ii)—Shields shall be installed in a manner to restrict lateral or other hazardous movement of the shield in the event of the application of sudden lateral loads.

Commentary
The shield can tip or slide sideways in the event of a collapse. Normally 6 in on each side is the maximum that should be allowed. If there is a larger void, soil should be replaced along the sides to prevent movement.

1926.652(g)(1)(iii)—Employees shall be protected from the hazard of cave-ins when entering or exiting the areas protected by shields.

1926.652(g)(1)(iv)—Employees shall not be allowed in shields when shields are being installed, removed, or moved vertically.

1926.652(g)(2)—Additional requirement for shield systems used in trench excavations. Excavations of earth material to a level not greater than 2 feet (.61 m) below the bottom of a shield shall be permitted, but only if the shield is designed to resist the forces calculated for the full depth of the trench, and there are no indications while the trench is open of a possible loss of soil from behind or below the bottom of the shield.

Commentary
See 1926.652(e)(2)—Additional requirements for support systems for trench excavations, above.

References

U.S. Department of Labor., *29 CFR 1926, Appendix A Soil Classification*, Occupational Safety & Health Administration, Washington, January 29, 2008, www.osha.gov.

APPENDIX 2

OSHA Appendix A, Soil Classification and Commentary

AP2.1 OSHA Appendix A Soil Classification

The following is federal OSHA 1926 Subpart P Appendix A with commentary. The commentary is strictly the author's interpretation and not to be interpreted as OSHA policy or consensus of opinion from any group or organization. The comments are intended to provide a clearer understanding of the OSHA text and to highlight topics that should be emphasized in competent person training seminars.

> **1926 Appendix A(a)** Scope and application
>
> **1926 Appendix A(a)(1) Scope.**
>
> This appendix describes a method of classifying soil and rock deposits based on site and environmental conditions, and on the structure and composition of the earth deposits. The appendix contains definitions, sets forth requirements, and describes acceptable visual and manual tests for use in classifying soils.

Commentary

In 1989 OSHA promulgated this current 29 CFR 1926 Subpart P Appendix A, Soil Classification System, that we use today. In developing standard practice in the field, it was evident that a method had to be adopted for determining soil conditions in the field as they are encountered. The basis for classification had to be clear, be easy to achieve, and not bring confusion to the courtroom during litigation. The basis of slope selection and the use of manufactured shoring equipment are at the heart of this soil type selection system. The use of this system does not require an education in geotechnical engineering, is easy to learn, can be performed on the spot, and does not necessarily require laboratory testing. It is self-contained in the sense that it provides definitions, requirements, and acceptable methods of field testing to meet the requirements. Everyone today with experience in excavation work is familiar with the basis of the system, classification of soil into type A, B, and C in a descending hierarchy of soil stability. To select any standard practice protective system, the soil must first be classified in accordance with this system.

1926 Appendix A(a)(2) **Application.**

This appendix applies when a sloping or benching system is designed in accordance with the requirements set forth in 1926.652(b)(2) as a method of protection for employees from cave-ins. This appendix also applies when timber shoring for excavations is designed as a method of protection from cave-ins in accordance with appendix C to subpart P of part 1926, and when aluminum hydraulic shoring is designed in accordance with appendix D. This appendix also applies if other protective systems are designed and selected for use from data prepared in accordance with the requirements set forth in 1926.652(c), and the use of the data is predicated on the use of the soil classification system set forth in this appendix.

Commentary

The stability requirements for sloping and strength requirements for shoring equipment are based on the soil type in which they are placed. Prior to use of any prescriptive worker protection system being utilized by the employer and her or his competent person, the soil must first be classified in accordance with Appendix A. The requirement is further stated in the following OSHA sections:

- 1926.652(b)(2)—Determination of slopes and configurations using OSHA Appendix B
- 1926.652(c)(1) Appendix C, Timber Shoring
- 1926.652(c)(1) Appendix D, Aluminum Hydraulic Shoring
- 1926.652(c)(2) Designs using Manufacturer's Tabulated Data

Also look in tabulated data issued by manufacturers of shoring equipment and engineers for requirement to first identify soil in accordance with Appendix A.

1926 Appendix A (b) Definitions.

The definitions and examples given below are based on, in whole or in part, the following; American Society for Testing Materials (ASTM) Standards D653-85 and D2488; The Unified Soils Classification System; The U.S. Department of Agriculture (USDA) Textural Classification Scheme; and The National Bureau of Standards Report BSS-121.

Commentary

In fact Appendix A was derived in part from these references; however, they should not be seen here as incorporated by reference and made a part of Appendix A. This soil classification system was intended to stand alone so that the person in the field could use and make decisions based on it without looking further than the document. From a legal standpoint, a judge and jury should judge whether the person who used it conformed to the requirements of Appendix A as it is written. A competent person does not necessarily have the education or experience to interpret the referenced standards and classification systems.

"Cemented soil" means a soil in which the particles are held together by a chemical agent, such as calcium carbonate, such that a hand-size sample cannot be crushed into powder or individual soil particles by finger pressure.

"Cohesive soil" means clay (fine grained soil), or soil with a high clay content, which has cohesive strength. Cohesive soil does not crumble, can be excavated with vertical side slopes, and is plastic when moist. Cohesive soil is hard to break up when dry, and exhibits significant cohesion when submerged. Cohesive soils include clayey silt, sandy clay, silty clay, clay and organic clay.

Commentary

See Article 5.3, Cohesive Soils, for more in-depth description.

Plastic means that something can be changed in shape, can be remolded, and can still look like the same material. Chewing gum is a good example of a plastic material. With cohesive soils think of modeling clay that can be molded into shapes. If it is too dry or has too much sand in it, the sample will crack and crumble.

Clay particles are flat and can only be seen individually through a microscope. They have an electronic interaction between the particles and also with water that is contained in the clay mixture. Silt particles are very fine particles of sand and soil that have no interparticle interaction. When silts and clays are mixed, the portion of clay particles causes cohesion and cohesive soil attributes.

Clayey silt has more silt than clay, and silty clay has more clay particles than silt. Organic clay has decomposed vegetation mixed in with clay.

Only cohesive soil can be type A.

"Dry soil" means soil that does not exhibit visible signs of moisture content.

"Fissured" means a soil material that has a tendency to break along definite planes of fracture with little resistance, or a material that exhibits open cracks, such as tension cracks, in an exposed surface.

"Granular soil" means gravel, sand, or silt (coarse grained soil) with little or no clay content. Granular soil has no cohesive strength. Some moist granular soils exhibit apparent cohesion. Granular soil cannot be molded when moist and crumbles easily when dry.

Commentary

Granular soil cannot be type A.

> **"Layered system"** means two or more distinctly different soil or rock types arranged in layers. Micaceous seams or weakened planes in rock or shale are considered layered.

Commentary

Excavations almost always consist of layered soils. The least stable soil layer controls the sloping and benching requirement for the soils above it.

Layers can be any thickness. In the case of layered soils that are flat, the thickness matters because a thick layer of soft soil can squeeze out or slough while a very thin layer would have much less tendency to do so. In the case of shored trenches, flat thin layers have a minimal lateral loading effect on shoring equipment compared to thicker layers. The thick or predominant soils will have the greatest effect on shoring loads and should be taken into consideration when selecting soil type for trench jack or shield use. For example, stiff clays with a thin 1-in thick layer of soft clay in between would exert a load on the shoring comparable to that of stiff clays.

In layered systems where the layers are sloping, generally greater than 4:1, there is a tendency for one layer to slide over another no matter how thick the layer. In this case the worse type of soil condition should be used for all soils above.

Micaceous seams are formed from mica and can separate or slide as sheets of paper do.

> **"Moist soil"** means a condition in which a soil looks and feels damp. Moist cohesive soil can easily be shaped into a ball and rolled into small diameter threads before crumbling. Moist granular soil that contains some cohesive material will exhibit signs of cohesion between particles.
>
> **"Plastic"** means a property of a soil which allows the soil to be deformed or molded without cracking, or appreciable volume change.
>
> **"Saturated soil"** means a soil in which the voids are filled with water. Saturation does not require flow. Saturation, or near saturation, is necessary for the proper use of instruments such as a pocket penetrometer or sheer vane.

Commentary

Cohesive clay soils have adsorbed water that is part of the bond force; see Article 6.2 for more on adsorption. If the clay is dried out to the point where the adsorbed water is not present, then the pocket penetrometer, thumb test, and sheer vane test are not correct and should not be used. There is also water that can flow or be squeezed out of the clay. The tests are still correct if this water is not present.

"Soil classification system" means, for the purpose of this subpart, a method of categorizing soil and rock deposits in a hierarchy of Stable Rock, Type A, Type B, and Type C, in decreasing order of stability. The categories are determined based on an analysis of the properties and performance characteristics of the deposits and the characteristics of the deposits and the environmental conditions of exposure.

Commentary

Manufacturers of shoring equipment have promoted and used in their tabulated data an intermediate type C-60 soil classification that in the hierarchy falls between type B and type C soils; see Sec. 5.8.1 and Table 5.7. This category is generally accepted within the shoring industry by contractors and those reviewing worker protection plans. Federal OSHA and state OSHA programs do not appear to be opposed to this; however, this appendix and the excavation safety standards do not recognize or address the type C-60 soil classification.

From a technical and legal standpoint, manufacturers of shoring equipment are entitled to utilize any type of soil classification system they want in their tabulated data as long as it is clear and understandable to those using it. The manufacturer and the user are responsible for getting it right, not OSHA.

The last sentence in the "Soil classification system" definition is confusing and could be more clearly written as:
The categories are determined based on an analysis of:

a. the properties and performance characteristics of the deposits; and
b. the characteristics of the deposits and the environmental conditions of exposure.

Properties are associated with possession. Basic properties of soil deposits are

- The unique property of rock is that it is solid, can be smashed but not smeared or molded.
- The unique property of cohesive soil is that it sticks together, has an electronic bond, and can be molded without falling apart.
- The unique property of noncohesive soil is that it does not stick together. It stands up during excavation due to friction between particles.

Characteristics are associated with behavior. Basic characteristics of soil deposits are

- A characteristic of rock is that it is solid like a chunk of concrete although it has joints and cracks in it. If the joints and cracks are oriented at steep angles, the rock can slide or tip over when lateral support is removed. If this is the case, then the rock is not stable. There needs to be some sort of factor of safety between stable and nonstable.

- Characteristic of cohesive soils is that they stick together. When lateral support is removed, they slough when soft and develop tension cracks at the surface prior to failure when they are not soft. They are more stable than noncohesive soils.
- Characteristic of noncohesive soils is that when lateral support is removed, they ravel, slide in, and settle at their angle of internal friction. They are less stable than cohesive soils.

Characteristics of the deposits and the environmental conditions of exposure refer to how the deposits behave over time and exposure to weather events such as

- Over time excavated soil and rock banks exposed to heat and air can dry out and become less stable.
- Over time excavated soil banks exposed to rain, a rising water table from rivers, and irrigation will cause them to be less stable.

"Stable rock" means natural solid mineral matter that can be excavated with vertical sides and remain intact while exposed.

"Submerged soil" means soil which is underwater or is free seeping.

"Type A" means cohesive soils with an unconfined, compressive strength of 1.5 ton per square foot (tsf) (144 kPa) or greater. Examples of cohesive soils are: clay, silty clay, sandy clay, clay loam and, in some cases, silty clay loam and sandy clay loam. Cemented soils such as caliche and hardpan are also considered Type A. However, no soil is Type A if:

(i) The soil is fissured; or

(ii) The soil is subject to vibration from heavy traffic, pile driving, or similar effects; or

(iii) The soil has been previously disturbed; or

(iv) The soil is part of a sloped, layered system where the layers dip into the excavation on a slope of four horizontal to one vertical (4H:1V) or greater; or

(v) The material is subject to other factors that would require it to be classified as a less stable material.

Commentary

The author has heard often that there cannot be type A soil because it is impossible for the requirements and exclusions to exist simultaneously. There are always vibrations on a construction site, and there are always factors that would cause it to be less stable. Many people have the impression that in typical governmental fashion they gave it and took it away at the same time. A deeper look at this shows that this is not the case, and the unintended consequence and impression should be avoided. There is plenty of type A soil out there, and the cases where there are exclusions, it is for good reason.

First it is important to understand the reasons for having type A soil because if there is no advantage, it is easier and best to use the type B classification. Advantages to type A soil are as follows:

- With open cut protective systems OSHA Appendix B allows seven different sloping configurations in type A soil where type B soils have four options and type C soils have two options.
- In type A soils slopes can be ¾:1 and in a short-term case ½:1
- Trench jacks can be spaced at maximum 8 ft on center to deeper depths than with B and C soils.
- The apparent soil loading on support systems is less than with B and C soils.

All these advantages translate into more cost-effective worker protection systems, faster production, less impact on surrounding streets and buried facilities, and better quality in the final product, especially on excavation-intensive projects. For these reasons it is important to understand the type A soil classification.

Type A soils are easy to understand and classify. Here it is step by step:

> **"Type A"** means cohesive soils with an unconfined, compressive strength of 1.5 tons per square foot (tsf) (144 kPa) or greater.

Type A soil must be cohesive and cannot be noncohesive. Proof of cohesion is found visually in the following:

> **1926 Appendix A (d)(1) Visual tests (ii)** Observe soil as it is excavated. Soil that remains in clumps when excavated is cohesive. Soil that breaks up easily and does not stay in clumps is granular.

Proof of cohesion is found manually in the following:

> **1926 Appendix A (d)(2) Manual tests (i)** Plasticity. Mold a moist or wet sample of soil into a ball and attempt to roll it into threads as thin as 1/8-inch in diameter. Cohesive material can be successfully rolled into threads without crumbling. For example, if at least a two inch (50 mm) length of 1/8-inch thread can be held on one end without tearing, the soil is cohesive.

Type A soil must have an unconfined compressive strength greater than 1.5 tons per square foot (tsf) (Fig. AP2.1). This is the same type of test that is performed on a concrete cylinder to determine concrete strength. Concrete strength tests are reported in pounds per square inch (psi).

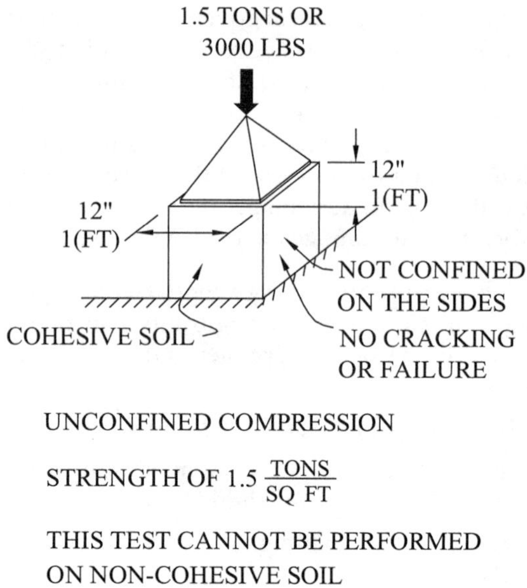

Figure AP2.1 Unconfined compressive strength, *qu*.

Proof of unconfined compressive strength is found manually in the field in the following:

> **1926 Appendix A (d)(2) Manual tests. (iii) Thumb penetration.** The thumb penetration test can be used to estimate the unconfined compressive strength of cohesive soils. (This test is based on the thumb penetration test described in American Society for Testing and Materials (ASTM) Standard designation D2488—"Standard Recommended Practice for Description of Soils (Visual—Manual Procedure).") **Type A soils with an unconfined compressive strength of 1.5 tsf can be readily indented by the thumb; however, they can be penetrated by the thumb only with very great effort. …**
>
> Examples of cohesive soils are: clay, silty clay, sandy clay, clay loam and, in some cases, silty clay loam and sandy clay loam.

These are Uniform Soil Classification System terms and are used in the description portion of soils report bore logs and other places in the report. Although use of the soils report is not a requirement of this classification system, it can be used to back up your conclusion or indicate on the boring log where to look in the field for type A soils.

> Cemented soils such as caliche and hardpan are also considered Type A.

Caliche is a hardened deposit of calcium carbonate, a material used in cement. The calcium carbonate cements together other materials, including gravel, sand, clay, and silt. It is found in semiarid, desert soils. Caliche is also known as hardpan. Caliche is generally light-colored but can range from white to light pink to reddish brown, depending on the impurities present. It is generally found on or near the surface, but it can be found in deeper subsoil deposits as well. The layers can vary from a few inches to feet thick, and multiple layers can exist in a single location.

Hardpan is a general term for a dense layer of soil, residing usually below the uppermost topsoil layer. There are different types of hardpan, all sharing the general characteristic of being a distinct soil layer that is largely impervious to water. Some hardpans are formed by deposits in the soil that fuse and bind the soil particles.

There are no visual or manual tests for caliche and hardpan listed in Appendix A; however, the definition suggests that visually they should be white to light pink to reddish brown. A manual test for cemented soils is to pour hydrochloric acid on a sample. If there is calcium carbonate in the sample, it will give off fumes.

As a practical matter, caliche and hardpan tend to behave more as rock and would be between rock and type A soil in a hierarchy.

However, no soil is Type A if:

(i) The soil is fissured;

Commentary

A fissure is a crack. The same as cracks in rock, cohesive soils can have cracks or lines of discontinuity. Geological fissures are created over time by events in the history of the deposit such as droughts, removal of overburden soil by erosion, removal of lateral support by rivers and wind erosion, and seismicity. Only cohesive soils that are affected by these events are fissured. After the cracks are generated, weathering from water and chemical intrusion makes the fissures permanent. Fissures become less prominent as depth in the cohesive layer increases. The presence of these types of fissures should be reported in the soils report.

For the purpose of this soil classification system, any crack in the soil is considered a fissure. Cohesive soils when excavated and subsequently exposed to air dry out and form cracks over time. Vertical banks form cracks at the surface due to drying and removal of the lateral support. Movement of heavy equipment near the excavation can cause movement and cracking. The persons planning the excavation work should consider the effect of these events in their overall planning, and the competent person at the site should look for evidence of fissures created by these events on a daily basis.

The reason that fissures are important to soil stability is that shear strength is dependent on having a continuous cohesive mixture and is reduced with the presence of fissures. Trench walls in geologically fissured soils look normal when first excavated but can quickly spall or completely fall within the first few minutes of being opened up.

The failures are rapid and unpredictable. Trench jacks can prevent bank failure; however, without the use of plywood they will not prevent spall failures. Drying and stress cracking due to construction activities occur over days and weeks, but the failure is still rapid with unpredictable timing.

A visual test for fissures is the following:

> **1926 Appendix A (d)(1) Visual tests (iii)** Observe the side of the opened excavation and the surface area adjacent to the excavation. Crack-like openings such as tension cracks could indicate fissured material. If chunks of soil spall off a vertical side, the soil could be fissured. Small spalls are evidence of moving ground and are indications of potentially hazardous situations.

Manual test for fissures are as follows:

> **1926 Appendix A (d)(2) Manual tests (ii)** Dry strength. If the soil is dry and crumbles on its own or with moderate pressure into individual grains or fine powder, it is granular (any combination of gravel, sand, or silt). If the soil is dry and falls into clumps which break up into smaller clumps, but the smaller clumps can only be broken up with difficulty, it may be clay in any combination with gravel, sand or silt. **If the dry soil breaks into clumps which do not break up into small clumps and which can only be broken with difficulty, and there is no visual indication the soil is fissured, the soil may be considered unfissured.**

And

> (v) Drying test. The basic purpose of the drying test is to differentiate between cohesive material with fissures, unfissured cohesive material, and granular material. The procedure for the drying test involves drying a sample of soil that is approximately one inch thick (2.54 cm) and six inches (15.24 cm) in diameter until it is thoroughly dry:
>
> (A) If the sample develops cracks as it dries, significant fissures are indicated.
>
> (B) Samples that dry without cracking are to be broken by hand. If considerable force is necessary to break a sample, the soil has significant cohesive material content. The soil can be classified as an unfissured cohesive material and the unconfined compressive strength should be determined.
>
> (C) If a sample breaks easily by hand, it is either a fissured cohesive material or a granular material. To distinguish between the two, pulverize the dried clumps of the sample by hand or by stepping on them. If the clumps do not pulverize easily, the material is cohesive with fissures. If they pulverize easily into very small fragments, the material is granular.

Note that only one manual test is required. Both of these tests are very similar, but the first one is easier to use.

> However, no soil is Type A if:
>
> (ii) The soil is subject to vibration from heavy traffic, pile driving, or similar effects;

Commentary

All soil on a construction project is subject to vibrations from surrounding activities. The only way for this statement to be an exclusion and not a prima facie eliminator of type A soil is to have to add the requirement that the "vibrations affect the excavation." The author's conclusion is that the intent of the statement is for vibrations that cause the excavation to ravel or slough. This conclusion is backed up further in the language of the qualifying visual test with the statement *sources of vibrations that may effect the stability of the excavation face*.

> **1926 Appendix A (d)(1) Visual test (vii)** Observe the area adjacent to the excavation and the area within the excavation for sources of vibration that may affect the stability of the excavation face.

Vibrations that do not affect the stability of the excavation face do not count. Observations of sloughing and cracking in cohesive soil excavation faces as a direct effect of local vibrations would cause the down rating of the soil from A to B.

In the author's experience the only time that vibrations have had an effect on cohesive soil stability is in soft clays, unconfined compression strength qu < 1000 psf, which is already type B soil by virtue of its strength. Vibrations can cause fissured soil to spall. Vibrations can cause noncohesive soils to ravel and even cause slope failure.

> However, no soil is Type A if:
>
> (iii) The soil has been previously disturbed;

Commentary

Previously disturbed soil is found primarily in soil that has been excavated and replaced. It is found in

- Existing pipeline and utility excavations
- Road and overpass fills
- Landfills

Naturally occurring previously disturbed soils would be

- Earthquake fissures
- Landslides
- Riverbeds and alluvial fans

Previously disturbed soils are considered less stable because they are not natural predictable soil deposits, there is no guarantee of degree of compaction, and there are usually subsurface facilities in the vicinity. A visual test for previously disturbed soil is

> **1926 Appendix A (d)(1) Visual tests (iv)** Observe the area adjacent to the excavation and the excavation itself for evidence of existing utility and other underground structures, and to identify previously disturbed soil.

Visual indications on the surface should be identified at the start of the project during the existing buried facilities investigation.

During excavation, if it is not certain that they do not exist, a competent person (this can be the excavator operator or grade checker) should always be there to identify changed soil characteristics such as different color, loose compaction, imported pipeline bedding and backfill material, and changed moisture condition due to water collection in permeable fill materials. Cohesive clay soils are generally impermeable.

There is no manual test for previously disturbed soils set out in Appendix A. Soils reports and bore logs will usually report fill areas, landslides, and earthquake fissures. The soil classification change only has to be made in the area where the previously disturbed soils exist. A pipeline excavation that is being sloped at ¾:1 would need to be sloped at 1:1 in the area of an intersecting utility and would change back to the original sloping configuration after the line was passed.

> However, no soil is Type A if:
>
> (iv) The soil is part of a sloped, layered system where the layers dip into the excavation on a slope of four horizontal to one vertical (4H:1V) or greater;

Commentary

This condition is usually found in hilly terrain. The danger is that one layer can slide over the top of another and start a slide or close a trench (Fig. AP2.2). A visual test for this is as follows:

> 1926 Appendix A (d)(1) Visual test (v) Observed the opened side of the excavation to identify layered systems. Examine layered systems to identify if the layers slope toward the excavation. Estimate the degree of slope of the layers.

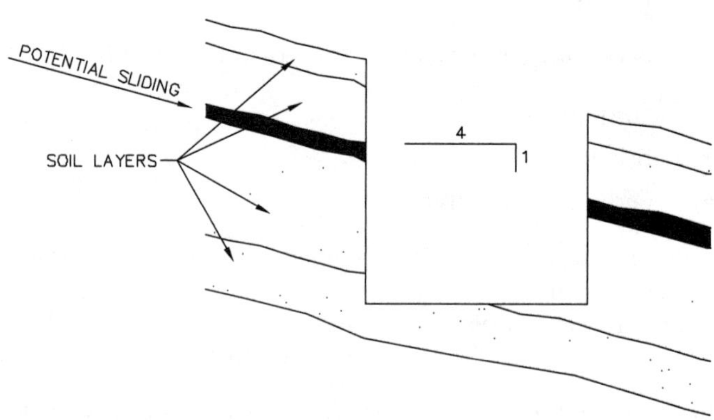

FIGURE AP2.2 Sloped layered system.

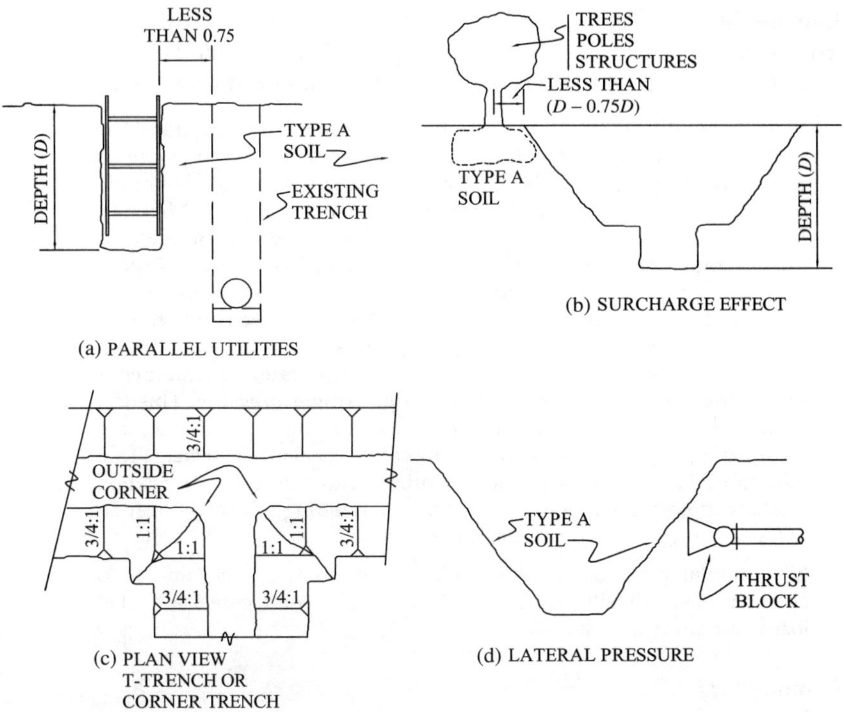

FIGURE AP2.3 Conditions affecting stability of type A soil.

A manual test for this would be to measure the slope of the layers. The horizontal distance divided by the vertical distance should be greater than 4.

However, no soil is Type A if:

(v) The material is subject to other factors that would require it to be classified as a less stable material.

Commentary

Some examples that would cause a type A soil to be less stable are shown in Figure AP2.3. Whether the trench is sloped or shored, these conditions still affect the stability of the soils. These conditions are also present with type B and C soils, and protective systems for these conditions should be thoroughly planned before starting the excavation. The competent person should also be on the lookout for conditions other than what is shown here that affect soil stability.

"Type B" means:

(i) Cohesive soil with an unconfined compressive strength greater than 0.5 tsf (48 kPa) but less than 1.5 tsf (144 kPa); or

Commentary

The deciding factor for type B cohesive soils is unconfined compressive strength *qu* (Fig AP2.1). The defining manual test is as follows:

> **1926 Appendix A (d)(2) Manual test (iii)** Thumb penetration. The thumb penetration test can be used to estimate the unconfined compressive strength of cohesive soils. (This test is based on the thumb penetration test described in American Society for Testing and Materials (ASTM) Standard designation D2488—"Standard Recommended Practice for Description of Soils (Visual—Manual Procedure).") **Type A soils with an unconfined compressive strength of 1.5 tsf can be readily indented by the thumb; however, they can be penetrated by the thumb only with very great effort. Type C soils with an unconfined compressive strength of 0.5 tsf can be easily penetrated several inches by the thumb, and can be molded by light finger pressure.** This test should be conducted on an undisturbed soil sample, such as a large clump of spoil, as soon as practicable after excavation to keep to a minimum the effects of exposure to drying influences. If the excavation is later exposed to wetting influences (rain, flooding), the classification of the soil must be changed accordingly.
>
> (ii) Granular cohesionless soils including: angular gravel (similar to crushed rock), silt, silt loam, sandy loam and, in some cases, silty clay loam and sandy clay loam.

Commentary

Cohesionless soils have more than 50 percent granular particles and less than 50 percent clay particles. The key to cohesionless granular soils is angularity. Marbles will not stack in a pile, but crush them and they can be poured into a pile. This is the material characteristic that divides type B and type C cohesionless soils.

Angular gravel includes quarried rock that is crushed such as aggregate base rock in roads, aggregate for concrete, crushed drain rock, chip seal rock for roads, crushed and natural angular sands for pipe bedding, broken rock for riprap, and large rock for jetties.

In the geological setting rock, gravel, and sand that have angular particles and are not rounded can be considered type B soils. Any type of jointed rock, shale, decomposed rock, and sandstone that is excavated comes out with angular particles. Be aware that jointed rock with bedding planes at steep angles will not necessarily stand up at typical 1:1 type B soil slopes.

Rounded particles are generally the product of the natural forces of gravity, water, and wind. River, streambed, and alluvial fan soils have been tumbled and smoothed due to their journey from the mountains toward the sea. Riverbeds and streambeds can be found anywhere since they have meandered through time and are often found buried in unpredictable places by other soil types. This is one of the reasons that a pipeline excavation production in cohesionless soils can go from hundreds of feet per day to zero or negative production in a matter of minutes. As the productivity goes down, the risk of soil collapse on workers goes up exponentially. The excavator operator and competent person should always be on the lookout for this type of changing condition.

Cobbles or "potato rocks" and pea rock would be another example of rounded particles.

Loam is a mixture of sand, silt, and clay with the sand and silt being predominant. In some proportions these soils can still exhibit plasticity and hold water.

A visual test for noncohesive soils is this:

> **1926 Appendix A (d)(1) Visual tests (i)** Observe samples of soil that are excavated and soil in the sides of the excavation. Estimate the range of particle sizes and the relative amounts of the particle sizes. Soil that is primarily composed of fine-grained material is cohesive material. Soil composed primarily of coarse-grained sand or gravel is granular material.

Fine-grained is generally referred to as soil that passes through a No. 20 (0.074-mm) sieve, individually undetectable by the eye. Clay is often used as another term for this; however, there are quartz grains and other particle grains that meet this criterion but do not have the characteristics of clay.

In the field if more than 50 percent of the soil by weight is judged to consist of grains that can be distinguished separately, then it is considered to be coarse-grained.

A visual test for angularity is not listed in the appendix; however, angularity is obvious in large sand and gravel deposits.

A manual test for angularity is also not listed in the appendix. Fine sand particles tend to be wind blown and generally round; however, if they feel abrasive when rubbed between the fingers, they could be considered angular. Also looking at them through a magnifying glass could help determine angularity. Angularity of fine particles of soil can be important because the behavior of the finest particles in a mixture of soil generally determines the characteristics of the soil. For example, the behavior of concrete is determined by the finest particles in the mixture, cement. Accordingly if there are clay particles present in noncohesive soils, they will have a strengthening affect.

Another manual test that is not listed in the appendix is to place a soil sample in a jar of water and completely shake it up. The different fractions of the soil will settle with the largest particles on the bottom and the lightest clay particles on the top. An estimate of the clay content is helpful in predicting the strength of the soil. If the clay content is greater than 50 percent, the soil is cohesive.

> (iii) Previously disturbed soils except those which would otherwise be classed as Type C soil.
>
> (iv) Soil that meets the unconfined compressive strength or cementation requirements for Type A, but is fissured or subject to vibration; or

Commentary

In the author's opinion this should read "… or subject to vibrations that affect the excavation."

(v) Dry rock that is not stable; or

(vi) Material that is part of a sloped, layered system where the layers dip into the excavation on a slope less steep than four horizontal to one vertical (4H:1V), but only if the material would otherwise be classified as Type B.

Commentary

There should be another description for type B soil here as follows:

(vii) Type A material that is subject to other factors that would require it to be classified as a less stable material.

These factors should again be evaluated. For example, outside trench corners, thrust blocks, and other lateral pressure forces close to the edge of a sloped or shored type B excavation would still have an effect on the stability of the excavation.

"Type C" means:

(i) Cohesive soil with an unconfined compressive strength of 0.5 tsf (48 kPa) or less; or

Commentary

The manual test for this is as follows:

1926 Appendix A (d)(2) Manual test (iii) Thumb penetration. The thumb penetration test can be used to estimate the unconfined compressive strength of cohesive soils. (This test is based on the thumb penetration test described in American Society for Testing and Materials (ASTM) Standard designation D2488—"Standard Recommended Practice for Description of Soils (Visual—Manual Procedure).") …**Type C soils with an unconfined compressive strength of 0.5 tsf can be easily penetrated several inches by the thumb, and can be molded by light finger pressure….**

(ii) Granular soils including gravel, sand, and loamy sand;

Commentary

Rounded particles are a determining physical property of type C soils. See discussion above regarding particle angularity.

(iii) Submerged soil or soil from which water is freely seeping;

Commentary

Submerged soil is obvious, but freely seeping is not quite so clear. The author has many times heard the statement that if there is water present in the excavation, it must be classified as type C soil. The presence of water is not sufficient evidence that a soil must be classified as type C.

Water can be present in the bottom of a trench for other reasons besides the fact that the soil is saturated with it, such as if it originated from below the bottom, flowed down the bottom of the trench from another location, or was drenched from surface level water sources such as rain and drainage ditches. The visual test for water is a little bit clearer on this:

> **1926 Appendix A (d)(1) Visual tests (vi)** Observe the area adjacent to the excavation and the sides of the opened excavation for evidence of surface water, water seeping from the sides of the excavation, or the location of the level of the water table.

The key words are "…water seeping from the sides of the excavation, or the location of the water table." If the water level is pumped down or just quits running from the sides of the excavation, it can then be classified as type A or B depending on other factors.

The presence of water is not to be taken lightly. Underwater the buoyant force of the water causes rock, sand, and gravel to be lighter, hence the resistant friction forces are less. Above water, flow through rock cracks and soil layers lubricates and causes sliding. Inundated noncohesive soils that are rapidly pumped down will develop seepage forces that will cause them to flow into the excavation. The apparent cohesive force in noncohesive soils due to moisture and capillary action is completely lost when the soil becomes wet. Stable excavations sloped at 1:1 during the dry summer months can become very unstable during wet winter months. The sloping or shore loading decision must take into consideration the length of time that it will be open and the environmental factors such as water that it will be exposed to. To change sloping or strengthen shoring after construction of a structure inside the excavation is started is extremely expensive, if not impossible. When a soil becomes type C later in the life of the excavation, it is very hard to ignore the 1½:1 sloping requirement which is what most contractors in that situation end up asking others to do. If there is a subsequent failure in a shoring system after it has been excavated and working properly, the most likely culprit is water.

> (iv) Submerged rock that is not stable, or
>
> (v) Material in a sloped, layered system where the layers dip into the excavation or a slope of four horizontal to one vertical (4H:1V) or steeper.

Commentary

This should also read "… or steeper when the soil does not meet the requirements of Type B soil."

> **"Unconfined compressive strength"** means the load per unit area at which a soil will fail in compression. It can be determined by laboratory testing, or estimated in the field using a pocket penetrometer, by thumb penetration tests, and other methods.

"Wet soil" means soil that contains significantly more moisture than moist soil, but in such a range of values that cohesive material will slump or begin to flow when vibrated. Granular material that would exhibit cohesive properties when moist will lose those cohesive properties when wet.

1926 Appendix A (c) Requirements

(1) Classification of soil and rock deposits.

Each soil and Rock deposit shall be classified by a competent person as Stable Rock, Type A, Type B, or Type C in accordance with the definitions set forth in paragraph (b) of this appendix.

Commentary

The cardinal sin in excavation safety is to not classify the soil prior to utilizing a worker protection system—getting the classification wrong does not even hold a close second to not classifying. In excavation-intensive projects being too conservative in classifying the soil can be extremely expensive or cost-prohibitive. Understanding and using Appendix A is an OSHA excavation safety requirement.

This requirement provides to the competent person in the field, at the time the soil condition is encountered, the information needed to make informed decisions about the sloping and strength requirements of worker protection systems. Paragraph (b) is the definitions above.

(2) Basis of classification.

The classification of the deposits shall be made based on the results of at least one visual and at least one manual analysis. Such analyses shall be conducted by a competent person using tests described in paragraph (d) below, or in other recognized methods of soil classification and testing such as those adopted by the American Society for Testing Materials, or the U.S. Department of Agriculture textural classification system.

Commentary

The conclusions drawn here are to be formed with understanding and good judgment. Others might come to a different conclusion; however, if the decisions by the competent person were made in good faith and conformance with this appendix, it would be extremely hard to make a legal case of negligence against the competent person or employer.

The Appendix A classification system is intended to stand alone, and although it may lack sufficient visual and manual tests, when combined with common sense and observation it is adequate to form worker safety decisions based on it. Knowledge of other classification systems is not a requirement of Appendix A; their use is optional. The Uniform Soil Classification System should also be listed here as it is seen as more applicable to construction, while the USDA system is applicable to soils in agricultural use.

(3) Visual and manual analyses.

The visual and manual analyses, such as those noted as being acceptable in paragraph (d) of this appendix, shall be designed and conducted to provide sufficient quantitative and qualitative information as may be necessary to identify properly the properties, factors, and conditions affecting the classification of the deposits.

Commentary

The point being made here is not to make decisions based on guesswork or lack of effort to find the information.

(4) Layered systems.

In a layered system, the system shall be classified in accordance with its weakest layer. However, each layer may be classified individually where a more stable layer lies under a less stable layer.

Commentary

The converse of this statement is that if the soil is weak on the bottom, it is to be considered that it is weak in the layers above it.

Very thin layers, say less than 4 in, present a classification problem because even though they may be weak, they do not present as much of an adverse soil condition as thick weak layers. In shored systems the effect can be minimal, while in sloped systems they can eventually present more problems.

(5) Reclassification.

If, after classifying a deposit, the properties, factors, or conditions affecting its classification change in any way, the changes shall be evaluated by a competent person. The deposit shall be reclassified as necessary to reflect the changed circumstances.

Commentary

Factors that change the properties of the soil include

- Rock drying and slaking
- Cohesive soils drying and cracking due to exposure to the atmosphere and heat
- Noncohesive soils that are exposed to water inundation, rain, and rising water tables from nearby streams, rivers, irrigation, and rainstorms
- The addition of large surcharge loads from heavy equipment and cranes
- Backfill and compactions operations

1926 Appendix A (d) Acceptable visual and manual tests.

Commentary

These tests can be carried out in the field. Unless otherwise stated, they should be taken on the soil as it is encountered. Tests can be performed on the trench walls if it is safe or from the undisturbed portion of excavator bucket scoops.

1926 Appendix A (d)(1) Visual tests.

Visual analysis is conducted to determine qualitative information regarding the excavation site in general, the soil adjacent to the excavation, the soil forming the sides of the open excavation, and the soil taken as samples from excavated material.

(i) Observe samples of soil that are excavated and soil in the sides of the excavation. Estimate the range of particle sizes and the relative amounts of the particle sizes. Soil that is primarily composed of fine-grained material is cohesive material. Soil composed primarily of coarse-grained sand or gravel is granular material.

(ii) Observe soil as it is excavated. Soil that remains in clumps when excavated is cohesive. Soil that breaks up easily and does not stay in clumps is granular.

(iii) Observe the side of the opened excavation and the surface area adjacent to the excavation. Crack-like openings such as tension cracks could indicate fissured material. If chunks of soil spall off a vertical side, the soil could be fissured. Small spalls are evidence of moving ground and are indications of potentially hazardous situations.

(iv) Observe the area adjacent to the excavation and the excavation itself for evidence of existing utility and other underground structures, and to identify previously disturbed soil.

(v) Observed the opened side of the excavation to identify layered systems. Examine layered systems to identify if the layers slope toward the excavation. Estimate the degree of slope of the layers.

(vi) Observe the area adjacent to the excavation and the sides of the opened excavation for evidence of surface water, water seeping from the sides of the excavation, or the location of the level of the water table.

(vii) Observe the area adjacent to the excavation and the area within the excavation for sources of vibration that may affect the stability of the excavation face.

1926 Appendix A (d)(2) Manual Tests.

Manual analysis of soil samples is conducted to determine quantitative as well as qualitative properties of soil and to provide more information in order to classify soil properly.

(i) Plasticity. Mold a moist or wet sample of soil into a ball and attempt to roll it into threads as thin as 1/8-inch in diameter. Cohesive material can be successfully rolled into threads without crumbling. For example, if at least a two inch (50 mm) length of 1/8-inch thread can be held on one end without tearing, the soil is cohesive.

(ii) Dry strength. If the soil is dry and crumbles on its own or with moderate pressure into individual grains or fine powder, it is granular (any combination of gravel, sand, or silt). If the soil is dry and falls into clumps which break up into smaller clumps, but the smaller clumps can only be broken up with difficulty, it may be clay in any combination with gravel, sand or silt. If the dry soil breaks into clumps which do not break up into small clumps and which can only be broken with difficulty, and there is no visual indication the soil is fissured, the soil may be considered unfissured.

(iii) Thumb penetration. The thumb penetration test can be used to estimate the unconfined compressive strength of cohesive soils. (This test is based on the thumb penetration test described in American Society for Testing and Materials (ASTM) Standard designation D2488—"Standard Recommended Practice for Description of Soils (Visual—Manual Procedure).") Type A soils with an unconfined compressive strength of 1.5 tsf can be readily indented by the thumb; however, they can be penetrated by the thumb only with very great effort. Type C soils with an unconfined compressive strength of 0.5 tsf can be easily penetrated several inches by the thumb, and can be molded by light finger pressure. This test should be conducted on an undisturbed soil sample, such as a large clump of spoil, as soon as practicable after excavation to keep to a minimum the effects of exposure to drying influences. If the excavation is later exposed to wetting influences (rain, flooding), the classification of the soil must be changed accordingly.

(iv) Other strength tests. Estimates of unconfined compressive strength of soils can also be obtained by use of a pocket penetrometer or by using a hand-operated shear vane.

(v) Drying test. The basic purpose of the drying test is to differentiate between cohesive material with fissures, unfissured cohesive material, and granular material. The procedure for the drying test involves drying a sample of soil that is approximately one inch thick (2.54 cm) and six inches (15.24 cm) in diameter until it is thoroughly dry:

(A) If the sample develops cracks as it dries, significant fissures are indicated.

(B) Samples that dry without cracking are to be broken by hand. If considerable force is necessary to break a sample, the soil has significant cohesive material content. The soil can be classified as an unfissured cohesive material and the unconfined compressive strength should be determined.

(C) If a sample breaks easily by hand, it is either a fissured cohesive material or a granular material. To distinguish between the two, pulverize the dried clumps of the sample by hand or by stepping on them. If the clumps do not pulverize easily, the material is cohesive with fissures. If they pulverize easily into very small fragments, the material is granular.

AP2.2 C-60 Soil Classification

Manufacturers of shoring equipment have promoted and used in their tabulated data an intermediate type C-60 soil classification that in the hierarchy falls between type B and type C soils (see Table 5.7). This category is generally accepted within the shoring industry by

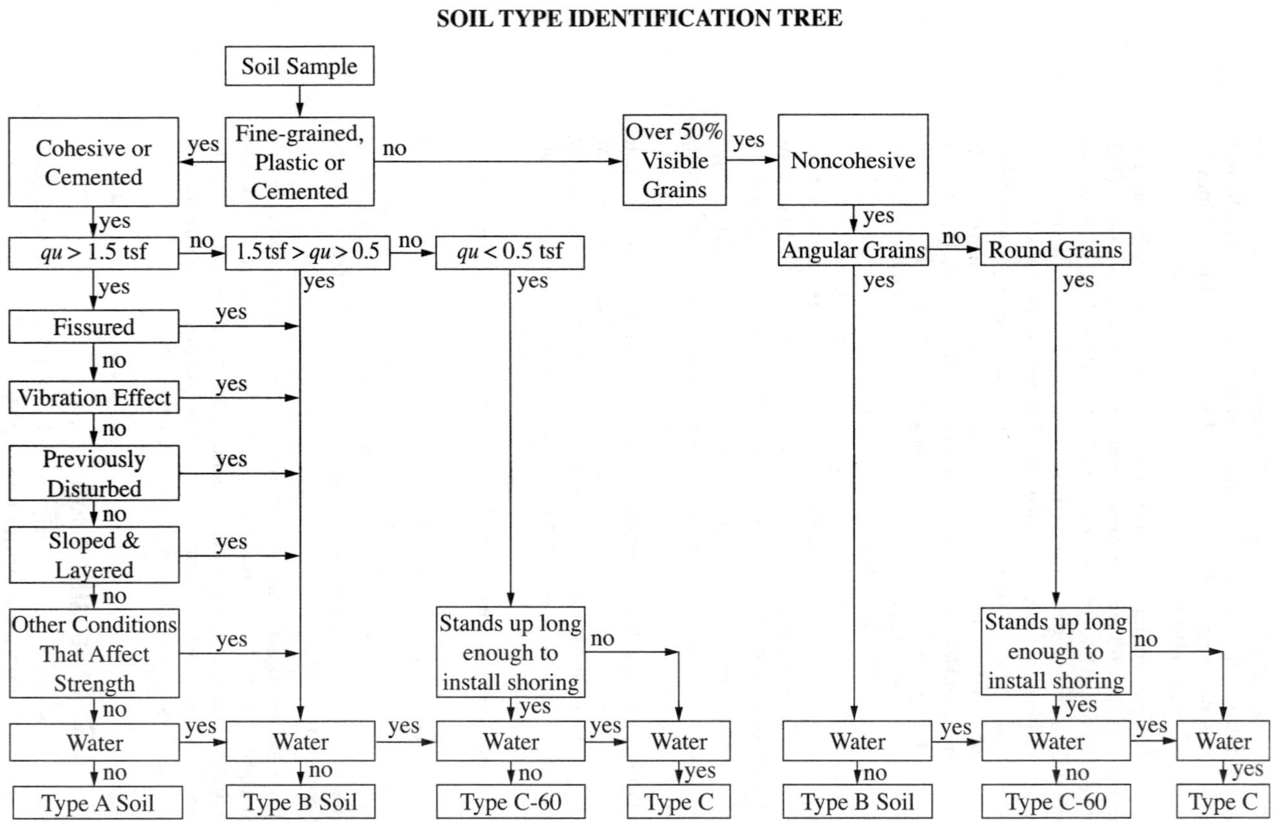

Note: *qu* = unconfined compressive strength in tons per square foot (tsf).

Figure AP2.4 Soil type identification tree.

contractors and those reviewing worker protection plans. Federal OSHA and state OSHA programs do not appear to be opposed to this; however, this appendix and the excavation safety standards do not recognize or address the type C-60 soil classification.

Prior to establishment of Appendix A in 1989, the original OSHA soil identification system also attached a soil lateral loading factor to the soil designations. They were often referred to as OSHA type A-20, type B-40, and type C-80 soils and indicated that, for instance, an excavation 10 ft deep in type B soil would deliver a 20 × 40 = 800 psf lateral load to the shoring. For shoring equipment such as shoring shields at 20 ft deep in C-80, the shield needed to be rated 20 × 80 = 1600 psf, twice as much as for type B soil and generally out of the range of normal shield construction at the time. In the case of trench jacks, some argued that cohesive C soil was too soft to stand long enough to get trench jacks installed and that the soil would slough in between even if one could get them installed. In noncohesive C soil the excavation walls would not stand up long enough to get the shores in. Consequently tabulated data developed by OSHA for hydraulic aluminum trench jacks only addressed soil types A and B.

In response to that assumption, there is plenty of soil that does not meet the type B soil requirements that stands up and acts more as type B than type C soil; the shoring manufacturers defined type C-60 soil and developed tabulated data for use of their equipment under this category. The very simple and easy to use defining test for C-60 soil is

> **Manufacturers' Definitive Test for Type C-60 Soil If the soil** stands up long enough to install the shoring it can be considered C-60.

In the design and tabulation of operating data of shoring equipment in C-60 soil, a 60 times excavation depth rectangular load is anticipated on the shoring. In Fig. AP2.4, the author has included the C-60 soil classification.

Reference

U.S. Department of Labor, 29 CFR 1926, *Appendix A Soil Classification*, Occupational Safety & Health Administration., Washington, Jan. 29, 2008, www.osha.gov.

Index

O

T

U

V

W